Virus — Host Cell Interactions

Virus — Host Cell Interactions

Editors

Thomas Hoenen
Allison Groseth

MDPI • Basel • Beijing • Wuhan • Barcelona • Belgrade • Manchester • Tokyo • Cluj • Tianjin

Editors

Thomas Hoenen
Institute of Molecular
Virology and Cell Biology
Friedrich-Loeffler-Institut
Greifswald - Insel Riems
Germany

Allison Groseth
Institute of Molecular
Virology and Cell Biology
Friedrich-Loeffler-Institut
Greifswald - Insel Riems
Germany

Editorial Office
MDPI
St. Alban-Anlage 66
4052 Basel, Switzerland

This is a reprint of articles from the Special Issue published online in the open access journal *Cells* (ISSN 2073-4409) (available at: www.mdpi.com/journal/cells/special_issues/HostCells_Interactions).

For citation purposes, cite each article independently as indicated on the article page online and as indicated below:

LastName, A.A.; LastName, B.B.; LastName, C.C. Article Title. *Journal Name* **Year**, *Volume Number*, Page Range.

ISBN 978-3-0365-6559-0 (Hbk)
ISBN 978-3-0365-6558-3 (PDF)

Contents

About the Editors

Thomas Hoenen

Thomas Hoenen is a Laboratory Head in the Institute of Molecular Virology and Cell Biology of the Friedrich-Loeffler-Institut (FLI) on the Isle of Riems near Greifswald, Germany. He studied human biology at the Philipps University in Marburg, Germany, and obtained his PhD in 2007 from the same university after performing his thesis work on the biological functions of ebolavirus matrix proteins. After postdoctoral training at the Public Health Agency of Canada in Winnipeg, MB, and at the Philipps University in Marburg he joined the National Institutes of Health (NIH) in Hamilton, MT, USA, where he continued his work on reverse genetics systems for ebolaviruses and other filoviruses. In 2015, he joined the FLI as the head of the Laboratory for Molecular Biology of Filoviruses, and additionally took on the role of head of the biosafety level 4 (BSL4) laboratory of the FLI following its opening a few years later. In 2021, he habilitated in Virology and obtained his *venia legendi* from the University of Greifswald. In the same year, he became a scientific director and head of the Laboratory for Integrative Cell and Infection Biology. His work focuses on the development and application of reverse genetics systems for highly pathogenic viruses, and, in particular, his ebolavirus transcription and replication-competent virus-like particle system, which allows the modelling of the virus life cycle without the need for a BSL4 laboratory, has been shared with more than 50 laboratories worldwide and has served as a template for the generation of similar systems for a number of other viruses. He is especially interested in virus–host cell interactions of ebolaviruses and other highly pathogenic viruses, in order to identify novel targets for antivirals with broad-spectrum activity, and to better understand viral and host virulence determinants to facilitate fact-based risk assessments for these viruses.

Allison Groseth

Allison Groseth is a Laboratory Head at the FLI. She performed both her undergraduate and graduate training in Canada with a BSc in Biochemistry from the University of Victoria and a PhD in Medical Microbiology from the University of Manitoba. Her PhD work was performed with the Special Pathogens Program of the Public Health Agency of Canada and focused on viral determinants underlying differences in virulence between filovirus species. Her postdoctoral work at the Philipps University in Marburg, Germany, focused on identifying host responses associated with arenavirus pathogenesis, particularly those regulating cell death and cytokine expression. Further work as a Staff Scientist in the Laboratory of Virology of the NIH at Rocky Mountain Laboratories in Hamilton, MT, USA, focused on similar questions in the context of the emergence of highly pathogenic orthobunyaviruses, while also looking to identify common host factors associated with infection by unrelated hemorrhagic fever-causing pathogens. In 2015, she was recruited to the FLI, first as a Junior Group Leader, and since 2020 as a Laboratory Head in the Institute of Molecular Virology and Cell Biology. She habilitated in Virology and obtained her *venia legendi* from the University of Greifswald in 2022. Her research group continues to focus on understanding the virus–host interface and particularly on developing a detailed mechanistic view of differences in host cell response regulation during infection with highly pathogenic and apathogenic arenaviruses in order to identify critical players in these processes that can serve as antiviral targets. In particular, recent areas of interest include studying the regulation of cell death and cytokine responses, and the role of kinase signaling in these processes, as well as dissecting mechanisms associated with induction (and evasion) of dsRNA detection.

Editorial

Virus–Host Cell Interactions

Thomas Hoenen [1],* and **Allison Groseth** [2],*

1   Laboratory for Integrative Cell and Infection Biology, Institute of Molecular Virology and Cell Biology, Friedrich-Loeffler-Institut, 17493 Greifswald, Germany
2   Laboratory for Arenavirus Biology, Institute of Molecular Virology and Cell Biology, Friedrich-Loeffler-Institut, 17493 Greifswald, Germany
*   Correspondence: thomas.hoenen@fli.de (T.H.); allison.groseth@fli.de (A.G.)

Citation: Hoenen, T.; Groseth, A. Virus–Host Cell Interactions. *Cells* **2022**, *11*, 804. https://doi.org/10.3390/cells11050804

Received: 16 February 2022
Accepted: 18 February 2022
Published: 25 February 2022

Publisher's Note: MDPI stays neutral with regard to jurisdictional claims in published maps and institutional affiliations.

As obligate intracellular parasites, viruses are intimately interconnected with their host cells. Virus–host cell interactions allow viruses to exploit cells for their own purposes, but they also provide a means for the host cell to combat virus infection. This close connection between host and viral processes means that the scientific fields of cell biology and virology have often inspired each other.

In particular, many discoveries in cell biology have been made possible by the study of viruses, while, at the same time, our fundamental understanding of the virus life cycle is inherently rooted in the principles of cell biology. By examining cellular responses to infection, we can gain insights regarding the mechanisms associated with the restriction of virus infection or, in cases where control is ineffective, pathogenesis. Such knowledge is a prerequisite for the successful modulation of these responses to develop host-directed therapies for the control of viral infections. Furthermore, from a practical point of view, virus–host cell interactions also provide important targets for the development of indirectly acting antivirals, which have a reduced likelihood to develop resistance due to their reliance on host cell components.

This Special Issue of *Cells* compiles both review papers discussing the current state of our knowledge regarding important emerging themes related to virus–host cell interactions, and research articles reporting new discoveries in this scientific area. These publications cover various aspects of the virus life cycle, from virus entry, uncoating, and virus replication in specialized replication compartments, all the way to virus particle production. They also touch on the complex interplay of viruses with the immune system, and current topics such as the modification of viral RNAs.

One of the first interactions of a virus with its host cell is during the entry process, which is generally facilitated by the interaction of a virus surface protein with a cellular receptor, and sometimes by additional interactions with cellular attachment factors, in order to facilitate virus uptake. Here, a mechanism that is increasingly recognized as being exploited by diverse virus families is apoptotic mimicry, which in its canonical form involves the exposure of phosphatidylserine in the outer leaflet of the viral membrane. This exposed phosphatidylserine can then be recognized by host cell proteins, such as those of the T-cell immunoglobulin and mucin domain (TIM) family. Kirui et al. [1] demonstrate that this mechanism is also used by Chikungunya virus. Their study also acts as an important reminder regarding the importance of working with authentic viruses, since their results with authentic Chikungunya virus show marked differences to previous results that were obtained based on experiments with pseudotyped particles alone.

In a second study on virus entry, but from the perspective of the role that virus entry receptor expression can play in directing infection outcome in different tissues, DeBuysscher et al. [2] show that Nipah virus efficiently replicates in human smooth muscle cells, even though these cells lack the canonical Nipah virus receptor ephrin B2. Furthermore, this lack of ephrin B2 appears to protect these cells from cell–cell fusion and cytopathic effects seen in other target cells that express this host factor, and suggests that smooth muscle cells might play an important role in pathogenesis by harboring and amplifying viruses that then infect and damage neighboring endothelial cells.

After entry, the next hurdle that many viruses have to overcome in order to establish a successful infection is the uncoating of the virus genetic material to facilitate its release. In a featured review on Influenza virus uncoating, Moreira at al. [3] discuss how the virus utilizes host cell pathways for this purpose. In particular, they highlight the role of a number of cellular factors, such as ubiquitin, histone deacetylase 6 (HDAC6), and transportin 1, and discuss potential contributions of these proteins for the uncoating process of other RNA viruses, as well as their potential as targets for broad-spectrum-antivirals.

For viruses that replicate in the cytoplasm, genome replication and viral transcription are increasingly being appreciated to take place in specialized replication organelles. For positive-strand RNA viruses, these are predominantly membranous structures. In contrast, for a growing number of negative-strand RNA viruses, viral RNA synthesis has been shown to be localized in inclusion bodies that are not delineated from their surroundings by membranes. Two reviews highlight the progress that is being made in understanding both of these types of replication organelles. Nguyen-Dingh and Herker [4] focus on the membranous replication organelles induced by positive-strand RNA viruses, whereas Dolnik et al. [5] discuss exciting recent progress in our understanding of negative-sense RNA virus replication structures as liquid organelles, i.e., compartments that are held together by liquid–liquid phase separation, rather than by a surrounding membrane.

Within these replication organelles, transcription leads to the generation of viral mRNAs. However, what has remained underappreciated in virology is that, in many cases, these mRNAs can be modified by the addition of methyl groups by host cell proteins. In the second featured review of this Special Issue, Courtney [6] provides a comprehensive overview regarding the current state of our knowledge with respect to this emerging topic, highlighting not only recent scientific advancements in understanding the functional implications of RNA modifications, but also giving an overview of the available methods for exploring them in a viral context.

Finally, in order to complete their life cycle, viruses have to exit their host cell, and this process is often intimately interlinked with the subversion of cellular host factors. In the case of non-segmented negative-sense RNA viruses, particle production is frequently driven by a dedicated viral matrix protein. Focusing on the Ebola virus matrix protein VP40, Paparisto et al. show that the process of particle production is inhibited by the cellular protein HECT and RCC1-like containing domain 5 (HERC5) [7]. However, interestingly, the mechanism for this does not appear to be an inhibition on the level of matrix protein function, but rather involves the depletion of mRNAs encoding for VP40. Consequently, this study emphasizes not only the role virus host–cell interactions play in supporting the virus life cycle, but also the role that antiviral factors can play in inhibiting key viral processes.

Of course, host–pathogen interactions involved in antiviral control not only occur at the level of specific antiviral cellular factors or cellular antiviral responses, but can also include a broader range of interactions with the immune system. This is highlighted in a comprehensive review by Muralidharan and Reid, in which they illuminate the complex roles of neutrophils during arbovirus infections [8]. Using examples from a wide range of arboviruses, including Zika virus, Dengue virus, West Nile virus, and various alphaviruses, they highlight not only the beneficial roles that neutrophils can play, but also their sometimes-detrimental roles in augmenting disease pathology.

An equally complex but currently underappreciated topic is that of pathogen–host interactions in the context of coinfections involving several disease agents, and particularly co-infections of viruses and bacteria. In this context, a study by Nickol et al. [9] demonstrates the impact of coinfection with Influenza virus and methicillin-resistant *Staphylococcus aureus* on the expression of bacterial virulence factors as well as on the infected host, particularly regarding its cytokine response and the integrity of the alveolar-capillary barrier. This work provides mechanistic support for the clinical observation that severe influenza infections are frequently complicated by bacterial coinfections.

Overall, this Special Issue provides examples of the crucial role that virus–host cell interactions play in the biology of a diverse range of viruses, and highlights the importance of better understanding such interactions in order to be better able to combat virus infections and the mechanisms that contribute to pathogenesis and disease.

Conflicts of Interest: The authors declare no conflict of interest.

References

1. Kirui, J.; Abidine, Y.; Lenman, A.; Islam, K.; Gwon, Y.D.; Lasswitz, L.; Evander, M.; Bally, M.; Gerold, G. The Phosphatidylserine Receptor TIM-1 Enhances Authentic Chikungunya Virus Cell Entry. *Cells* **2021**, *10*, 1828. [CrossRef] [PubMed]
2. DeBuysscher, B.L.; Scott, D.P.; Rosenke, R.; Wahl, V.; Feldmann, H.; Prescott, J. Nipah Virus Efficiently Replicates in Human Smooth Muscle Cells without Cytopathic Effect. *Cells* **2021**, *10*, 1319. [CrossRef] [PubMed]
3. Moreira, E.A.; Yamauchi, Y.; Matthias, P. How Influenza Virus Uses Host Cell Pathways during Uncoating. *Cells* **2021**, *10*, 1722. [CrossRef] [PubMed]
4. Nguyen-Dinh, V.; Herker, E. Ultrastructural Features of Membranous Replication Organelles Induced by Positive-Stranded RNA Viruses. *Cells* **2021**, *10*, 2407. [CrossRef] [PubMed]
5. Dolnik, O.; Gerresheim, G.K.; Biedenkopf, N. New Perspectives on the Biogenesis of Viral Inclusion Bodies in Negative-Sense RNA Virus Infections. *Cells* **2021**, *10*, 1460. [CrossRef] [PubMed]
6. Courtney, D.G. Post-Transcriptional Regulation of Viral RNA through Epitranscriptional Modification. *Cells* **2021**, *10*, 1129. [CrossRef] [PubMed]
7. Paparisto, E.; Hunt, N.R.; Labach, D.S.; Coleman, M.D.; Di Gravio, E.J.; Dodge, M.J.; Friesen, N.J.; Cote, M.; Muller, A.; Hoenen, T.; et al. Interferon-Induced HERC5 Inhibits Ebola Virus Particle Production and Is Antagonized by Ebola Glycoprotein. *Cells* **2021**, *10*, 2399. [CrossRef] [PubMed]
8. Muralidharan, A.; Reid, S.P. Complex Roles of Neutrophils during Arboviral Infections. *Cells* **2021**, *10*, 1324. [CrossRef] [PubMed]
9. Nickol, M.E.; Lyle, S.M.; Dennehy, B.; Kindrachuk, J. Dysregulated Host Responses Underlie 2009 Pandemic Influenza-Methicillin Resistant Staphylococcus aureus Coinfection Pathogenesis at the Alveolar-Capillary Barrier. *Cells* **2020**, *9*, 2472. [CrossRef] [PubMed]

Article

The Phosphatidylserine Receptor TIM-1 Enhances Authentic Chikungunya Virus Cell Entry

Jared Kirui [1,2], Yara Abidine [3,4], Annasara Lenman [1,3], Koushikul Islam [3], Yong-Dae Gwon [3], Lisa Lasswitz [1,2], Magnus Evander [3], Marta Bally [3,4] and Gisa Gerold [1,2,3,4,*]

1. Centre for Experimental and Clinical Infection Research, TWINCORE, Institute for Experimental Virology, a Joint Venture between the Medical School Hannover and the Helmholtz Centre for Infection Research, 30625 Hannover, Germany; Jared.Kirui@tiho-hannover.de (J.K.); anna-sara.lenman@umu.se (A.L.); Lisa.Lasswitz@tiho-hannover.de (L.L.)
2. Department of Biochemistry & Research Center for Emerging Infections and Zoonoses (RIZ), University of Veterinary Medicine Hannover, 30559 Hannover, Germany
3. Department of Clinical Microbiology, Umeå University, 90185 Umeå, Sweden; yara.abidine@umu.se (Y.A.); islam.koushikul@umu.se (K.I.); kwon.yongdae@umu.se (Y.-D.G.); magnus.evander@umu.se (M.E.); marta.bally@umu.se (M.B.)
4. Wallenberg Centre for Molecular Medicine (WCMM), Umeå University, 90185 Umeå, Sweden
* Correspondence: Gisa.Gerold@tiho-hannover.de

Abstract: Chikungunya virus (CHIKV) is a re-emerging, mosquito-transmitted, enveloped positive stranded RNA virus. Chikungunya fever is characterized by acute and chronic debilitating arthritis. Although multiple host factors have been shown to enhance CHIKV infection, the molecular mechanisms of cell entry and entry factors remain poorly understood. The phosphatidylserine-dependent receptors, T-cell immunoglobulin and mucin domain 1 (TIM-1) and Axl receptor tyrosine kinase (Axl), are transmembrane proteins that can serve as entry factors for enveloped viruses. Previous studies used pseudoviruses to delineate the role of TIM-1 and Axl in CHIKV entry. Conversely, here, we use the authentic CHIKV and cells ectopically expressing TIM-1 or Axl and demonstrate a role for TIM-1 in CHIKV infection. To further characterize TIM-1-dependent CHIKV infection, we generated cells expressing domain mutants of TIM-1. We show that point mutations in the phosphatidylserine binding site of TIM-1 lead to reduced cell binding, entry, and infection of CHIKV. Ectopic expression of TIM-1 renders immortalized keratinocytes permissive to CHIKV, whereas silencing of endogenously expressed TIM-1 in human hepatoma cells reduces CHIKV infection. Altogether, our findings indicate that, unlike Axl, TIM-1 readily promotes the productive entry of authentic CHIKV into target cells.

Keywords: Chikungunya virus; CHIKV; alphavirus; enveloped virus; phosphatidylserine; T-cell immunoglobulin and mucin domain 1; TIM-1; Axl receptor tyrosine kinase; Axl; entry

Citation: Kirui, J.; Abidine, Y.; Lenman, A.; Islam, K.; Gwon, Y.-D.; Lasswitz, L.; Evander, M.; Bally, M.; Gerold, G. The Phosphatidylserine Receptor TIM-1 Enhances Authentic Chikungunya Virus Cell Entry. *Cells* **2021**, *10*, 1828. https://doi.org/10.3390/cells10071828

Academic Editors: Thomas Hoenen and Allison Groseth

Received: 2 June 2021
Accepted: 15 July 2021
Published: 20 July 2021

Publisher's Note: MDPI stays neutral with regard to jurisdictional claims in published maps and institutional affiliations.

1. Introduction

Chikungunya fever, caused by chikungunya virus (CHIKV), has emerged as a global health problem in the last seven decades [1,2]. CHIKV is an arbovirus and member of the Togaviridae family, genus *Alphavirus* transmitted to humans mainly by *Aedes* (*Ae.*) *aegypti* and *Ae. albopictus* mosquitoes [3]. The species CHIKV consists of three main genotypes, namely East-Central-South-African (ECSA), West African, and Asian [4]. It is estimated that about 75–95% of infected individuals develop chikungunya fever, with symptoms such as high fever, intense asthenia, myalgia, rash, and debilitating joint pain that turns chronic in 12–49% of patients [5,6]. Therapeutic options for CHIKV are limited since there are currently no specific antivirals and no licensed vaccines.

CHIKV has a wide cellular and tissue tropism which may be attributed to use of ubiquitously expressed molecules or several cell specific factors for entry. These molecules likely determine CHIKV pathogenesis and represent promising targets for antiviral strategies [7–11]. Multiple attachment factors and putative receptors for CHIKV and other alphaviruses have been documented [12,13]. For instance, ATP synthase β subunit (ATPSβ) is a host factor in mosquito cells [14] and prohibitins [15], glycosaminoglycans [16,17], phosphatidylserine (PtdSer)-mediated virus entry-enhancing receptors (PVEERs) [18,19], and MXRA8 [20] are host factors in mammalian cells. Interaction with the cell surface molecules is mediated by the viral E2 glycoprotein, whose domain B contains receptor binding sites [17,21]. For MXRA8, a recently identified receptor for several alphaviruses [20] E1 is additionally important as MXRA8 engages amino acid residues at the E1 and E2 glycoprotein heterodimer interface [22,23]. Phagocytic cells express PVEERs through which they bind PtdSer present on the outer leaflets of apoptotic bodies [24–26]. Similarly, epithelial cells expressing T-cell immunoglobulin and mucin domain 1 (TIM-1) act as semi-professional phagocytes and are involved in the clearance of apoptotic bodies [27,28], a process mediated by phosphorylation of residues in the cytoplasmic domain of TIM-1 [29]. Some enveloped viruses have evolved to incorporate PtdSer in the viral membrane, hence disguised as apoptotic bodies, a phenomenon termed as apoptotic mimicry [19,30,31]. TIM-1 and Axl receptor tyrosine kinase (Axl) are PVEERs associated with enhanced cell entry by enveloped viruses. This includes alphaviruses, filoviruses, and flaviviruses, among others [18,19,32–34]. Using CHIKV glycoprotein based pseudoviruses, the TIM family of proteins and Axl were shown to enhance infection [19,33]. TIM-1 and Axl are single pass transmembrane proteins with distinct ectodomains and cytoplasmic domains. TIM-1 interacts with PtdSer through a binding pocket known as metal ion ligand binding site (MILIBS) on its extracellular immunoglobulin-like variable (Ig-V) domain [33]. Axl indirectly binds phosphatidylserine through ligands, namely growth arrest-specific factor 6 (Gas6) [35] or protein S1 (ProS1) [36]. In the skin, TIM-1 and Axl are predominantly expressed by keratinocytes in the basal layer of the epidermis [37,38]. HaCat cells derived from spontaneously immortalized keratinocytes serve as a relevant model to study keratinocytes in vitro [39]. However, they hardly express TIM-1 and Axl. The role of TIM-1 and Axl expression in permissiveness of keratinocytes is yet to be characterized.

After binding to a receptor on the plasma membrane, CHIKV primarily enters human host cells by clathrin-mediated endocytosis [40,41]. However, clathrin-independent pathways have also been reported [8,42]. Upon endocytosis, CHIKV particles are delivered to early endosomes in mammalian cells [42], whereas in mosquito cells, the complexes traffic further to maturing or late endosomes before membrane fusion occurs [43]. The discrepancy in the endosomal fusion compartments may be due to the variability of endosomal cues between cells. Nonetheless, general molecular mechanisms involved in fusion are highly conserved between alphaviruses. Specifically, the acidic endosomal environment triggers a class II membrane fusion mechanism [44,45] and release of the nucleocapsid into the cytosol [46,47].

In the current study, we have examined the role of TIM-1 and Axl in CHIKV infection using different genotypes of authentic virus. Our experiments show that TIM-1 unlike Axl is functional as an entry factor for CHIKV and that the PtdSer binding site as well as the cytoplasmic domain are essential for infection. These results indicate that CHIKV exploits the apoptotic cell clearance pathway to facilitate the rapid and efficient infection of human cells.

2. Materials and Methods

2.1. Cells and Viruses

Human embryonic kidney 293T (HEK293T) cells [48] obtained from American Type Culture Collection (ATCC, CRL-3612), baby hamster kidney cells (BHK-21, ATCC CCL-10), human hepatoma derived (Huh7.5) cells [49] (kindly provided by Charles Rice, Rockefeller University, New York, NY, USA), spontaneously immortalized human skin keratinocytes (HaCat cells) and dermal fibroblasts were kindly provided by PD Dr. F. Pessler, Twincore, Hannover and Vero cells (ATCC CRL-1586) were cultured in Dulbecco's modified essential medium (DMEM, Gibco™, Paisley, Scotland, UK) supplemented with 10% fetal calf serum (FCS, Gibco™), 100 U/mL penicillin, 100 µg/ml streptomycin, 1% non-essential amino acids and 2 mM L-glutamine, at 37 °C in 5% CO_2 humidified incubator. Chinese hamster ovary (CHOK1 and CHO745) cells obtained from ATCC were cultured in RPMI-1640 (Gibco™, Paisley, Scotland, UK) supplemented with 10% fetal calf serum (FCS, Gibco™), 100 U/mL penicillin, 100 µg/ml streptomycin, 1% non-essential amino acids, and 2 mM L-Glutamine.

The chikungunya virus strains; East Central South African (ECSA) LR2006-OPY1 strain (3'GFP-CHIKV) [50] and West African (WA) 37997 strain (5'GFP-CHIKV) [51], both encoding green fluorescent protein (GFP) gene, and Asian (181/25) vaccine strain [52] encoding either mCherry-fluorescent protein (mc-CHIKV) or nano-luciferase gene (nLuc-CHIKV) fused to the N-terminus of E2 glycoprotein (Supplementary Figure S1). The mCherry or nano-luciferase proteins are expressed on the viral envelope as previously described [53]. The plasmids used for the production of CHIKV were kindly provided by Graham Simmons, San Francisco, CA, USA. The live attenuated Venezuelan equine encephalitis virus (VEEV) TC-83 strain expressing a GFP reporter gene [54] was generated by Mike Diamond, Saint Louis, MO, USA. Sindbis virus, Lövånger (KF737350.1) strain [55] was provided by Magnus Evander, Umeå, Sweden. The GFP expressing human adenovirus type HAdV-C5 (HAdV-5) was obtained from Vector Development Laboratory, Houston, TX, USA.

2.2. Plasmids and Antibodies

Gene fragments (gBlocks) encoding wild type TIM-1/Axl and respective mutant open reading frames were commercially synthesized by Integrated DNA Technologies (IDT, Inc., Coralville, IA, USA). The gene fragments were amplified by PCR and cloned into pWPI_BLR vector using the Gibson assembly method according to the manufacturer's instructions (New England Biolabs, Ipswich, MA, USA) and direct sequencing used to confirm inserts.

TIM-1 polyclonal (TIM-1 pAb, AF1750) and Axl polyclonal (Axl pAb) antibodies were purchased from R&D systems while TIM-1 monoclonal antibody (TIM-1 mAb) was purchased from BioLegend®, San Diego, CA, USA.

2.3. RNA Transfection by Electroporation

HEK293T cells (1×10^6) were resuspended in 400 µL of cytomix electroporation buffer (2 mM ATP, 5 mM glutathione, 120 mM KCl, 0.15 mM $CaCl_2$, 10 mM K_2HPO_4/KH_2PO_4 (pH 7.6), 25 mM HEPES, 2 mM of EGTA and 5 mM $MgCl_2$). Either CHIKV sub-genomic replicon (SGR) or full-length CHIKV genome encoding nanoluciferase gene were added at a concentration of 1 µg and 1.7 µg respectively. The mixture was transferred to a 0.4 cm sterile cuvette and electroporated at 240 V and 975 Ω using a Bio-Rad electroporator system. After electroporation, cells were gently resuspended in 2 mL of prewarmed DMEM supplemented with 10% FCS, 2 mM L-Glutamine. Cells were seeded in duplicates in 96 and 24-well plates for SGR and the full-length RNA assays respectively. The cells were cultured at 37 °C and 5% CO_2. At the indicated timepoints, the cells were lysed for luciferase assay.

2.4. Generation of Lentiviral Vectors and Transduction of Cells

To generate lentiviral pseudoparticles, HEK293T cells were co-transfected with three plasmids. For pseudoparticles used to generate cells stably expressing a protein of interest, the cells were transfected with pVSV-G encoding the G protein for the Vesicular stomatitis virus (VSV), the lentiviral packaging plasmid pCMV_ΔR8-74 and the pWPI (from Didier Trono (Addgene plasmid # 12254) encoding either the wild type or mutant (TIM-1 or Axl) and a blasticidin resistance gene. Sodium butyrate was added 24 h post transfection in order to boost plasmid transcription [56]. For cell entry experiments, the pseudoparticles were generated using a plasmid encoding for a glycoprotein of the virus of interest (CHIKV, EBOV or VSV), the lentiviral packaging plasmid pCMV_ΔR8-74, and a pWPI plasmid encoding a luciferase gene as a reporter protein. At 48 and 72 h post transfection, lentiviral particles were harvested by filtering the supernatant through a 0.45 μm pore size filter. The lentiviral particles were stabilized by adding HEPES and polybrene was added to improve the efficiency of gene transfer [57].

To generate cells stably expressing TIM-1 or Axl, the particles were added to a monolayer of cells for five hours of transduction then replenished with fresh media. Selection for positively transduced cells with blasticidin (5 μg/mL) commenced 48 h post-transduction. To determine the role of surface proteins in virus cell entry experiments, cells were transduced with lentiviral pseudoparticles for 4 h and incubated with fresh media for 24 h.

2.5. RNA Interference

Huh7.5 cells pre-seeded in 6-well plates for five hours were transiently transfected with a pool of three siRNAs (Ambion™ Silencer™ Select) for TIM-1 (s230290, s230291, s25632) and MXRA8 (s29242, s29241, s29240) and a control non-targeting (NT) siRNA (AM4637) (ThermoFisher) using Lipofectamine RNAiMAX Reagent protocol (ThermoFisher, Waltham, MA, USA). At 48 h post-transfection, cells were assessed for expression, then seeded for infection and viability testing. The cells were infected with CHIKV at the indicated MOI and susceptibility determined by flow cytometry at 24 h post infection.

2.6. Cell Viability and Proliferation Assay

The cellular metabolic activity was measured using the MTT assay as previously described [58]. Cells were seeded in 96-well plates at a density of 2×10^4 cells/well. Medium was replaced with 50 μL of 0.5 mg/mL MTT in media and incubated for 2 h at 37 °C and 5% CO_2. Afterwards, 50 μL per well of Dimethyl sulfoxide (DMSO) was added to solubilize the crystals. After 30 min at room temperature, the absorbance was measured at a wavelength of 560 nm on a spectrophotometer microplate-reader (BioTek™ Synergy™ 2).

2.7. Cell Culture Derived CHIKV Stock Production and Titration

Plasmid DNA (20 μg) encoding CHIKV genome was linearized using NotI (New England Biolabs) endonuclease restriction digestion. Complete linearization was confirmed by agarose gel electrophoresis and linearized DNA purified using the QIAprep Spin Miniprep kit (QIAgen, Hilden, Germany) following the manufacturer protocol. A 100 μL in vitro transcription reaction was prepared using 2 μg of the DNA mixed with nuclease free water, 10 μL of RNA polymerase buffer (10×), 10 μL of rNTP-mix (25 nM each, Roche, Basel, Switzerland), 5 μL of 5′cap Analog, 2.5 μL of RNAse inhibitor (Promega, 40 U/μL, Madison, WI, USA), and 6 μL of SP6 polymerase (New England BioLabs, Ipswich, MA, USA). The mixture was incubated for 2 h at 37 °C after which 4 μL of SP6 polymerase was added and incubated for a further 2 h. The reaction was stopped by the addition of 7.5 μL of DNAse (Promega, 1 U/μL) and incubated for 30 min at 37 °C to digest the DNA template. The synthesized RNA was purified using NucleoSpin RNA Clean-up kit (Macherey-Nagel, Düren, Germany) and analyzed by agarose gel electrophoresis. Afterwards, the concentration was determined on a spectrophotometer and aliquots of 20 μg frozen at −80 °C.

To produce the CHIKV reporter viruses, 20 μg of the in vitro transcribed RNA was electroporated into 1×10^7 BHK-21 cells in Opti-Mem® (Gibco™). Electroporation was performed using a Gene Pulser (Bio-Rad, Hercules, CA, USA) at 250 V, two pulses at an interval of one second and 15 ms pulse length. Cells were immediately transferred into 10 mL of complete DMEM and seeded on 10-cm dishes. The supernatant containing CHIKV was collected 48 h post electroporation and cellular debris removed using a 0.45-μm pore size filter. The virus was concentrated by either ultracentrifugation through a 20% sucrose gradient or by use of 100 MW amicon tubes (Merck, Darmstadt, Germany). The supernatant was subsequently stored in small aliquots at −80 °C. Virus titers were assessed by flow cytometry (FACS) and luciferase assay on HEK293T cells and expressed as median Tissue Culture Infectious Dose (TCID$_{50}$).

2.8. Infection and Antibody Inhibition Assay

To determine susceptibility to CHIKV, cells were seeded on cell culture plates coated with poly-*L*-lysine (HEK293T) or uncoated (Huh7.5, HaCats, fibroblasts and CHO) at the densities indicated below. After an overnight culture, cells were transferred to biosafety level three lab and incubated with CHIKV at indicated MOI for four hours. The inoculum was replaced with fresh medium and cells incubated for the specified time points. Infection was determined by flow cytometry or luciferase assay depending on the virus used.

For antibody inhibition of CHIKV, HEK293T cells stably expressing TIM-1 WT were seeded in a 96-well plate at a density of 2×10^4 cells/well 24 h prior to the experiment. Cells were preincubated in media with the indicated concentration of anti-TIM-1 polyclonal antibody in DMEM complete. Identical concentrations of IgG isotype were used as control. After 30 min, the cells were inoculated with GFP tagged CHIKV at MOI of 0.01 for 4 h in the presence of the antibody, washed and incubated with culture medium. Susceptibility was analyzed by flow cytometry 24 h post infection.

To determine susceptibility to VEEV, parental HEK293T cells and cells expressing TIM-1 WT or Axl WT were seeded in 96-well plates (2×10^4 cells/well) coated with poly-*L*-lysine. After 24 h, cells were inoculated with virus at MOI of 0.001, 0.01, 0.1, and 1 and incubated for 4 h at 37 °C. The inoculum was removed and fresh DMEM supplemented with 10% FCS added. After 16 h of infection, cells were detached and GFP expression determined by flow cytometry.

To determine susceptibility to SINV, parental HEK293T cells and cells expressing TIM-1 WT or Axl WT were seeded in 6-well plates (1.3×10^6 cells/well) coated with poly-*L*-lysine. After 24 h, cells were inoculated with virus at MOI of 0.01 and 0.1 and incubated for 1 h at 37 °C. The inoculum was removed and fresh DMEM supplemented with 10% FCS added. The supernatant from the wells was collected after 24 h of infection and titrated at a 10-fold dilution on Vero cells seeded in 12-well plates (5×10^5 cells/well). After 1 h at 37 °C the virus was removed and carboxymethyl cellulose overlay added. After 48 h the wells were fixed with 4% PFA and stained with crystal violet solution. The titer was determined by counting visible plaques.

To determine susceptibility to HAdV-C5-GFP, parental HEK293T cells and cells expressing TIM-1 WT or Axl WT were seeded in black 96-well plates with transparent bottom (3×10^4 cells/well). Cells were washed twice with DMEM before addition of serial dilutions of HAdV-C5-GFP. After 1 h at 37 °C, the virus was removed and fresh DMEM supplemented with 2% FBS was added to the cells. Then, 24 h post infection the plates were fixed with 4% PFA for 10 min and GFP expression was imaged using a Trophos system (Luminy Biotech Enterprises, Marseille, France).

2.9. Virus Binding and Endosomal Escape Assays

Cells were suspended in binding buffer (DMEM supplemented with 20 mM HEPES, 1 mM calcium chloride and 0.2% human serum albumin, pH 7.4) with mCherry-fluorescent CHIKV (mc-CHIKV) at MOI of 50 and incubated at 4 °C on a shaker for 2 h. The cells were washed 3 times in binding buffer and fixed using 4% PFA. The cells suspended in FACS buffer were then analyzed for binding by flow cytometry.

For endosomal escape, cells inoculated with nano-luciferase CHIKV (nLuc-CHIKV) in DMEM complete medium were incubated on ice for one hour to allow maximum attachment and synchronized entry of virus. Cells were washed three times with DMEM to remove unbound nLuc-CHIKV. Binding efficiency between parental and TIM-1 WT/mutant expressing HEK293T cells was determined by luciferase assay. After removing unbound virus, fresh DMEM complete was added and cells were transferred to 37 °C to initiate particle uptake. At the indicated time points, medium supplemented with 20 mM ammonium chloride was added in order to prevent endosomal acidification and escape of the virus from endosomes. After 10 h of continuous incubation with ammonium chloride at 37 °C, the cells were lysed by freeze-thawing and productive infection assessed by determining the enzymatic activity of the newly translated luciferase after initial replication of the incoming viral genomes.

2.10. Confocal Microscopy and Live Cell Imaging

Live-cell imaging was carried out using a laser confocal spinning-disc coupled to a motorized Ti-E inverted microscope (Nikon, Tokyo, Japan) and equipped with a Yokogawa CSU-X1 5000 Spinning Disk Unit and an EMCCD camera iXon Ultra DU-888 (Andor Technologies, Belfast, Northern Ireland). Time-lapse movies were acquired using 60× and 100× objectives (NA = 1.49) and NIS-Elements AR DUO software and were recorded at an acquisition rate of 3 frames per second for 2 min.

Prior to imaging, cells were stained using a membrane permeable dye for living cells, Calcein AM (C3099, Invitrogen, Carlsbad, CA, USA) at a concentration of 0.2 μM for 10 min at 37 °C. While Calcein AM is usually used to probe cell viability, here it was used to identify the cells and make it possible to count them using confocal imaging. Cells were kept in a growth chamber (37 °C, 5% CO_2) for the entire acquisition time. Labeled mc-CHIKV viruses were immediately imaged after they were added to the cells and were excited using the 561 nm laser. Before and after each time-lapse recording, the cells and viruses were imaged using a 488 nm and 561 nm laser excitation to check for autofluorescence.

Internalized viral particles were quantified by quenching of the extracellular viral particles using Trypan Blue (Gibco). Twenty minutes after addition of mc-CHIKV to the cells, Trypan Blue (Gibco™) was added at the concentration of 0.4% to quench the extracellular viral particles. Zstack images of the cells after addition of Trypan Blue were recorded using a 488 nm and 561 nm laser excitation. The number of internalized particles per cell was then counted using Fiji and the multipoint tool. The number of cells used for the quantification of the bound and the internalized particles was determined using images taken in the green TIM-1 channel. Finally, the total number of virus particles and of intracellular virus particles was each divided by the number of cells.

2.11. Single Particle Tracking

Recorded movies of mc-CHIKV diffusion at the cell surface were processed and analyzed using TrackMate [59] and Matlab DC-MSS (Divide-and-Conquer Moment Scaling Spectrum) transient diffusion analysis [60]. First, the movies were pre-processed using Fiji by correcting uneven background using a rolling ball of 50 and by filtering the noise (despeckle). The virus trajectories were then reconstructed using TrackMate (ImageJ, 1.53j, University of Wisconsin, Madison, WI, USA) where the virus particles were detected with sub-pixel localization and linking of frame-to-frame displacement of 1 μm and a maximum gap of 2 μm and 20 frames. Aggregates and large particles were manually excluded from the analysis. Trajectories longer than 60 frames were then segmented and classified in Matlab using a built-in script and DC-MSS. Briefly, diffusion classification was done using the moment scaling spectrum (MSS) where high order moments of the displacement distribution are considered and the slope of the MSS reflects the motion type: a slope of 0.5 implies free normal diffusion, a slope between 0 and 0.5 yields anomalous motion, and a slope of 0 represents immobile particles. Trajectories were segmented depending on the motion type with a rolling-window of 21 frames. For each segment, diffusion properties such as the diffusion coefficient and confinement radius were extracted as detailed by Vega et al. [60]. Moreover, the time spent in immobile, anomalous, or free motion type was calculated by dividing the sum of the time spent in one motion type by the total time spent by all segments in all motion types.

2.12. Western Blot Analysis

Cells were washed three times using PBS and suspended for 30 min on ice in lysis buffer (1% Nonidet P40, 10% glycerol, 1 mM $CaCl_2$ in HEPES/NaCl) supplemented with

1% protease inhibitor (Sigma-Aldrich #P8340, Burlington, MA, USA). Supernatants were collected after centrifugation and total protein concentration determined by Bradford assay. Then, 25 µg of total protein was separated in reducing conditions by sodium dodecyl sulphate-polyacrylamide gel electrophoresis (SDS-PAGE). The proteins were transferred to polyvinylidene difluoride (PVDF) membrane (Bio-Rad) followed by blocking with 5% skimmed milk in PBS supplemented with 0.5% Tween 20 (PBS-T). The membranes were incubated with their respective primary antibody for one hour at room temperature. After washing three times using PBS-T, the membranes were incubated with the indicated horse radish peroxidase (HRP)-conjugated secondary antibodies. Following extensive washing, protein levels were detected using ECL Prime Western blot detection system (GE Healthcare, Chicago, IL, USA) and visualized using the ChemoStar Professional Imager System (Intas, Göttingen, Germany).

2.13. Luciferase Assay

Luciferase activity was determined as previously described [61] in cells inoculated with lentiviral pseudoparticles or authentic CHIKV and in cells electroporated with CHIKV subgenomic or full-length RNA. Firefly luciferase activity was measured by mixing 20 µL cell lysate with 72 µL firefly luciferase assay buffer [25 mM glycyl-glycine (pH 7.8), 15 mM KPO_4 (pH 7.8), 15 mM $MgSO_4$, 4 mM EGTA, 1 mM DTT and 2 mM ATP (pH 7.6)] and 40 µL of firefly luciferase substrate (0.2 mM D-luciferin in 25 mM glycyl-glycine). Nano-luciferase activity was measured by adding 80 µL of 1:1000 coelenterazine solution (0.42 mg/mL in methanol) to 20 µL of the lysate. Luciferase activity was measured in a plate luminometer (LB960 CentroXS3, Berthold technologies) in white luminometer 96-well plate.

2.14. Surface Staining and Flow Cytometry

The expression of TIM-1, Axl, and MXRA8 was analyzed by staining cells with anti-TIM-1, anti-Axl (R&D Systems, Minnneapolis, MN, USA), and anti-MXRA8 (JSR life sciences, Sunnyvale, CA, USA) monoclonal antibodies without prior fixation. The primary staining with unconjugated antibody was followed by secondary staining with either Alexa 488- or 647-conjugated anti-mouse/goat IgG antibody (ThermoFisher, Waltham, MA, USA). The respective IgG isotype was used as control. The expression of TIM-1 and Axl was analyzed by APC-conjugated monoclonal anti-TIM-1 (BioLegend®) and anti-Axl (R&D Systems, Minnneapolis, MN, USA) antibodies respectively. Appropriate APC-conjugated isotype control antibodies from BioLegend® and R&D Biosystems were used. All flow cytometry analyses in this study were performed using Sony Spectral Cell Analyzer (Sony Biotechnology, San Jose, CA, USA) and data analyzed by FlowJo V10 Software.

2.15. Statistical Analysis

Experiments were performed in at least three biological replicates, each carried out in technical triplicates unless otherwise specified. Results are plotted as mean ± standard error of mean (SEM) of three biological replicates unless otherwise indicated. Statistical analyses were performed in GraphPad Prism 8 (GraphPad Software, Inc., San Diego, CA, USA) using analysis of variance (ANOVA) followed by Dunnett's multiple comparison test. Statistical relevance for binding and internalization of CHIKV was calculated using Welch t-test. Statistical relevance was reached for $p \leq 0.05$ (*), $p \leq 0.01$ (**), $p \leq 0.001$ (***), and $p \leq 0.0001$ (****); $p > 0.05$ (ns) was considered non-significant.

3. Results

3.1. Ectopic TIM-1 Expression Enhances CHIKV Infection in HEK293T Cells

In order to investigate the role of wild type TIM-1 (TIM-1 WT) and Axl (Axl WT) in mediating CHIKV infection, we generated HEK293T cells stably expressing TIM-1 WT and Axl WT (Figure 1A and Supplementary Figure S2A). The immunoblot of TIM-1 (predicted molecular weight = 39.3 kDa) and Axl (predicted molecular weight = 98.3 kDa) confirmed expression of both proteins, however also revealed additional bands at higher molecular weight, which are most likely attributed to post-translational modifications. TIM-1 has several Ser, Thr, and Pro as well as Asn residues in the mucin domain that can be modified by O-linked or N-linked glycans, respectively [62]. Moreover, TIM-1 forms homodimers due to high affinity between residues in the IgV domains [63,64]. Axl mainly undergoes N-linked glycosylation [65]. The predicted Axl molecular weight of ~98.3 kDa may explain the lower band. However, the band with molecular weight of ~130 kDa is considered to be fully glycosylated and functional [66]. We inoculated cells with increasing multiplicities of infection (MOI) and analyzed susceptibility to the green fluorescent protein encoding CHIKV (3'GFP-CHIKV) by flow cytometry after 24 h. The cells that ectopically express TIM-1 WT were more susceptible to CHIKV, especially at MOI $\leq$ 0.1. At MOI of 0.01, TIM-1 WT expression increased susceptibility by 12-fold whereas Axl WT expression resulted in a two-fold increase (Figure 1B,C). Notably, TIM-1 WT and Axl WT expressing HEK293T cells were more susceptible to CHIKV (33-fold and 10-fold respectively) and Ebola virus (18-fold and five-fold respectively) glycoprotein-based lentiviral pseudoparticles demonstrating that the proteins are functional as previously reported [18,67–69] (Supplementary Figure S2B). Next, we analyzed the susceptibility of the cells to authentic CHIKV at different hours post infection (hpi). At 24 hpi, TIM-1 WT expressing cells were 69% positive for CHIKV while parental cells and Axl WT expressing cells were 11% and 21% respectively positive for CHIKV (Figure 1C). To evaluate the dependence of different CHIKV genotypes on TIM-1 WT, we inoculated cells with strains of ECSA (MOI = 0.01), West African (WA, MOI = 0.01), and Asian (181/25, MOI = 0.1) genotypes. In comparison to the control cells, TIM-1 WT expression in HEK293T cells consistently enhanced the infection with all tested CHIKV strains (ECSA: 28-fold, WA: 14-fold, and 181/25: 10-fold) (Figure 1D). Moreover, we observed a dose-dependent inhibition of 3'GFP-CHIKV infection of TIM-1 expressing cells using an anti-TIM-1 polyclonal antibody (α-TIM-1 Ab) while an isotype IgG control antibody slightly affected CHIKV infection, but not in a dose-dependent manner (Figure 1E).

In order to establish the role of TIM-1 in the presence or absence of glycosaminoglycans (GAGs), we expressed human TIM-1 in Chinese hamster ovary (CHO) cells with (CHOK1) and without GAGs (CHO745) [70,71]. We inoculated CHO cells with strains of ECSA (MOI = 0.01), WA (MOI = 0.01), and Asian (MOI = 0.1) genotypes. In comparison to control cells, the expression of TIM-1 in CHOK1 cells increased susceptibility by 11-fold (ECSA), 21-fold (WA), and 15-fold (181/25) whereas a ~7-fold increase was observed in CHO745 for all tested genotypes. CHO745 cells expressing TIM-1 were generally approximately two-fold less susceptible to CHIKV than CHOK1 expressing TIM-1 (Supplementary Figure S3A) despite similar susceptibility of parental cells lacking TIM-1 (Supplementary Table S1). This suggests that the observed TIM-1-dependent enhancement of CHIKV infection is slightly modulated by the expression of GAGs.

To test if other alphaviruses use TIM-1 and/or Axl as host factor, we inoculated HEK293T cells expressing TIM-1 and Axl with Venezuelan eastern equine encephalitis virus (VEEV) and with Sindbis virus (SINV) at increasing MOI. In comparison to the parental cells, TIM-1 expression enhanced VEEV infection by 1.4-fold at a MOI of 0.1 (Figure S3B). SINV infection increased four-fold upon TIM-1 expression as measured by infectious particle release (Supplementary Figure S3C). As a control, we challenged the cells with serially diluted human adenovirus-5 (HAdV-5), a non-enveloped virus. Infection of the cells was independent of TIM-1 and Axl (Supplementary Figure S3D). Taken together, these observations indicate that unlike Axl, TIM-1 expression enhances infection of different CHIKV genotypes and alphaviruses. However, in comparison to CHIKV, VEEV and SINV appear to be less dependent on TIM-1.

In order to establish if TIM-1 has a role in the replication of CHIKV, we generated HEK293T cells expressing TIM-1 wild type (TIM-1 WT), TIM-1 with a double mutation in the PtdSer-binding pocket (TIM-1ΔMIL, N114A, and D115A), and TIM-1 lacking the cytoplasmic domain (TIM-1ΔCyt). We then compared the replication in parental cells to TIM-1 WT, TIM-1ΔMIL, and TIM-1ΔCyt by electroporating CHIKV subgenomic RNA encoding for nano-luciferase. Our findings showed similar RNA replication between the cells as determined by luciferase assay (Supplementary Figure S4A). Furthermore, we electroporated the cells with the full length CHIKV RNA encoding nano-luciferase gene (nLuc-CHIKV) to confirm the observation made by electroporating subgenomic RNA. Infection was stopped at 4, 6, 8, and 10 h timepoints to avoid re-infection by de novo virus. By use of the luciferase assay, we observed luciferase activity across the cell variants (Supplementary Figure S4B), suggesting that TIM-1 has no influence on the replication of CHIKV in HEK293T cells. Collectively, this finding implies that the role of TIM-1 in CHIKV infection may be at the level of binding and entry.

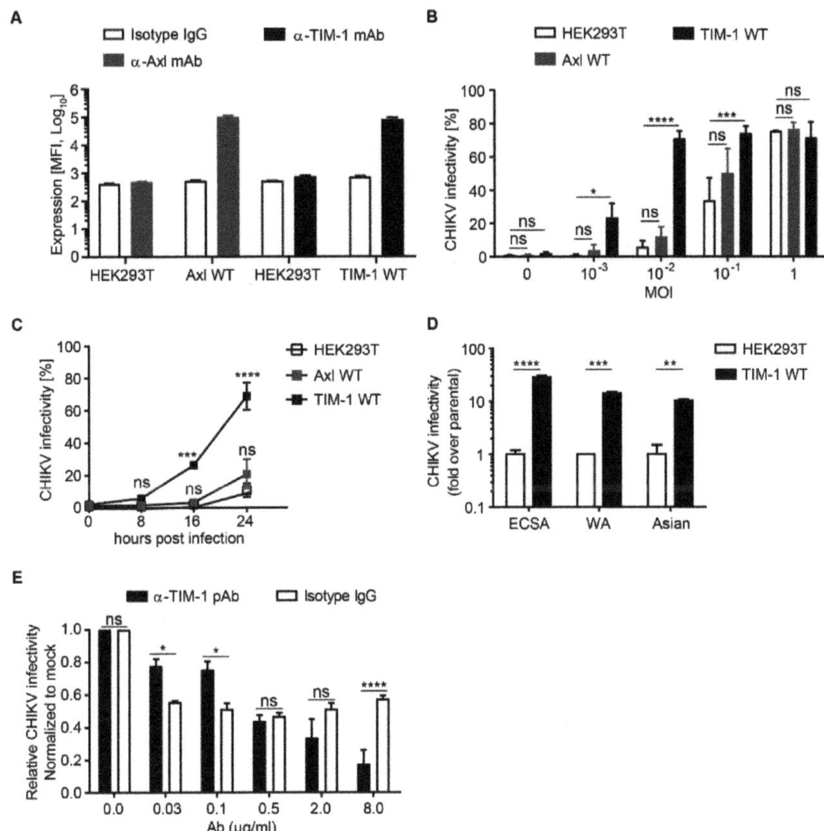

Figure 1. Ectopic expression of TIM-1 enhances CHIKV infection (**A**) Surface expression of Axl and TIM-1 wild type (WT) proteins on parental HEK293T and stably transduced cells was evaluated by monoclonal antibody staining and flow cytometry. Cells stained with an isotype IgG were used as control. (**B**) Parental, Axl WT and TIM-1 WT expressing cells were challenged with ECSA 3'GFP-CHIKV at indicated MOI and (**C**) for different infection durations at MOI of 0.01. Infection levels were assessed by flow cytometry and plotted as percentage of GFP positive cells. (**D**) Parental and TIM-1 WT expressing cells were challenged with CHIKV strains of ECSA 3'GFP-CHIKV (MOI = 0.01), WA 5'GFP-CHIKV (MOI = 0.01) and Asian mc-CHIKV (MOI = 0.1) genotypes and infection assessed by flow cytometry. (**E**) HEK293T cells expressing TIM-1 were pre-incubated for 30 min with increasing concentrations of TIM-1 polyclonal antibody (black bars) or isotype control antibody (mock, white bars) before inoculation with ECSA 3'GFP-CHIKV at MOI of 0.01. After 4 h the cells were washed and infection levels analyzed by flow cytometry 20 h later as in (**B**) and (**C**). Error bars represent standard error of the mean (SEM) of three biological replicates. Statistical significance was calculated using a Dunnet's multiple comparisons test (2way ANOVA) ns > 0.05, * $p < 0.05$, ** $p < 0.01$, *** $p < 0.001$ and **** $p < 0.0001$.

3.2. The TIM-1 Phosphatidylserine-Binding Domain Is Crucial for TIM-1-Dependent Infection

Next, we characterized the role of intra- and extracellular domains of TIM-1 in enhancing CHIKV binding, uptake, and infection using HEK293T cells expressing TIM-1 WT, TIM-1ΔMIL, and TIM-1ΔCyt (Figure 2A). We stained TIM-1 WT and TIM-1 deletion mutant expressing cells with antibodies against the ectodomain and sorted cell populations with similar TIM-1 surface expression levels (mean fluorescent intensity, MFI) (Figure 2B). We challenged the parental HEK293T cells and cells expressing TIM-1 WT, TIM-1ΔMIL, or TIM-1ΔCyt with ECSA (MOI 0.01), WA (MOI 0.01), and Asian (MOI 0.1) strains of CHIKV. After 24 h, infection levels in cells expressing the ectodomain mutant (TIM-1ΔMIL) were similar to those of parental cells (Figure 2C). Conversely, we observed increased infection levels in cells expressing TIM-1 WT and TIM-1ΔCyt. Expression of TIM-1 WT increased infection with ECSA, WA and Asian (181/25) strains by 25-fold, 21-fold and five-fold, respectively while expression of TIM-1ΔCyt increased infection by 21-fold, 17-fold and seven-fold, respectively (Figure 2C). Notably, infectivity of the Asian vaccine strain (181/25) was lower in comparison to ECSA and WA stains. The lower infectivity is attributed to attenuation due to substitution of two amino acids at positions 12 and 82 in the E2 envelope glycoprotein responsible for receptor binding [52,72]. We observed similar infectivity between cells expressing TIM-1 WT and TIM-1ΔCyt, implying that the cytoplasmic domain of TIM-1 is dispensable for CHIKV infection of HEK293T cells. To further characterize the role of TIM-1 ectodomain in CHIKV infection, we determined the competence of parental cells and cells expressing TIM-1 WT or TIM-1ΔMIL to bind and internalize mCherry-fluorescent CHIKV (mc-CHIKV, Asian genotype). The mc-CHIKV presents the mCherry on the virion surface due to a mCherry-E2 fusion and can be detected by flow cytometry or confocal microscopy. After two hours of CHIKV binding on ice to avoid internalization, we assessed cell bound virus particles by flow cytometry. In comparison to the parental cells, the expression of TIM-1 WT or TIM-1ΔMIL resulted in increased binding of CHIKV (Figure 2D). This increase was more pronounced for TIM-1 WT than for the ectodomain mutant, however this observation did not reach statistical significance in this assay. Imaging of CHIKV binding within the first 20 min of a live cell confocal experiment at 37 °C further confirmed that cells expressing TIM-1 WT bind 2-fold more CHIKV particles compared to TIM-1ΔMIL and this observation reached statistical significance (Figure 2E). Collectively, our findings here indicate that CHIKV depends on the phosphatidylserine binding domain of TIM-1 for efficient binding and infection.

3.3. Single Particle Tracking of CHIKV Confirms PtdSer Domain Requirement

To characterize the effect of the PtdSer binding site on the diffusive behavior of CHIKV on the cell membrane, we used single particle tracking of live cells. The mc-CHIKV particles were added to the cells immediately prior to imaging and movies were recorded at three frames per second for two minutes. A total of N_{virus} = 1523 and 472 virus particles for TIM-1 WT and TIM-1ΔMIL respectively were analyzed and a representative trajectory of CHIKV on TIM-1 WT cells is shown in Figure 3A, illustrating the lateral diffusion of the virus at the cell surface (movies. SA-B). The extracted trajectories were further analyzed by segmenting each track depending on the three-diffusion types: immobile and the mobile motions, anomalous confined and free motion (Figure 3B,C corresponding to the track shown in Figure 3A). The average segment length ranged between 12 and 30 seconds with the shortest being six seconds. While in all cases the virus spent 8% and 7% (for WT and mutant respectively) of the time being immobile, for mobile particles, the characteristics of virus diffusion at the cell surface were affected by the mutation in the PtdSer binding site. Indeed, upon mutation of the TIM-1 MILIBS, CHIKV spent less time diffusing freely (35% vs. 48% for the TIM-1 WT cells) (Figure 3D), albeit with a higher diffusion coefficient (Figure 3E). The confinement radius also increased slightly (Figure 3F), in spite of the fact that the anomalous diffusion coefficient remained unaffected. Together, these results reveal that the PtdSer binding site of TIM-1 contributes to modulating the lateral virus Brownian movement at the plasma membrane by reducing the diffusion coefficient of the free motion and the area of virus diffusion at the cell surface, leading to a more confined motion, which could benefit virus internalization.

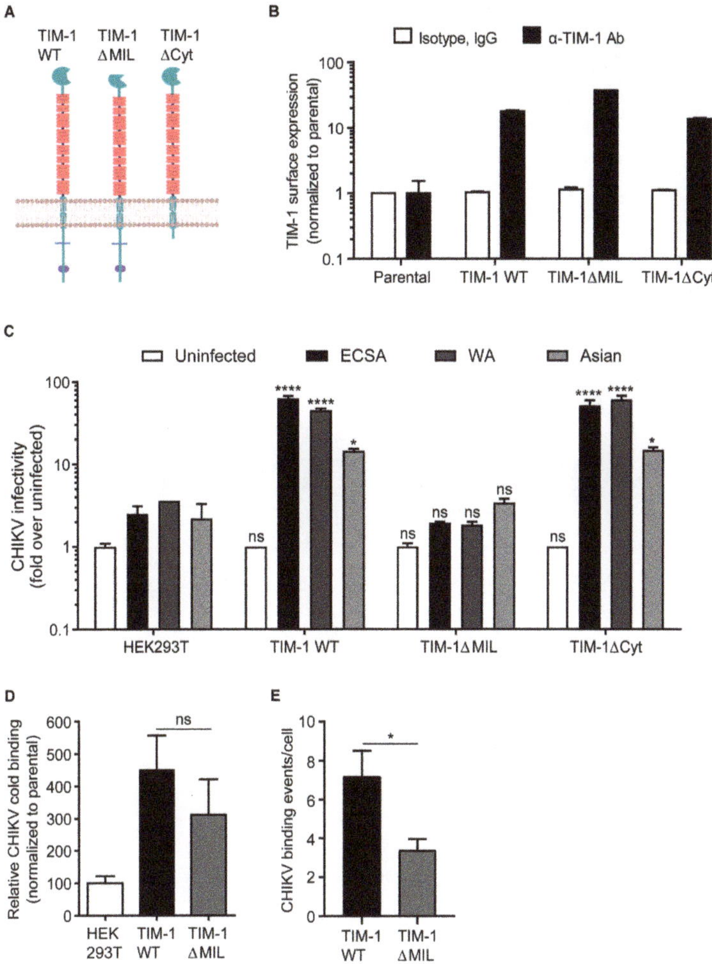

Figure 2. Phosphatidylserine-binding domain of TIM-1 is crucial for CHIKV entry and infection (**A**) Schematic representation of TIM-1 WT, TIM-1ΔMIL with a mutation in the phosphatidylserine binding site and TIM-1ΔCyt that lacks part of the cytoplasmic domain (made with BioRender. com). (**B**) Expression levels of TIM-1 in parental and HEK293T cells transduced with lentiviral pseudoparticles to stably express TIM-1 WT, TIM-1ΔMIL or TIM-1ΔCyt assessed as in Figure 1. (**C**) Parental HEK293T cells expressing TIM-1 WT, TIM-1ΔMIL and TIM-1ΔCyt were challenged with CHIKV strains of ECSA 3′GFP-CHIKV (MOI = 0.01), WA 5′GFP-CHIKV (MOI = 0.01) or Asian mc-CHIKV genotypes (MOI = 0.1) and infection assessed by flow cytometry at 24 hpi. (**D**) Cold binding of fluorescent mc-CHIKV at MOI of 50. After two hours and extensive washes, cells were fixed and analyzed by flow cytometry. (**E**) Live cell imaging of CHIKV binding. Cells were inoculated with fluorescent Asian mc-CHIKV at MOI of 50 and monitored by confocal microscopy. Number of CHIKV binding events to the plasma membrane within 20 min were counted. The error bars represent standard errors of the mean (SEM). Statistical significance was calculated using Dunnet's multiple comparisons test (2way ANOVA) (**C**) and Welch *t*-test (**E**) with ns > 0.05, * $p < 0.05$ and **** $p < 0.0001$.

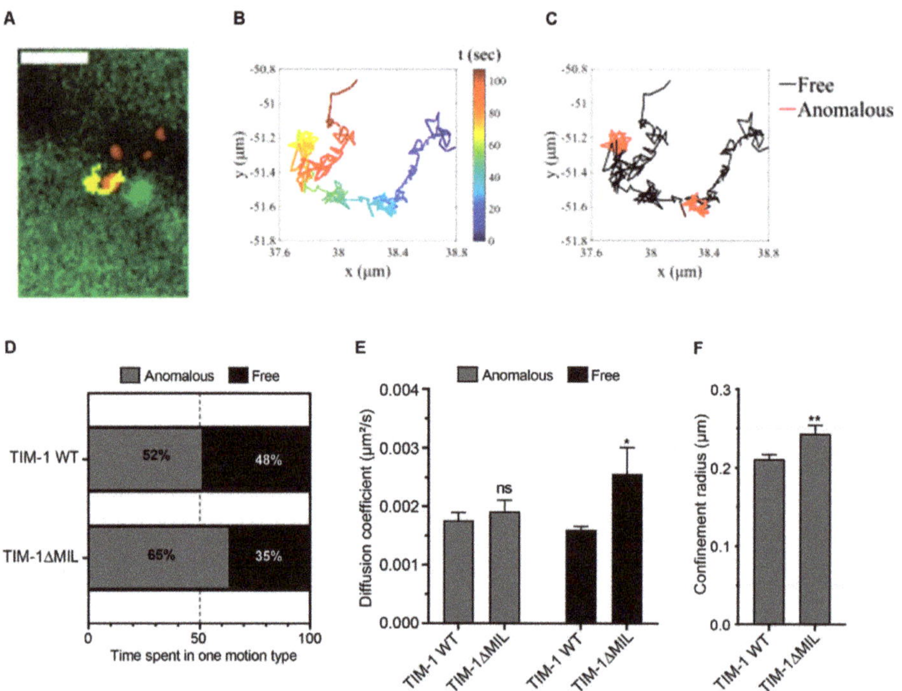

Figure 3. Single particle tracking of CHIKV confirms PtdSer domain requirement. (**A**) Overlay of the fluorescence image of CalceinAM stained TIM-1 WT cells (green) and labeled Asian mc-CHIKV particles (red) with the virus diffusion trajectory (yellow) at the cell surface. The bar represents 2 µm. (**B**) The CHIKV trajectory shown in (**A**) as a time-lapse with the time in seconds presented as a color bar (**C**) Segmentation of the trajectory shown in (**B**) and classification of the segments using moment scaling spectrum (red: anomalous confined motion—black: Brownian free motion). (**D**) The fraction of time spent by the mobile particles either in confined anomalous or Brownian free motion. The total number of viruses analyzed here is $N_{virus} = 1523$, 472 for TIM-1 WT and TIM-1 ΔMIL respectively. (**E**) The mean of the diffusion coefficient for anomalous and free motion of each segment is calculated using MSS as described in methods. (**F**) The confinement radius of the anomalous motion is presented for each cell type. The error bars represent standard errors of the mean (SEM) with $N_{segments} = 1296$, 414 for TIM-1 WT and TIM-1ΔMIL respectively. Statistical significance was calculated using a Welch *t*-test with ns > 0.05, * $p < 0.05$ and ** $p < 0.01$. For D, the percentage of fraction was obtained by counting, so no statistical analysis is provided.

3.4. Entry Kinetics of CHIKV Are Altered by TIM-1

Next, we aimed to dissect the role of TIM-1 protein domains in the initial steps of the CHIKV infection cycle. After binding to the cell surface, CHIKV is thought to be primarily internalized by clathrin-mediated endocytosis and reach endosomes where low pH-dependent membrane fusion occurs and the nucleocapsid is released (endosomal escape) [40,41]. Eventually, the capsid dissociates and the released genome undergoes translation and replication [9]. Addition of medium supplemented with 20 mM of lysoso-motropic ammonium chloride (NH_4Cl) raises the pH in the endosomes hence blocking the endosomal escape of viruses [73–75]. To that end, we compared the CHIKV endosomal escape kinetics in HEK293T cells expressing either TIM-1 WT, TIM-1ΔMIL, or TIM-1ΔCyt.

We inoculated cells with CHIKV carrying a nano-luciferase fused E2 glycoprotein (nLuc-CHIKV) and synchronized virus binding for one hour in the cold. After extensive washes, we determined binding efficiency by luciferase assay, making use of the virion incorporated nano-luciferase E2 fusion protein (Supplementary Figure S5). To initiate CHIKV internalization and endosomal escape, we then shifted the cells to 37 °C. At the indicated timepoints, we exposed the cells to 20 mM NH_4Cl to raise the endosomal pH and prevent membrane fusion. After 10 h at 37 °C, we quantified the enzymatic activity of the luciferase E2 fusion protein translated from newly replicated CHIKV genomes as a measure for productive infection (Figure 4A). In comparison to the parental cells, TIM-1 WT and TIM-1ΔCyt expressing cells more efficiently bound nLuc-CHIKV by three-fold (Supplementary Figure S5). To determine TIM-1-dependent entry/endosomal escape kinetics, and to exclude the role of other cellular factors, we normalized the data to that of the parental cells at each time point. We observed a TIM-1 dependent enhancement of entry kinetics after 20 min and for all subsequent timepoints. We also observed enhanced endosomal escape as compared to parental cells in cells expressing TIM-1ΔCyt and this became apparent at one hour post temperature shift. In contrast, cells expressing TIM-1ΔMIL showed no significant enhancement of endosomal escape as compared to parental cells (Figure 4B and Supplementary Table S2). Thus, although cells expressing either TIM-1 WT or TIM-1ΔCyt bound equal numbers of CHIKV particles (Supplementary Figure S5), TIM-1ΔCyt expressing cells needed more time to internalize CHIKV in comparison to TIM-1 WT expressing cells. Similarly, after 20 min of incubation at 37 °C, we observed by confocal microscopy that the number of internalized mc-CHIKV virions in TIM-1 WT expressing cells was three-fold higher than the number of virions in cells expressing TIM-1ΔMIL (Figure 4C). Altogether, these results indicate that, in addition to the TIM-1 PtdSer binding domain, the cytoplasmic domain may modulate TIM-1-dependent CHIKV entry kinetics.

3.5. TIM-1 Expression Renders Keratinocyte Derived HaCat Cells Permissive to CHIKV

The skin is the primary entry point of CHIKV. Human epidermal keratinocytes express Axl and TIM-1 in the stratum basale layer [37,38] and are susceptible to CHIKV [20]. In order to determine the role of Axl and TIM-1 in CHIKV infection of the skin, we used HaCat cells, a derivative of immortalized keratinocytes that acts as a relevant model to study keratinocytes in vitro [39]. However, surface staining using monoclonal antibodies and flow cytometry revealed that HaCat cells endogenously express modest levels of Axl (Figure 5A), but no detectable levels of TIM-1 (Figure 5B). Hence, we used lentiviral pseudoparticles to generate cells stably expressing Axl WT, Gas6 binding site mutant (Axl E66R_T84R), tyrosine kinase domain mutant (Axl K567A), Axl lacking the cytoplasmic domain (AxlΔCyt), and Axl with a naturally occurring single nucleotide polymorphism (SNP, Axl R295W) (Figure 5A and Supplementary Figure S6A). We also generated cells expressing TIM-1 wild type (TIM-1 WT), TIM-1 with a mutation in the PtdSer-binding pocket (TIM-1ΔMIL including N114A and D115A), TIM-1 lacking the cytoplasmic domain (TIM-1ΔCyt), TIM-1 with single and double mutations (K338A, K346A, and TIM-1ΔUbi) in the cytoplasmic ubiquitination motif, TIM-1 lacking the cytoplasmic domain (TIM-1ΔCyt), and TIM-1 with a naturally occurring single nucleotide polymorphism (SNP) in the Ig-V domain (TIM-1 S51L) (Figure 5B and Supplementary Figure S6B). We detected expression of all Axl and TIM-1 variants on the surface of HaCat cells. Expression levels were comparable with the exception of the Axl E66R_T84R and the TIM-1-1ΔCyt, which displayed slightly reduced expression as compared to the respective WT protein.

In order to establish if Axl and TIM-1 have a role in CHIKV entry into HaCat cells, we transduced the cells with lentiviral pseudoparticles (pp) decorated with glycoproteins of CHIKV (CHIKVpp) and VSV (VSVGpp). In comparison to parental HaCat cells, which did not support CHIKV pseudoparticle entry, Axl WT enhanced CHIKVpp and VSVGpp entry into HaCat cells by ~two-fold while all the Axl mutant variants did not enhance entry (Supplementary Figure S7A). This was in contrast to the ten-fold CHIKVpp increase in entry observed in HEK293T cells (Supplementary Figure S2B). TIM-1 WT expression enhanced pp entry (five-fold and nine-fold for CHIKVpp and VSVGpp, respectively), as did the TIM-1 mutants TIM-1ΔUbi, TIM-1 K338R and TIM-1 K346R (five-fold and six-fold for CHIKVpp and VSVpp respectively), TIM-1ΔCyt (two-fold for both CHIKVpp and VSVGpp) and TIM-1 S51L (three-fold and five-fold for CHIKVpp and VSVGpp respectively). There was no entry enhancement of CHIKVpp or VSVGpp entry into cells expressing the PtdSer-binding pocket mutants; TIM-1ΔMIL, N114A and D115A (Supplementary Figure S7B). Unlike expression in HEK293T cells, TIM-1 WT expression in HaCat cells enhanced VSVGpp entry implying that VSV dependence of TIM-1 is cell-type specific. Our results suggest that the ubiquitination motif of TIM-1 is dispensable for entry into HaCat cells whereas the cytoplasmic domain in general, the PtdSer-binding pocket and the naturally occurring ectodomain SNP (TIM-1 S51L) are required for the full CHIKV entry factor function of TIM-1 in HaCat cells.

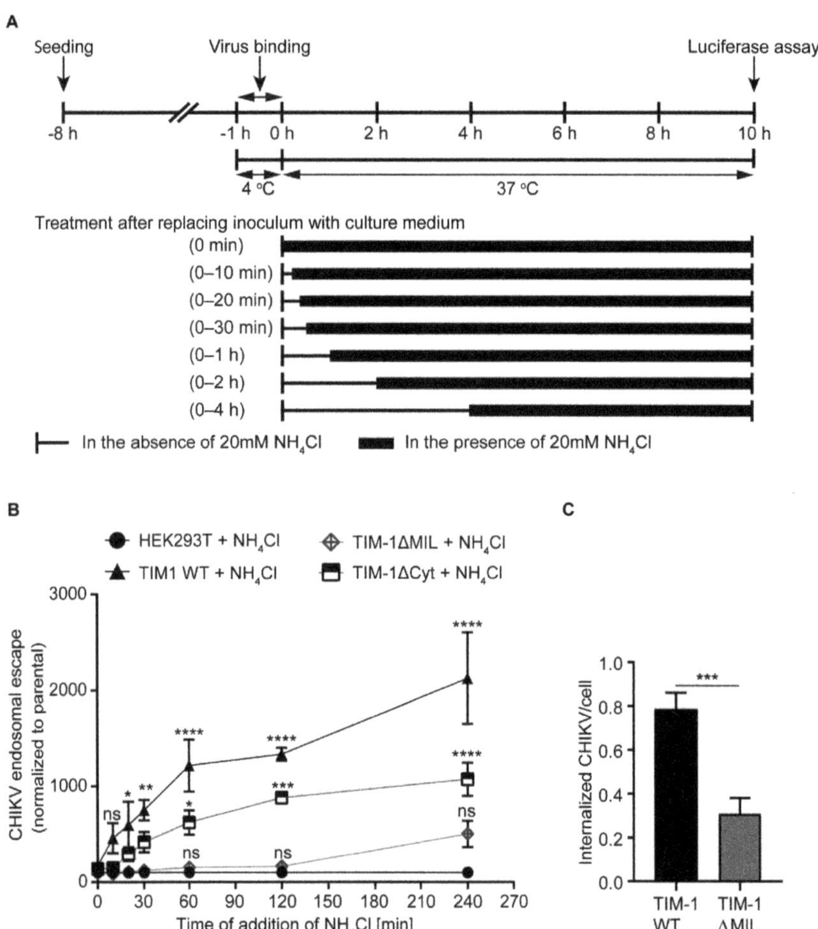

Figure 4. Entry kinetics of CHIKV are altered by TIM-1. (**A**) Scheme showing binding and entry assay. CHIKV (Asian genotype) encoding nano-luciferase fused to E2 glycoprotein (nLuc-CHIKV) was added to parental and TIM-1 expressing HEK293T cells, which were subsequently incubated at 4 °C for 1 h to synchronize binding. After washing, cells were transferred to 37 °C and medium with 20 mM NH_4Cl was added at indicated time points. Assay was stopped after 10 h of incubation at 37 °C and relative endosomal escape determined by luciferase assay. (**B**) Entry kinetics of nLuc-CHIKV in the indicated cell lines normalized to parental HEK293T cells at each time point. (**C**) Live cell imaging of internalized CHIKV. Cells were inoculated with fluorescent Asian mc-CHIKV at MOI of 50 and monitored by confocal microscopy. After 20 min of live imaging of fluorescent mc-CHIKV (Asian genotype) and cells at 37 °C, trypan blue was added to quench extracellular particles and only internalized viruses were imaged and counted using ImageJ. The error bars represent standard errors of the mean (SEM). Statistical significance was calculated using Dunnet's multiple comparisons test (2way ANOVA) (B) and Welch *t*-test (C) with * $p < 0.05$, ** $p < 0.01$, *** $p < 0.001$ and **** $p < 0.0001$.

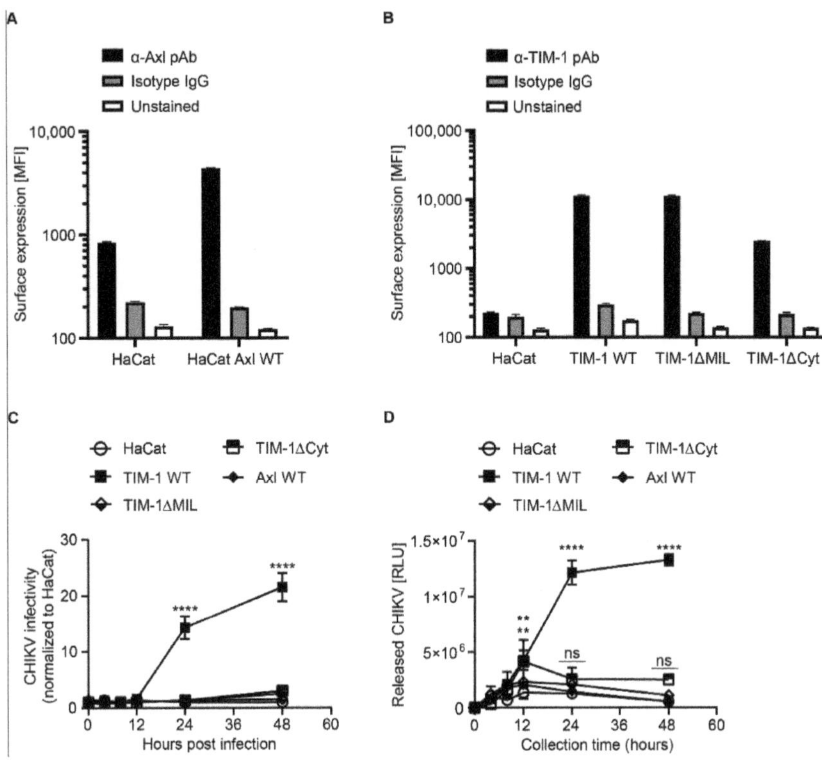

Figure 5. TIM-1 renders HaCat cells permissive to CHIKV. (**A**) Cell surface expression of Axl in HaCat cells with and without ectopic Axl expression (Axl WT) analyzed by antibody staining and flow cytometry. (**B**) Cell surface expression of ectopic TIM-1 WT and TIM-1 mutants in HaCat cells analyzed by antibody staining and flow cytometry. (**C**) Parental HaCat immortalized keratinocytes and HaCat cells expressing either TIM-1 WT, TIM-1ΔMIL, TIM-1ΔCyt or Axl WT were inoculated with Asian genotype nano-luciferase CHIKV reporter virus. After four hours, the cells were washed extensively to remove unbound virus. Expression of nano-luciferase attesting for viral replication was monitored over time using luciferase assay. Permissive fold change relative to parental HaCat cells at the indicated time points post infection is shown. (**D**) Released progeny virions in culture supernatants from (**C**) were used to inoculate human dermal fibroblasts and infection levels determined at 24 h post infection. The error bars represent mean ± SEM of three independent experiments. Statistical significance was calculated using a Dunnet's multiple comparisons test (2way ANOVA) ns > 0.05, ** $p < 0.01$ and **** $p < 0.0001$.

We next challenged the Axl and TIM-1 expressing HaCat cells with authentic nLuc-CHIKV and determined infectivity by luciferase assay. In comparison to the parental cells, which were refractory to CHIKV infection, HaCat cells expressing TIM-1 WT were 14-fold and 22-fold more permissive to nLuc-CHIKV at 24 hpi and at 48 hpi, respectively (Figure 5C and Supplementary Table S1). Interestingly, neither the expression of TIM-1ΔMIL, TIM-1ΔCyt, nor Axl rendered HaCat cells susceptible. In contrast to HEK293T cells, where the cytoplasmic domain of TIM-1 was dispensable for CHIKV infection (Figure 2C), the cytoplasmic domain of TIM-1 was required for infection of HaCat cells (Figure 5C and Supplementary Table S1). In order to establish if infection of HaCat cells was productive, we collected HaCat cell supernatants at the indicated time points post infection, inoculated human dermal fibroblasts with the supernatants for 24 h and determined CHIKV infection of the fibroblasts by luciferase assay. Fibroblasts are known to be permissive to CHIKV and the infection is MXRA8-dependent [20], however our data cannot exclude a role for Axl in this cell line. In comparison to the parental cells, expression of TIM-1 WT, resulted in a 10-fold and 24-fold higher release of infectious CHIKV particles at 24 hpi and 48 hpi, respectively. Expression of TIM-1ΔCyt and TIM-1ΔMIL in HaCat cells resulted in a five-fold and two-fold higher release of CHIKV compared to parental cells. The expression of Axl did not yield infectious CHIKV particle release (Figure 5D). Altogether, these findings underpin that both the ectodomain and the cytoplasmic domain of TIM-1 complement each other and play a role in CHIKV infection. The findings also demonstrate the cell type specific dependence of CHIKV on TIM-1.

3.6. Endogenous TIM-1 Mediates CHIKV Infection of Hepatoma Cells

Chikungunya virus has a wide tissue and cellular tropism and previous reports indicate that it infects the liver [76,77]. Hence, we determined the expression levels of TIM-1 and Axl in Huh7.5 cells—a human hepatoma derived cell line. We found that Huh7.5 cells predominantly express TIM-1 while Axl expression is negligible (Figure 6A and Supplementary Figure S8). To analyze the role of TIM-1 in CHIKV infection of Huh7.5 cells, we used a pool of three siRNAs to specifically knockdown TIM-1, MXRA8, a known factor for CHIKV entry (as positive control) [20,22], or both (Supplementary Figure S8D). The viability of cells transfected with a pool of siRNA was similar to control cells transfected with a non-targeting (NT) siRNA (Figure 6B). 48 h post silencing, the surface expression (mean fluorescent intensity, MFI) of TIM-1 and MXRA8 was reduced by four-fold and two-fold, respectively (Figure 6C). In comparison to cells transfected with NT siRNA, we observed that siRNA mediated silencing of TIM-1 reduced susceptibility to CHIKV by ~25%. Knockdown of MXRA8 resulted in ~60% reduction in susceptibility to CHIKV similar to simultaneous silencing of TIM-1 and MXRA8 (Figure 6D). Together, these findings show that, in the presence of MXRA8, endogenous TIM-1 plays a role in CHIKV infection of Huh7.5 cells.

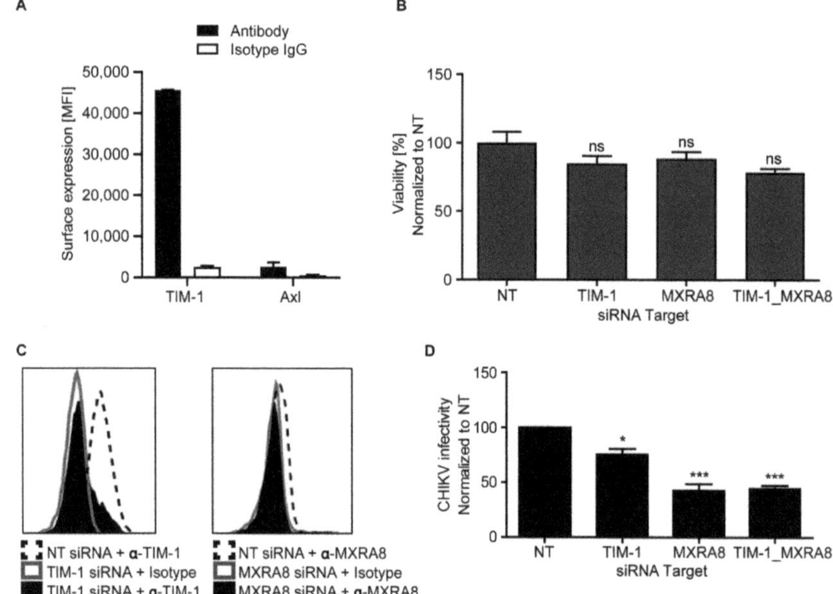

Figure 6. Endogenous expression of TIM-1 facilitates CHIKV infection of hepatocytes. (**A**) Endogenous surface expression of TIM-1 and Axl in Huh7.5 cells. (**B**) MTT assay-based viability of Huh7.5 cells. Cells were re-seeded in 96-well plate 48 h after treatment with targeting and non-targeting (NT) siRNA and incubated overnight. Afterwards, MTT assay was performed to compare proliferation of cells. (**C**) TIM-1 and MXRA8 surface expression 48 h after siRNA treatment measured by antibody staining followed by flow cytometry. One representative dataset shown. (**D**) Huh7.5 cell susceptibility to CHIKV after treatment with targeting and NT siRNA. Cells were inoculated with 5′GFP CHIKV (WA genotype) and infectivity determined

by flow cytometry 24 hpi. The error bars represent mean ± SEM of at least three independent experiments. Statistical significance was calculated using a Dunnet's multiple comparisons test (2way ANOVA) ns > 0.05, * $p < 0.05$ and *** $p < 0.001$.

4. Discussion

Phosphatidylserine-binding proteins, including TIM-1 and Axl receptor tyrosine kinase, are important host factors for a number of viruses [18,33,34,78–80]. In the present study, we demonstrated that human TIM-1 plays a key role in CHIKV infection of human cells. In contrast to findings in previous studies that used CHIKV glycoprotein pseudotyped viruses, we observed negligible enhancement of susceptibility to authentic CHIKV in cells expressing Axl [19,80]. TIM-1 associates with PtdSer on virions through the immunoglobulin-like variable (Ig-V) domain whereas Axl interacts through a growth arrest-specific 6 ligand (Gas6) or protein-S to enhance viral entry, which may explain the differential use of both proteins observed in this study [18,33,34,79].

HEK293T cells expressing TIM-1 exhibited a dose-dependent reduction in susceptibility to CHIKV when pre-treated with a TIM-1 specific antibody. This is an indication that the ectodomain of TIM-1 is critical for its function in CHIKV infection. Furthermore, ectopic expression of TIM-1 in HEK293T cells enhanced binding and subsequent internalization of CHIKV particles. The PtdSer-binding pocket also known as metal ion ligand-binding site (MILIBS) is conserved across all TIMs [81]. We mutated the MILIBS in human TIM-1 by replacing Asn and Asp residues (N114A and D115A) or both (TIM-1ΔMIL) with Ala at positions 114 and 115 respectively. We observed that the HEK293T cells expressing TIM-1ΔMIL similarly bind CHIKV as cells expressing TIM-1 WT when incubated with the mCherry-fluorescent virus on ice and detected by flow cytometry. This suggests that apart from the MILIBS, other residues within TIM-1 or other factors on the cell surface contribute to CHIKV binding [82]. However, when CHIKV was added to the cells in normal medium and immediately observed by confocal microscopy at 37 °C, CHIKV bound more efficiently to cells expressing TIM-1 WT in comparison to TIM-1ΔMIL. Consistently, we observed that subsequent internalization and infection was completely hampered in cells expressing TIM-1ΔMIL. According to our findings, the residues in the MILIBS are essential for TIM-1-mediated CHIKV infection and this is in line with previous reports [18,33,34,80]. Ectodomain residues outside the MILIBS were found to be less important in Dengue virus (DENV) infection [34]. Conversely, EBOV infection is additionally mediated by the direct interaction between viral glycoprotein (GP) and TIM-1 ectodomain residues outside the MILIBS but within the Ig-V domain [34,83,84]. Our study shows that TIM-1 residues in the Ig-V domain other than in the MILIBS are also needed for the entry factor function of TIM-1 in the context of CHIKV infection. Specifically, we observed reduced CHIKV pseudoparticle entry in HaCat cells expressing TIM-1-S51L in comparison to cells expressing wild type TIM-1. This observation suggests that TIM-1 residues outside the MILIBS may play a role in CHIKV infection implying that different viruses use a distinct set of TIM-1 residues for infection. A better understanding of the molecular interaction between amino acid residues in the IgV domain of TIM-1, PtdSer and viral glycoproteins may help in the development of antiviral factors. For instance, Song et al. recently developed a reagent that specifically binds PtdSer and/or phosphatidylethanolamine and could inhibit ZIKV infection [85]. Analysis of the diffusive behavior of CHIKV upon binding to the cell surface not only confirms that the PtdSer binding site promotes virus binding but also influences its diffusive behavior. Specifically, expression of TIM-1 with an intact PtdSer binding site leads to a decrease in the diffusion coefficient of the virus as well as the area of diffusion, indicating that the virus binds to TIM-1 WT directly or indirectly within membrane protein complexes. CHIKV diffusion may be slowed down by assembly of a higher molecular weight complex. This is in line with a TIRF microscopy study which reported that 76% of TIM-1-GFP spots on HeLa cells are confined, 19% transported, and 5% diffusive [37]. The majority of TIM-1-GFP tracked for Dengue virus internalization displayed confined displacement at the plasma membrane [37]. As CHIKV diffusion coefficient and confinement radius increased for TIM-1ΔMIL expressing cells, one can hypothesize that the

PtdSer binding site stabilizes the interaction of CHIKV with TIM-1 protein complexes. This further highlights the role of the PtdSer binding site in the attachment of CHIKV to the cell membrane.

The hampered entry and infection in cells expressing TIM-1ΔMIL may signify that the MILIBS is involved in signaling for internalization of CHIKV, possibly by mediating interaction with signaling receptors. Our data suggest that TIM-1 signaling through the cytoplasmic tail is dispensable for CHIKV entry into HEK293T cells as the cytoplasmic tail deletion of TIM-1 (TIM-1ΔCyt) does not impact CHIKV infection at a later time point. A similar observation was seen in DENV infection of HEK293T cells expressing TIM-1ΔCyt [34]. However, the endosomal escape assay showed that CHIKV entry into cells expressing TIM-1ΔCyt was slower in comparison to TIM-1 WT cells, suggesting that the cytoplasmic domain of TIM-1 may support efficient internalization. The importance of the cytoplasmic domain in TIM-1-dependent CHIKV infection was apparent in HaCat cells expressing TIM-1ΔCyt. Here, only TIM-1 WT enhanced CHIKV infection while cells expressing TIM-1ΔCyt remained refractory to CHIKV infection similar to parental HaCat cells. The two lysine residues at positions K-338 and K-346 in the cytoplasmic domain of TIM-1 are targets of ubiquitin ligases [37]. Since ubiquitin chains are internalization signals, the lysine residues have the potential to initiate internalization of TIM-1 upon ubiquitination [86]. In this study, HaCat cells expressing TIM-1 ubiquitination motif mutants were still susceptible to CHIKV pseudoparticles implying that CHIKV infection of HaCat cells is independent of TIM-1 ubiquitination. Taken together, these findings indicate that TIM-1 mediates CHIKV infection by enhancing particle attachment and uptake into cells. Our data further show that the role of the ectodomain and cytoplasmic domain of TIM-1 may be cell type specific and presumably depends on the presence or absence of other attachment factors.

Previous studies on cell entry of alphaviruses have pointed out different entry pathways, including macropinocytosis [87] as well as clathrin-dependent and -independent endocytosis [8,42,88,89], which suggests cell type specific variations. Application of the lysosomotropic agent ammonium chloride led to reduced CHIKV infectivity confirming that TIM-1-mediated entry of CHIKV occurs via an endocytic pathway. Since viruses are obligate intracellular pathogens, fast cell entry benefits the maintenance of virus structural integrity for effective intra-cellular delivery of its genomic material. Delayed entry may lead to virus inactivation in the extracellular milieu for instance due to variations in pH [90]. The endosomal escape assay further emphasized the importance of the MILIBS and cytoplasmic domain of TIM-1 in rapid CHIKV entry into cells.

Li et al. observed that expression of TIM-1 inhibits HIV-1 release due to the association of PtdSer-binding domain with the PtdSer on the membranes of the budding virions leading to diminished virus production and replication [91]. Interestingly, TIM-1 was shown to increase replication and virus production of Japanese encephalitis virus (JEV) [32]. In the current study, we did not observe an increase in CHIKV replication upon expression of TIM-1 in HEK293T cells. Moreover, infected HaCat cells expressing TIM-1 WT produced more infectious CHIKV particles, likely due to enhanced initial virus entry. These results argue that, unlike for HIV-1, the enhanced entry of CHIKV due to the expression of TIM-1 WT is dominant over possible inhibition during release, resulting in increased viral production. Additionally, TIM-1-dependent inhibition of HIV-1 release may be attributed to the lower density of envelope glycoproteins in comparison to other viruses [92], hence PtdSer in the budding virus particles is readily accessible for binding TIM-1. Altogether, these observations suggest that TIM-1 may have different roles in specific virus families and these may yet be cell type dependent.

In the skin, which is the primary entry point of the mosquito transmitted CHIKV, TIM-1 and Axl are expressed by human epidermal keratinocytes [37,38]. HaCat cells derived from spontaneously immortalized keratinocytes act as a relevant model to study keratinocytes in vitro [39]. However, HaCat cells endogenously express Axl but not TIM-1 and they are refractory to CHIKV [93]. In our gain-of-function study, the ectopic expression of TIM-1 WT rendered HaCat cells permissive to authentic CHIKV. Conversely, expression of TIM-1ΔMIL, TIM-1ΔCyt and Axl in HaCat cells did not support CHIKV infection. Bernard et al. demonstrated that HaCat cells are refractory to CHIKV due to induction of interferon [93]. However, in the presence of TIM-1 WT we observed increased CHIKV susceptibility and premissiveness. Since the basal layer of epidermal keratinocytes expresses TIM-1 and HaCat cells become permissive to CHIKV upon ectopic expression of TIM-1, TIM-1 may play a physiological role in CHIKV infection of the skin. However, future experiments using primary keratinocytes will need to test this hypothesis. Our findings emphasize the role of the PtdSer binding domain and the cytoplasmic domain of TIM-1 in HaCat cells and demonstrate a cell type specific dependency of CHIKV on TIM-1. The requirement for the cytoplasmic domain of TIM-1 in CHIKV infection was more evident in HaCat cells, which remained non-permissive unlike HEK293T cells. This observation may be attributed to presence of alternative attachment factors in HEK293T cells, which render them readily susceptible even in the absence of TIM-1.

In order to analyze the role of endogenously expressed TIM-1 in CHIKV infection, we used the hepatoma derived Huh7.5 cell line [49], physiologically relevant cells since the virus infects the liver [76]. Huh7.5 cells predominantly express TIM-1 while Axl expression is negligible. Additionally, Huh7.5 cells express MXRA8, a known entry factor for CHIKV and other alphaviruses [20]. TIM-1 was shown to support binding and infection of hepatitis C virus (HCV) in Huh7.5 cells [94]. In a TIM-1 loss-of-function study, we silenced TIM-1 and MXRA8 and observed a reduction in CHIKV susceptibility for single and double knock downs. This suggests that TIM-1 plays a role in CHIKV infection in the presence of MXRA8. The observed residual susceptibility may be due to incomplete knock down. Additionally, hepatocytes are known to express TIM-1 splice-forms, which lack cytoplasmic tyrosine phosphorylation motif [95]. Hence, TIM-1 entry factor function may be partially impaired in these cells. Our results corroborate the observation made by Jemielity et al., that antibody blocking of TIM-1 barely inhibits entry of CHIKV pseudovirus in Huh7 cells [33]. Overall, these observations indicate that TIM-1 is not the only internalization factor in hepatocytes, however it may serve to concentrate CHIKV on the plasma membrane for subsequent internalization by alternative cellular factors. Glycosaminoglycans (GAGs) may act as co-receptors and additionally concentrate the virus [96]. Here, we observed that the expression of TIM-1 in cells with and without GAGs resulted in enhanced CHIKV infection, although GAGs led to an additional slight increase in susceptibility to CHIKV. This finding further suggests that CHIKV uses a number of cellular factors to broaden its tropism.

Wang et al. used CRISPR/Cas9-mediated gene editing to knockout TIM-1 in Huh7 cells and demonstrated that HCV genome replication was not dependent on expression of TIM-1 [94]. In this study, we investigated the role of TIM-1 in CHIKV replication by electroporating CHIKV subgenomic replicon RNA and full length CHIKV RNA into HEK293T cells expressing wild type TIM-1 and mutant variants. We found that expressing TIM-1 WT did not alter CHIKV replication compared to parental cells. This suggests that TIM-1 does not influence CHIKV genome replication but instead facilitates binding of viral particles on the cell surface to promote subsequent uptake and infection.

Axl is thought to enhance infection of Zika virus by antagonizing immune response [97,98]. In the current study, we did not observe a significant Axl-dependent enhancement of authentic CHIKV infection. HaCat cells which moderately express Axl were refractory to CHIKV infection even after transducing the cells to ectopically express more Axl. Ectopic expression of mutant Axl in HaCat cells served as additional control since Axl is known to dimerize upon activation and the mutant acts as a dominant negative [99]. The presence of mutant forms of Axl would disrupt the activity of the dimers formed and reduce the function of endogenous Axl. In line with the findings for WT Axl, we observed no reduction of infection with mutant Axl. We believe that the ectopically expressed WT Axl was functional as it enhanced the entry of CHIKV and Ebola virus glycoprotein-based lentiviral pseudoparticles. We speculate that the enhanced pseudovirus entry is due to exposure of PtdSer by lentivirus-based CHIKV pseudoparticles and hence binding of the particles to either Axl or TIM-1. The glycoproteins on lentiviral pseudoparticles are likely less densely packed than on authentic alphaviruses and hence PtdSer may be better accessible [92]. Hence, Axl may not efficiently bind authentic CHIKV membrane lipids through its ligands Gas6. Interestingly, mutating the TIM-1 PtdSer binding site abrogated CHIKV infection implying that either TIM-1 unlike Axl is able to access PtdSer in CHIKV particles for binding or that the TIM-1 MILIBS is required for a secondary function of TIM-1 necessary for its role as entry factor. Additionally, based on our observation that the TIM-1 ectodomain residue S51, which is outside the MILIBS, is required for CHIKV entry and a previous findings that TIM-1 interacts with EBOV glycoproteins [84], we speculate that in addition to PtdSer, CHIKV glycoproteins may interact with TIM-1. However, future experiments are required for clarification. Overall, the role of Axl as a virus cell entry factor is inconclusive as conflicting results have been published [75,97,100,101]. We postulate that authentic CHIKV infection is less dependent on Axl. However, additional studies on cells with relevant endogenous Axl expression are needed to confirm this notion.

Like other alphaviruses, CHIKV induces apoptosis [102–104], which is associated with PdtSer exposure on the outer leaflet of the plasma membrane where budding of virions occurs [105,106]. Consequently [107], CHIKV particles may acquire an apoptotic bleb-like membrane during egress and become a target for cells that express TIM-1 or other proteins of the TIM family. Apart from epithelial cells and T-helper cells (Th2), macrophages express TIM proteins which interact with PdtSer in the process of clearing apoptotic bodies [81]. The mouse ortholog of TIM-1 is preferentially expressed in Th2 cells and modulates T-cell activation and proliferation by signal transduction, increasing airway inflammation and allergy [108,109]. The extent to which TIM-1-mediated uptake of CHIKV into immune cells plays a role in antiviral immune responses requires further investigation.

In conclusion, our findings show that TIM-1 enhances CHIKV cell binding and entry, which may have implications for virus propagation and spread. The role of TIM-1 and its domains is cell line dependent and since it is endogenously expressed by primary keratinocytes, HaCat cells that ectopically express TIM-1 could act as a suitable model system. Ultimately, a better understanding of the interaction of CHIKV and cellular factors such as TIM-1 may inform the development of antiviral strategies to combat chikungunya fever.

Supplementary Materials: The following are available online at https://www.mdpi.com/article/10.3390/cells10071828/s1, Figure S1: Schematic representation of CHIKV strains, Figure S2: Expression of Axl and TIM-1 in HEK293T cells enhances entry of CHIKV glycoprotein-based pseudovirus, Figure S3: TIM-1-dependent infectivity of CHIKV genotypes and other *Alphavirus* species, Figure S4: CHIKV replicates independently of TIM-1, Figure S5: Relative binding of nLuc-CHIKV virus, Figure S6: Expression of Axl and TIM-1 in HaCat cells, Figure S7: TIM-1 enhances CHIKV and VSV pseudoparticle entry in HaCat cells, Figure S8: Surface expression of TIM-1, Axl and MXRA8 in Huh7.5 cells, Table S1: Baseline values after infection with authentic CHIKV, Table S2: Raw data of the endosomal escape assay in relative light units (RLU) and Movies: Lateral diffusion of the virus at the cell surface.

Author Contributions: All authors substantially contributed to this work. Conceptualization, J.K., Y.A., M.B., and G.G.; methodology, J.K., Y.A., K.I., Y.-D.G., A.L., and L.L.; validation, J.K., Y.A., M.B., and G.G.; writing—original draft preparation, J.K., writing—review and editing, J.K., Y.A., K.I., Y.-D.G., A.L., L.L., M.E., M.B., and G.G. All authors have read and agreed to the published version of the manuscript.

Funding: This work was funded by the German Academic Exchange Service (DAAD) to J.K., the Knut and Alice Wallenberg Foundation to G.G. and M.B., the Federal Ministry of Education and Research together with the Ministry of Science and Culture of Lower Saxony through the Professorinnen programm III to G.G., Swedish Research Council grant no. 2017-05607 to M.E., Wenner-Grens foundations postdoctoral fellowship to Y.A. (UPD2018-0193) and the international Infection Biology Ph.D. program of Hannover Biomedical Research School to J.K. This publication was supported by Deutsche Forschungsgemeinschaft and University of Veterinary Medicine Hannover, Foundation within the funding programme Open Access Publishing.

Institutional Review Board Statement: Not applicable.

Informed Consent Statement: Not applicable.

Data Availability Statement: Not applicable.

Acknowledgments: We thank Charles M. Rice for Huh-7.5 cells, Dirk Lindemann for the pc.Z.VSV-G plasmid and Didier Trono for pWPI and pCMVΔR8.74 constructs.

Conflicts of Interest: The authors declare no conflict of interest.

References

1. Morrison, T.E. Reemergence of chikungunya virus. *J. Virol.* **2014**, *88*, 11644–11647. [CrossRef] [PubMed]
2. Wahid, B.; Ali, A.; Rafique, S.; Idrees, M. Global expansion of chikungunya virus: Mapping the 64-year history. *Int. J. Infect. Dis.* **2017**, *58*, 69–76. [CrossRef]
3. Tsetsarkin, K.A.; Vanlandingham, D.L.; McGee, C.E.; Higgs, S. A single mutation in chikungunya virus affects vector specificity and epidemic potential. *PLoS Pathog.* **2007**, *3*, e201. [CrossRef] [PubMed]
4. Powers, A.M.; Brault, A.C.; Tesh, R.B.; Weaver, S.C. Re-emergence of Chikungunya and O'nyong-nyong viruses: Evidence for distinct geographical lineages and distant evolutionary relationships. *J. Gen. Virol.* **2000**, *81*, 471–479. [CrossRef]
5. Thiberville, S.-D.; Moyen, N.; Dupuis-Maguiraga, L.; Nougairede, A.; Gould, E.A.; Roques, P.; de Lamballerie, X. Chikungunya fever: Epidemiology, clinical syndrome, pathogenesis and therapy. *Antivir. Res.* **2013**, *99*, 345–370. [CrossRef]
6. Dupuis-Maguiraga, L.; Noret, M.; Brun, S.; Le Grand, R.; Gras, G.; Roques, P. Chikungunya disease: Infection-associated markers from the acute to the chronic phase of arbovirus-induced arthralgia. *PLoS Negl. Trop. Dis.* **2012**, *6*, e1446. [CrossRef] [PubMed]
7. van Duijl-Richter, M.K.S.; Hoornweg, T.E.; Rodenhuis-Zybert, I.A.; Smit, J.M. Early Events in Chikungunya Virus Infection-From Virus Cell Binding to Membrane Fusion. *Viruses* **2015**, *7*, 3647–3674. [CrossRef]
8. Sourisseau, M.; Schilte, C.; Casartelli, N.; Trouillet, C.; Guivel-Benhassine, F.; Rudnicka, D.; Sol-Foulon, N.; Le Roux, K.; Prevost, M.-C.; Fsihi, H.; et al. Characterization of reemerging chikungunya virus. *PLoS Pathog.* **2007**, *3*, e89. [CrossRef]
9. Solignat, M.; Gay, B.; Higgs, S.; Briant, L.; Devaux, C. Replication cycle of chikungunya: A re-emerging arbovirus. *Virology* **2009**, *393*, 183–197. [CrossRef]
10. Wikan, N.; Sakoonwatanyoo, P.; Ubol, S.; Yoksan, S.; Smith, D.R. Chikungunya virus infection of cell lines: Analysis of the East, Central and South African lineage. *PLoS ONE* **2012**, *7*, e31102. [CrossRef] [PubMed]
11. Weaver, S.C.; Osorio, J.E.; Livengood, J.A.; Chen, R.; Stinchcomb, D.T. Chikungunya virus and prospects for a vaccine. *Expert Rev. Vaccines* **2012**, *11*, 1087–1101. [CrossRef]
12. Schnierle, B.S. Cellular attachment and entry factors for chikungunya virus. *Viruses* **2019**, *11*, 1078. [CrossRef]
13. Holmes, A.C.; Basore, K.; Fremont, D.H.; Diamond, M.S. A molecular understanding of alphavirus entry. *PLoS Pathog.* **2020**, *16*, e1008876. [CrossRef]
14. Fongsaran, C.; Jirakanwisal, K.; Kuadkitkan, A.; Wikan, N.; Wintachai, P.; Thepparit, C.; Ubol, S.; Phaonakrop, N.; Roytrakul, S.; Smith, D.R. Involvement of ATP synthase β subunit in chikungunya virus entry into insect cells. *Arch. Virol.* **2014**, *159*, 3353–3364. [CrossRef] [PubMed]
15. Wintachai, P.; Wikan, N.; Kuadkitkan, A.; Jaimipuk, T.; Ubol, S.; Pulmanausahakul, R.; Auewarakul, P.; Kasinrerk, W.; Weng, W.-Y.; Panyasrivanit, M.; et al. Identification of prohibitin as a Chikungunya virus receptor protein. *J. Med. Virol.* **2012**, *84*, 1757–1770. [CrossRef]
16. Gardner, C.L.; Hritz, J.; Sun, C.; Vanlandingham, D.L.; Song, T.Y.; Ghedin, E.; Higgs, S.; Klimstra, W.B.; Ryman, K.D. Deliberate attenuation of chikungunya virus by adaptation to heparan sulfate-dependent infectivity: A model for rational arboviral vaccine design. *PLoS Negl. Trop. Dis.* **2014**, *8*, e2719. [CrossRef]
17. Silva, L.A.; Khomandiak, S.; Ashbrook, A.W.; Weller, R.; Heise, M.T.; Morrison, T.E.; Dermody, T.S. A single-amino-acid polymorphism in Chikungunya virus E2 glycoprotein influences glycosaminoglycan utilization. *J. Virol.* **2014**, *88*, 2385–2397. [CrossRef] [PubMed]
18. Moller-Tank, S.; Kondratowicz, A.S.; Davey, R.A.; Rennert, P.D.; Maury, W. Role of the phosphatidylserine receptor TIM-1 in enveloped-virus entry. *J. Virol.* **2013**, *87*, 8327–8341. [CrossRef]
19. Moller-Tank, S.; Maury, W. Phosphatidylserine receptors: Enhancers of enveloped virus entry and infection. *Virology* **2014**, *468*, 565–580. [CrossRef] [PubMed]
20. Zhang, R.; Kim, A.S.; Fox, J.M.; Nair, S.; Basore, K.; Klimstra, W.B.; Rimkunas, R.; Fong, R.H.; Lin, H.; Poddar, S.; et al. Mxra8 is a receptor for multiple arthritogenic alphaviruses. *Nature* **2018**, *557*, 570–574. [CrossRef]
21. Weber, C.; Berberich, E.; von Rhein, C.; Henß, L.; Hildt, E.; Schnierle, B.S. Identification of functional determinants in the chikungunya virus E2 protein. *PLoS Negl. Trop. Dis.* **2017**, *11*, e0005318. [CrossRef] [PubMed]
22. Basore, K.; Kim, A.S.; Nelson, C.A.; Zhang, R.; Smith, B.K.; Uranga, C.; Vang, L.; Cheng, M.; Gross, M.L.; Smith, J.; et al. Cryo-EM Structure of Chikungunya Virus in Complex with the Mxra8 Receptor. *Cell* **2019**, *177*, 1725–1737. [CrossRef]
23. Song, H.; Zhao, Z.; Chai, Y.; Jin, X.; Li, C.; Yuan, F.; Liu, S.; Gao, Z.; Wang, H.; Song, J.; et al. Molecular basis of arthritogenic alphavirus receptor MXRA8 binding to chikungunya virus envelope protein. *Cell* **2019**, *177*, 1714–1724. [CrossRef] [PubMed]

24. Ravichandran, K.S. Beginnings of a good apoptotic meal: The find-me and eat-me signaling pathways. *Immunity* **2011**, *35*, 445–455. [CrossRef]
25. Wang, Q.; Imamura, R.; Motani, K.; Kushiyama, H.; Nagata, S.; Suda, T. Pyroptotic cells externalize eat-me and release find-me signals and are efficiently engulfed by macrophages. *Int. Immunol.* **2013**, *25*, 363–372. [CrossRef]
26. Pietkiewicz, S.; Schmidt, J.H.; Lavrik, I.N. Quantification of apoptosis and necroptosis at the single cell level by a combination of Imaging Flow Cytometry with classical Annexin V/propidium iodide staining. *J. Immunol. Methods* **2015**, *423*, 99–103. [CrossRef]
27. Erwig, L.-P.; Henson, P.M. Immunological consequences of apoptotic cell phagocytosis. *Am. J. Pathol.* **2007**, *171*, 2–8. [CrossRef] [PubMed]
28. Ichimura, T.; Asseldonk, E.J.P.V.; Humphreys, B.D.; Gunaratnam, L.; Duffield, J.S.; Bonventre, J.V. Kidney injury molecule-1 is a phosphatidylserine receptor that confers a phagocytic phenotype on epithelial cells. *J. Clin. Investig.* **2008**, *118*, 1657–1668. [CrossRef] [PubMed]
29. Brooks, C.R.; Yeung, M.Y.; Brooks, Y.S.; Chen, H.; Ichimura, T.; Henderson, J.M.; Bonventre, J.V. KIM-1-/TIM-1-mediated phagocytosis links ATG5-/ULK1-dependent clearance of apoptotic cells to antigen presentation. *EMBO J.* **2015**, *34*, 2441–2464. [CrossRef]
30. Mercer, J.; Helenius, A. Apoptotic mimicry: Phosphatidylserine-mediated macropinocytosis of vaccinia virus. *Ann. N. Y. Acad. Sci.* **2010**, *1209*, 49–55. [CrossRef]
31. Amara, A.; Mercer, J. Viral apoptotic mimicry. *Nat. Rev. Microbiol.* **2015**, *13*, 461–469. [CrossRef]
32. Niu, J.; Jiang, Y.; Xu, H.; Zhao, C.; Zhou, G.; Chen, P.; Cao, R. TIM-1 Promotes Japanese Encephalitis Virus Entry and Infection. *Viruses* **2018**, *10*, 630. [CrossRef] [PubMed]
33. Jemielity, S.; Wang, J.J.; Chan, Y.K.; Ahmed, A.A.; Li, W.; Monahan, S.; Bu, X.; Farzan, M.; Freeman, G.J.; Umetsu, D.T.; et al. TIM-family proteins promote infection of multiple enveloped viruses through virion-associated phosphatidylserine. *PLoS Pathog.* **2013**, *9*, e1003232. [CrossRef]
34. Meertens, L.; Carnec, X.; Lecoin, M.P.; Ramdasi, R.; Guivel-Benhassine, F.; Lew, E.; Lemke, G.; Schwartz, O.; Amara, A. The TIM and TAM families of phosphatidylserine receptors mediate dengue virus entry. *Cell Host Microbe* **2012**, *12*, 544–557. [CrossRef]
35. Nagata, K.; Ohashi, K.; Nakano, T.; Arita, H.; Zong, C.; Hanafusa, H.; Mizuno, K. Identification of the product of growth arrest-specific gene 6 as a common ligand for Axl, Sky, and Mer receptor tyrosine kinases. *J. Biol. Chem.* **1996**, *271*, 30022–30027. [CrossRef] [PubMed]
36. Anderson, H.A.; Maylock, C.A.; Williams, J.A.; Paweletz, C.P.; Shu, H.; Shacter, E. Serum-derived protein S binds to phosphatidylserine and stimulates the phagocytosis of apoptotic cells. *Nat. Immunol.* **2003**, *4*, 87–91. [CrossRef] [PubMed]
37. Dejarnac, O.; Hafirassou, M.L.; Chazal, M.; Versapuech, M.; Gaillard, J.; Perera-Lecoin, M.; Umana-Diaz, C.; Bonnet-Madin, L.; Carnec, X.; Tinevez, J.-Y.; et al. TIM-1 Ubiquitination Mediates Dengue Virus Entry. *Cell Rep.* **2018**, *23*, 1779–1793. [CrossRef] [PubMed]
38. Bauer, T.; Zagórska, A.; Jurkin, J.; Yasmin, N.; Köffel, R.; Richter, S.; Gesslbauer, B.; Lemke, G.; Strobl, H. Identification of Axl as a downstream effector of TGF-β1 during Langerhans cell differentiation and epidermal homeostasis. *J. Exp. Med.* **2012**, *209*, 2033–2047. [CrossRef] [PubMed]
39. Boukamp, P.; Petrussevska, R.T.; Breitkreutz, D.; Hornung, J.; Markham, A.; Fusenig, N.E. Normal keratinization in a spontaneously immortalized aneuploid human keratinocyte cell line. *J. Cell Biol.* **1988**, *106*, 761–771. [CrossRef]
40. Hoornweg, T.E.; van Duijl-Richter, M.K.S.; Ayala Nuñez, N.V.; Albulescu, I.C.; van Hemert, M.J.; Smit, J.M. Dynamics of Chikungunya Virus Cell Entry Unraveled by Single-Virus Tracking in Living Cells. *J. Virol.* **2016**, *90*, 4745–4756. [CrossRef] [PubMed]
41. Ooi, Y.S.; Stiles, K.M.; Liu, C.Y.; Taylor, G.M.; Kielian, M. Genome-wide RNAi screen identifies novel host proteins required for alphavirus entry. *PLoS Pathog.* **2013**, *9*, e1003835. [CrossRef]
42. Bernard, E.; Solignat, M.; Gay, B.; Chazal, N.; Higgs, S.; Devaux, C.; Briant, L. Endocytosis of chikungunya virus into mammalian cells: Role of clathrin and early endosomal compartments. *PLoS ONE* **2010**, *5*, e11479. [CrossRef]
43. Lee, R.C.H.; Hapuarachchi, H.C.; Chen, K.C.; Hussain, K.M.; Chen, H.; Low, S.L.; Ng, L.C.; Lin, R.; Ng, M.M.-L.; Chu, J.J.H. Mosquito cellular factors and functions in mediating the infectious entry of chikungunya virus. *PLoS Negl. Trop. Dis.* **2013**, *7*, e2050. [CrossRef] [PubMed]
44. Kuo, S.-C.; Chen, Y.-J.; Wang, Y.-M.; Tsui, P.-Y.; Kuo, M.-D.; Wu, T.-Y.; Lo, S.J. Cell-based analysis of Chikungunya virus E1 protein in membrane fusion. *J. Biomed. Sci.* **2012**, *19*, 44. [CrossRef] [PubMed]
45. Kuo, S.-C.; Chen, Y.-J.; Wang, Y.-M.; Kuo, M.-D.; Jinn, T.-R.; Chen, W.-S.; Chang, Y.-C.; Tung, K.-L.; Wu, T.-Y.; Lo, S.J. Cell-based analysis of Chikungunya virus membrane fusion using baculovirus-expression vectors. *J. Virol. Methods* **2011**, *175*, 206–215. [CrossRef] [PubMed]
46. van Duijl-Richter, M.K.S.; Blijleven, J.S.; van Oijen, A.M.; Smit, J.M. Chikungunya virus fusion properties elucidated by single-particle and bulk approaches. *J. Gen. Virol.* **2015**, *96*, 2122–2132. [CrossRef]
47. Zeng, X.; Mukhopadhyay, S.; Brooks, C.L. Residue-level resolution of alphavirus envelope protein interactions in pH-dependent fusion. *Proc. Natl. Acad. Sci. USA* **2015**, *112*, 2034–2039. [CrossRef]
48. DuBridge, R.B.; Tang, P.; Hsia, H.C.; Leong, P.M.; Miller, J.H.; Calos, M.P. Analysis of mutation in human cells by using an Epstein-Barr virus shuttle system. *Mol. Cell. Biol.* **1987**, *7*, 379–387. [CrossRef]

49. Blight, K.J.; McKeating, J.A.; Rice, C.M. Highly permissive cell lines for subgenomic and genomic hepatitis C virus RNA replication. *J. Virol.* **2002**, *76*, 13001–13014. [CrossRef] [PubMed]

50. Tsetsarkin, K.; Higgs, S.; McGee, C.E.; De Lamballerie, X.; Charrel, R.N.; Vanlandingham, D.L. Infectious clones of Chikungunya virus (La Réunion isolate) for vector competence studies. *Vector Borne Zoonotic Dis.* **2006**, *6*, 325–337. [CrossRef]

51. Vanlandingham, D.L.; Tsetsarkin, K.; Hong, C.; Klingler, K.; McElroy, K.L.; Lehane, M.J.; Higgs, S. Development and characterization of a double subgenomic chikungunya virus infectious clone to express heterologous genes in Aedes aegypti mosquitoes. *Insect Biochem. Mol. Biol.* **2005**, *35*, 1162–1170. [CrossRef]

52. Levitt, N.H.; Ramsburg, H.H.; Hasty, S.E.; Repik, P.M.; Cole, F.E.; Lupton, H.W. Development of an attenuated strain of chikungunya virus for use in vaccine production. *Vaccine* **1986**, *4*, 157–162. [CrossRef]

53. Jin, J.; Sherman, M.B.; Chafets, D.; Dinglasan, N.; Lu, K.; Lee, T.-H.; Carlson, L.-A.; Muench, M.O.; Simmons, G. An attenuated replication-competent chikungunya virus with a fluorescently tagged envelope. *PLoS Negl. Trop. Dis.* **2018**, *12*, e0006693. [CrossRef] [PubMed]

54. Atasheva, S.; Krendelchtchikova, V.; Liopo, A.; Frolova, E.; Frolov, I. Interplay of acute and persistent infections caused by Venezuelan equine encephalitis virus encoding mutated capsid protein. *J. Virol.* **2010**, *84*, 10004–10015. [CrossRef]

55. Bergqvist, J.; Forsman, O.; Larsson, P.; Näslund, J.; Lilja, T.; Engdahl, C.; Lindström, A.; Gylfe, Å.; Ahlm, C.; Evander, M.; et al. Detection and isolation of Sindbis virus from mosquitoes captured during an outbreak in Sweden, 2013. *Vector Borne Zoonotic Dis.* **2015**, *15*, 133–140. [CrossRef]

56. Jaalouk, D.E.; Crosato, M.; Brodt, P.; Galipeau, J. Inhibition of histone deacetylation in 293GPG packaging cell line improves the production of self-inactivating MLV-derived retroviral vectors. *Virol. J.* **2006**, *3*, 27. [CrossRef] [PubMed]

57. Palsson, B.; Andreadis, S. The physico-chemical factors that govern retrovirus-mediated gene transfer. *Exp. Hematol.* **1997**, *25*, 94–102.

58. Mosmann, T. Rapid colorimetric assay for cellular growth and survival: Application to proliferation and cytotoxicity assays. *J. Immunol. Methods* **1983**, *65*, 55–63. [CrossRef]

59. Tinevez, J.-Y.; Perry, N.; Schindelin, J.; Hoopes, G.M.; Reynolds, G.D.; Laplantine, E.; Bednarek, S.Y.; Shorte, S.L.; Eliceiri, K.W. TrackMate: An open and extensible platform for single-particle tracking. *Methods* **2017**, *115*, 80–90. [CrossRef]

60. Vega, A.R.; Freeman, S.A.; Grinstein, S.; Jaqaman, K. Multistep track segmentation and motion classification for transient mobility analysis. *Biophys. J.* **2018**, *114*, 1018–1025. [CrossRef]

61. Alberione, M.P.; Moeller, R.; Kirui, J.; Ginkel, C.; Doepke, M.; Ströh, L.J.; Machtens, J.-P.; Pietschmann, T.; Gerold, G. Single-nucleotide variants in human CD81 influence hepatitis C virus infection of hepatoma cells. *Med. Microbiol. Immunol.* **2020**, *209*, 499–514. [CrossRef]

62. Kuchroo, V.K.; Umetsu, D.T.; DeKruyff, R.H.; Freeman, G.J. The TIM gene family: Emerging roles in immunity and disease. *Nat. Rev. Immunol.* **2003**, *3*, 454–462. [CrossRef]

63. Santiago, C.; Ballesteros, A.; Tami, C.; Martínez-Muñoz, L.; Kaplan, G.G.; Casasnovas, J.M. Structures of T Cell immunoglobulin mucin receptors 1 and 2 reveal mechanisms for regulation of immune responses by the TIM receptor family. *Immunity* **2007**, *26*, 299–310. [CrossRef]

64. Kane, L.P. T cell Ig and mucin domain proteins and immunity. *J. Immunol.* **2010**, *184*, 2743–2749. [CrossRef] [PubMed]

65. Li, J.; Jia, L.; Ma, Z.-H.; Ma, Q.-H.; Yang, X.-H.; Zhao, Y.-F. Axl glycosylation mediates tumor cell proliferation, invasion and lymphatic metastasis in murine hepatocellular carcinoma. *World J. Gastroenterol.* **2012**, *18*, 5369–5376. [CrossRef] [PubMed]

66. Lauter, M.; Weber, A.; Torka, R. Targeting of the AXL receptor tyrosine kinase by small molecule inhibitor leads to AXL cell surface accumulation by impairing the ubiquitin-dependent receptor degradation. *Cell Commun. Signal.* **2019**, *17*, 59. [CrossRef]

67. Brindley, M.A.; Hunt, C.L.; Kondratowicz, A.S.; Bowman, J.; Sinn, P.L.; McCray, P.B.; Quinn, K.; Weller, M.L.; Chiorini, J.A.; Maury, W. Tyrosine kinase receptor Axl enhances entry of Zaire ebolavirus without direct interactions with the viral glycoprotein. *Virology* **2011**, *415*, 83–94. [CrossRef] [PubMed]

68. Hunt, C.L.; Kolokoltsov, A.A.; Davey, R.A.; Maury, W. The Tyro3 receptor kinase Axl enhances macropinocytosis of Zaire ebolavirus. *J. Virol.* **2011**, *85*, 334–347. [CrossRef]

69. Zapatero-Belinchón, F.J.; Dietzel, E.; Dolnik, O.; Döhner, K.; Costa, R.; Hertel, B.; Veselkova, B.; Kirui, J.; Klintworth, A.; Manns, M.P.; et al. Characterization of the Filovirus-Resistant Cell Line SH-SY5Y Reveals Redundant Role of Cell Surface Entry Factors. *Viruses* **2019**, *11*, 275. [CrossRef]

70. Kao, F.T.; Puck, T.T. Genetics of somatic mammalian cells. IV. Properties of Chinese hamster cell mutants with respect to the requirement for proline. *Genetics* **1967**, *55*, 513–524. [CrossRef]

71. Esko, J.D.; Stewart, T.E.; Taylor, W.H. Animal cell mutants defective in glycosaminoglycan biosynthesis. *Proc. Natl. Acad. Sci. USA* **1985**, *82*, 3197–3201. [CrossRef] [PubMed]

72. Gorchakov, R.; Wang, E.; Leal, G.; Forrester, N.L.; Plante, K.; Rossi, S.L.; Partidos, C.D.; Adams, A.P.; Seymour, R.L.; Weger, J.; et al. Attenuation of Chikungunya virus vaccine strain 181/clone 25 is determined by two amino acid substitutions in the E2 envelope glycoprotein. *J. Virol.* **2012**, *86*, 6084–6096. [CrossRef] [PubMed]

73. Ohkuma, S.; Poole, B. Fluorescence probe measurement of the intralysosomal pH in living cells and the perturbation of pH by various agents. *Proc. Natl. Acad. Sci. USA* **1978**, *75*, 3327–3331. [CrossRef] [PubMed]

74. Ohkuma, S.; Poole, B. Cytoplasmic vacuolation of mouse peritoneal macrophages and the uptake into lysosomes of weakly basic substances. *J. Cell Biol.* **1981**, *90*, 656–664. [CrossRef] [PubMed]

75. Fedeli, C.; Torriani, G.; Galan-Navarro, C.; Moraz, M.-L.; Moreno, H.; Gerold, G.; Kunz, S. Axl can serve as entry factor for lassa virus depending on the functional glycosylation of dystroglycan. *J. Virol.* **2018**, *92*, e01613–e01617. [CrossRef]

76. Chua, H.H.; Abdul Rashid, K.; Law, W.C.; Hamizah, A.; Chem, Y.K.; Khairul, A.H.; Chua, K.B. A fatal case of chikungunya virus infection with liver involvement. *Med. J. Malays.* **2010**, *65*, 83–84.

77. Matusali, G.; Colavita, F.; Bordi, L.; Lalle, E.; Ippolito, G.; Capobianchi, M.R.; Castilletti, C. Tropism of the chikungunya virus. *Viruses* **2019**, *11*, 175. [CrossRef]

78. Bhattacharyya, S.; Zagórska, A.; Lew, E.D.; Shrestha, B.; Rothlin, C.V.; Naughton, J.; Diamond, M.S.; Lemke, G.; Young, J.A.T. Enveloped viruses disable innate immune responses in dendritic cells by direct activation of TAM receptors. *Cell Host Microbe* **2013**, *14*, 136–147. [CrossRef]

79. Morizono, K.; Xie, Y.; Olafsen, T.; Lee, B.; Dasgupta, A.; Wu, A.M.; Chen, I.S.Y. The soluble serum protein Gas6 bridges virion envelope phosphatidylserine to the TAM receptor tyrosine kinase Axl to mediate viral entry. *Cell Host Microbe* **2011**, *9*, 286–298. [CrossRef]

80. Morizono, K.; Chen, I.S.Y. Role of phosphatidylserine receptors in enveloped virus infection. *J. Virol.* **2014**, *88*, 4275–4290. [CrossRef]

81. Freeman, G.J.; Casasnovas, J.M.; Umetsu, D.T.; DeKruyff, R.H. TIM genes: A family of cell surface phosphatidylserine receptors that regulate innate and adaptive immunity. *Immunol. Rev.* **2010**, *235*, 172–189. [CrossRef]

82. Bishop, N.E.; Anderson, D.A. Early interactions of hepatitis A virus with cultured cells: Viral elution and the effect of pH and calcium ions. *Arch. Virol.* **1997**, *142*, 2161–2178. [CrossRef] [PubMed]

83. Kondratowicz, A.S.; Lennemann, N.J.; Sinn, P.L.; Davey, R.A.; Hunt, C.L.; Moller-Tank, S.; Meyerholz, D.K.; Rennert, P.; Mullins, R.F.; Brindley, M.; et al. T-cell immunoglobulin and mucin domain 1 (TIM-1) is a receptor for Zaire Ebolavirus and Lake Victoria Marburgvirus. *Proc. Natl. Acad. Sci. USA* **2011**, *108*, 8426–8431. [CrossRef]

84. Yuan, S.; Cao, L.; Ling, H.; Dang, M.; Sun, Y.; Zhang, X. TIM-1 acts a dual-attachment receptor for ebolavirus by interacting directly with viral GP and the PS on the viral envelope. *Protein Cell* **2015**, *6*, 814–824. [CrossRef] [PubMed]

85. Song, D.-H.; Garcia, G.; Situ, K.; Chua, B.A.; Hong, M.L.O.; Do, E.A.; Ramirez, C.M.; Harui, A.; Arumugaswami, V.; Morizono, K. Development of a blocker of the universal phosphatidylserine- and phosphatidylethanolamine-dependent viral entry pathways. *Virology* **2021**, *560*, 17–33. [CrossRef]

86. Shih, S.C.; Sloper-Mould, K.E.; Hicke, L. Monoubiquitin carries a novel internalization signal that is appended to activated receptors. *EMBO J.* **2000**, *19*, 187–198. [CrossRef] [PubMed]

87. Lee, C.H.R.; Mohamed Hussain, K.; Chu, J.J.H. Macropinocytosis dependent entry of Chikungunya virus into human muscle cells. *PLoS Negl. Trop. Dis.* **2019**, *13*, e0007610. [CrossRef]

88. Marsh, M.; Kielian, M.C.; Helenius, A. Semliki forest virus entry and the endocytic pathway. *Biochem. Soc. Trans.* **1984**, *12*, 981–983. [CrossRef]

89. Kielian, M.; Chanel-Vos, C.; Liao, M. Alphavirus Entry and Membrane Fusion. *Viruses* **2010**, *2*, 796–825. [CrossRef]

90. Pirtle, E.C.; Beran, G.W. Virus survival in the environment. *Rev. Sci. Tech.* **1991**, *10*, 733–748. [CrossRef] [PubMed]

91. Li, M.; Ablan, S.D.; Miao, C.; Zheng, Y.-M.; Fuller, M.S.; Rennert, P.D.; Maury, W.; Johnson, M.C.; Freed, E.O.; Liu, S.-L. TIM-family proteins inhibit HIV-1 release. *Proc. Natl. Acad. Sci. USA* **2014**, *111*, E3699–E3707. [CrossRef]

92. Amitai, A.; Chakraborty, A.K.; Kardar, M. The low spike density of HIV may have evolved because of the effects of T helper cell depletion on affinity maturation. *PLoS Comput. Biol.* **2018**, *14*, e1006408. [CrossRef]

93. Bernard, E.; Hamel, R.; Neyret, A.; Ekchariyawat, P.; Molès, J.-P.; Simmons, G.; Chazal, N.; Desprès, P.; Missé, D.; Briant, L. Human keratinocytes restrict chikungunya virus replication at a post-fusion step. *Virology* **2015**, *476*, 1–10. [CrossRef] [PubMed]

94. Wang, J.; Qiao, L.; Hou, Z.; Luo, G. TIM-1 Promotes Hepatitis C Virus Cell Attachment and Infection. *J. Virol.* **2017**, *91*, e01583–e01616. [CrossRef]

95. Bailly, V.; Zhang, Z.; Meier, W.; Cate, R.; Sanicola, M.; Bonventre, J.V. Shedding of kidney injury molecule-1, a putative adhesion protein involved in renal regeneration. *J. Biol. Chem.* **2002**, *277*, 39739–39748. [CrossRef]

96. Aquino, R.S.; Park, P.W. Glycosaminoglycans and infection. *Front. Biosci. (Landmark Ed.)* **2016**, *21*, 1260–1277. [CrossRef] [PubMed]

97. Chen, J.; Yang, Y.-F.; Yang, Y.; Zou, P.; Chen, J.; He, Y.; Shui, S.-L.; Cui, Y.-R.; Bai, R.; Liang, Y.-J.; et al. AXL promotes Zika virus infection in astrocytes by antagonizing type I interferon signalling. *Nat. Microbiol.* **2018**, *3*, 302–309. [CrossRef] [PubMed]

98. Meertens, L.; Labeau, A.; Dejarnac, O.; Cipriani, S.; Sinigaglia, L.; Bonnet-Madin, L.; Le Charpentier, T.; Hafirassou, M.L.; Zamborlini, A.; Cao-Lormeau, V.-M.; et al. Axl mediates ZIKA virus entry in human glial cells and modulates innate immune responses. *Cell Rep.* **2017**, *18*, 324–333. [CrossRef]

99. Sasaki, T.; Knyazev, P.G.; Clout, N.J.; Cheburkin, Y.; Göhring, W.; Ullrich, A.; Timpl, R.; Hohenester, E. Structural basis for Gas6-Axl signalling. *EMBO J.* **2006**, *25*, 80–87. [CrossRef]

100. Wang, Z.-Y.; Wang, Z.; Zhen, Z.-D.; Feng, K.-H.; Guo, J.; Gao, N.; Fan, D.-Y.; Han, D.-S.; Wang, P.-G.; An, J. Axl is not an indispensable factor for Zika virus infection in mice. *J. Gen. Virol.* **2017**, *98*, 2061–2068. [CrossRef]

101. Strange, D.P.; Jiyarom, B.; Pourhabibi Zarandi, N.; Xie, X.; Baker, C.; Sadri-Ardekani, H.; Shi, P.-Y.; Verma, S. Axl promotes zika virus entry and modulates the antiviral state of human sertoli cells. *MBio* **2019**, *10*, e01372–e01419. [CrossRef]

102. Khan, M.; Dhanwani, R.; Patro, I.K.; Rao, P.V.L.; Parida, M.M. Cellular IMPDH enzyme activity is a potential target for the inhibition of Chikungunya virus replication and virus induced apoptosis in cultured mammalian cells. *Antivir. Res.* **2011**, *89*, 1–8. [CrossRef]
103. Baer, A.; Lundberg, L.; Swales, D.; Waybright, N.; Pinkham, C.; Dinman, J.D.; Jacobs, J.L.; Kehn-Hall, K. Venezuelan Equine Encephalitis Virus Induces Apoptosis through the Unfolded Protein Response Activation of EGR1. *J. Virol.* **2016**, *90*, 3558–3572. [CrossRef] [PubMed]
104. Dhanwani, R.; Khan, M.; Bhaskar, A.S.B.; Singh, R.; Patro, I.K.; Rao, P.V.L.; Parida, M.M. Characterization of Chikungunya virus infection in human neuroblastoma SH-SY5Y cells: Role of apoptosis in neuronal cell death. *Virus Res.* **2012**, *163*, 563–572. [CrossRef]
105. Nayak, T.K.; Mamidi, P.; Kumar, A.; Singh, L.P.K.; Sahoo, S.S.; Chattopadhyay, S.; Chattopadhyay, S. Regulation of Viral Replication, Apoptosis and Pro-Inflammatory Responses by 17-AAG during Chikungunya Virus Infection in Macrophages. *Viruses* **2017**, *9*, 3. [CrossRef] [PubMed]
106. Krejbich-Trotot, P.; Denizot, M.; Hoarau, J.-J.; Jaffar-Bandjee, M.-C.; Das, T.; Gasque, P. Chikungunya virus mobilizes the apoptotic machinery to invade host cell defenses. *FASEB J.* **2011**, *25*, 314–325. [CrossRef]
107. Pialoux, G.; Gaüzère, B.-A.; Jauréguiberry, S.; Strobel, M. Chikungunya, an epidemic arbovirosis. *Lancet Infect. Dis.* **2007**, *7*, 319–327. [CrossRef]
108. Meyers, J.H.; Chakravarti, S.; Schlesinger, D.; Illes, Z.; Waldner, H.; Umetsu, S.E.; Kenny, J.; Zheng, X.X.; Umetsu, D.T.; DeKruyff, R.H.; et al. TIM-4 is the ligand for TIM-1, and the TIM-1-TIM-4 interaction regulates T cell proliferation. *Nat. Immunol.* **2005**, *6*, 455–464. [CrossRef]
109. Umetsu, S.E.; Lee, W.-L.; McIntire, J.J.; Downey, L.; Sanjanwala, B.; Akbari, O.; Berry, G.J.; Nagumo, H.; Freeman, G.J.; Umetsu, D.T.; et al. TIM-1 induces T cell activation and inhibits the development of peripheral tolerance. *Nat. Immunol.* **2005**, *6*, 447–454. [CrossRef]

Article

Nipah Virus Efficiently Replicates in Human Smooth Muscle Cells without Cytopathic Effect

Blair L. DeBuysscher [1,2], Dana P. Scott [3], Rebecca Rosenke [3], Victoria Wahl [4], Heinz Feldmann [1,*] and Joseph Prescott [1,5,*]

1   Laboratory of Virology, Division of Intramural Research, NIAID, NIH, Hamilton, MT 59840, USA; blair.debuy@gmail.com

2   Fred Hutchinson Cancer Research Center, Vaccine and Infectious Disease Division, Seattle, WA 98109-1024, USA

3   Rocky Mountain Veterinary Branch, Division of Intramural Research, NIAID, NIH, Hamilton, MT 59840, USA; dana.scott@nih.gov (D.P.S.); rebecca.rosenke@nih.gov (R.R.)

4   National Biodefense Analysis and Countermeasures Center, Department of Homeland Security, Frederick, MD 21702, USA; Victoria.Wahl@ST.DHS.GOV

5   Center for Biological Threats and Special Pathogens, Robert Koch Institute, 13353 Berlin, Germany

*   Correspondence: feldmannh@niaid.nih.gov (H.F.); prescottj@rki.de (J.P.)

Abstract: Nipah virus (NiV) is a highly pathogenic zoonotic virus with a broad species tropism, originating in pteropid bats. Human outbreaks of NiV disease occur almost annually, often with high case-fatality rates. The specific events that lead to pathogenesis are not well defined, but the disease has both respiratory and encephalitic components, with relapsing encephalitis occurring in some cases more than a year after initial infection. Several cell types are targets of NiV, dictated by the expression of the ephrin-B2/3 ligand on the cell's outer membrane, which interact with the NiV surface proteins. Vascular endothelial cells (ECs) are major targets of infection. Cytopathic effects (CPE), characterized by syncytia formation and cell death, and an ensuing vasculitis, are a major feature of the disease. Smooth muscle cells (SMCs) of the tunica media that line small blood vessels are infected in humans and animal models of NiV disease, although pathology or histologic changes associated with antigen-positive SMCs have not been reported. To gain an understanding of the possible contributions that SMCs might have in the development of NiV disease, we investigated the susceptibility and potential cytopathogenic changes of human SMCs to NiV infection in vitro. SMCs were permissive for NiV infection and resulted in high titers and prolonged NiV production, despite a lack of cytopathogenicity, and in the absence of detectable ephrin-B2/3. These results indicate that SMC might be important contributors to disease by producing progeny NiV during an infection, without suffering cytopathogenic consequences.

Keywords: Nipah virus; endothelial cells; smooth muscle cells; henipavirus; paramyxovirus; bat virus; fusion; syncytia

Citation: DeBuysscher, B.L.; Scott, D.P.; Rosenke, R.; Wahl, V.; Feldmann, H.; Prescott, J. Nipah Virus Efficiently Replicates in Human Smooth Muscle Cells without Cytopathic Effect. *Cells* **2021**, *10*, 1319. https://doi.org/10.3390/cells10061319

Academic Editors: Thomas Hoenen, Allison Groseth and Alexander E. Kalyuzhny

Received: 6 April 2021
Accepted: 18 May 2021
Published: 25 May 2021

Publisher's Note: MDPI stays neutral with regard to jurisdictional claims in published maps and institutional affiliations.

1. Introduction

Nipah virus (NiV) (family; *Paramyxoviridae*, genus; *henipavirus*; species *Nipah henipavirus*) was first recognized as a zoonotic pathogen in 1998–1999 in Malaysia, when it infected 265 people and caused 105 deaths. NiV was subsequently found to originate from pteropid bats (*Chiroptera: Pteropodidae*), its natural reservoir [1,2]. Since its discovery, outbreaks of NiV disease have occurred on an almost annual basis in India and Bangladesh, with resultant high case-fatality rates, often as high as 100% during small isolated outbreaks [3–5]. Clinically, human infection is characterized by fever, cough, dyspnea, headache, and loss of consciousness. The average duration of illness is approximately 9 days [6]. Autopsies of patients that have succumbed to the NiV infection have highlighted hallmarks of the disease, including systemic vasculitis and extensive endothelium destruction, and central nervous system (CNS) involvement [7,8]. The small blood vessels of the lungs and CNS, along with the heart and kidneys, are primary targets of infection and histologic changes. The endothelium of medium and large vessels are less involved, compared to the microvasculature, although infection of larger blood vessels can occur [9,10]. Blood vessels infected with NiV display marked inflammation, with leukocyte infiltration, thrombosis, necrosis, and often hemorrhaging is noted [8]. Multinucleated giant cells, resulting from the fusion and syncytia formation of infected endothelial cells (ECs) are a prominent disease feature and are directly involved in the development of vasculitis [11–14]. Syncytium involving ECs are observed in alveolar spaces, causing pulmonary edema, and viral antigen is seen in cells in the endothelium, tunica media, and in the alveolar spaces [15].

The tropism of NiV is thought to be primarily dictated by the expression of ephrin-B2 and ephrin-B3 on cells, the only identified receptors for NiV [16–18]. These receptors are highly conserved across Mammalia, likely accounting for the broad species tropism of NiV. These receptors normally function in cell-cell signaling, angiogenesis, and neuronal axon guidance [19–21]. Ephrin-B2 expression is observed in arterial ECs and neurons, with a high expression in the lung and brain, while ephrin-B3 expression is restricted to the brain stem and heart [9,16–18,22].

In vitro, various cell lines and primary cells have been evaluated for NiV permissibility, and some for their ephrin-B2/3 expression. Susceptibility to NiV infection has been linked to ephrin-B2 or B3 expression, and even in cases where cytoplasmic entry of NiV through macropinocytosis was noted, ephrin-B2 was still required for cytoplasmic entry [23]. In vivo, it has been described that smooth muscle cells (SMCs) of the tunica media often contain the NiV antigen, but little or no pathology has been associated directly with SMCs, unlike the extensive cytopathology observed in the endothelium. SMCs form the tunica media and are tightly linked physiologically to the endothelium for control of vascular stability and function during and after blood vessel formation. SMCs and ECs are developmentally linked as well, with both originating from the lateral plate mesoderm. Although ECs express various ephrins and Eph receptors in abundance, SMCs also express ephrins and their Eph receptors spatiotemporally, and receptor and membrane-bound ligand interactions between these two cell types are important for adhesion and cell motility during angiogenesis [24].

To gain an understanding of how these two in vivo targets for infection might contribute to NiV replication and cytopathogenicity, we investigated the susceptibility of human primary SMCs to NiV infection, and the consequence of exposure to NiV, compared to ECs in vitro. We show that human SMCs are permissive for infection and produce high and sustained titers of NiV, without the cytopathic effects that are prominent in ECs, such as fusion, syncytia formation, and cell death. Ephrin-B2/B3 was not detectable on the cell surface of SMCs, however, upon exogenous expression of ephrin-B2, NiV infected SMCs gained the ability to fuse and formed syncytia similar to what was observed in EC cultures. Infection in vitro mimics the in vivo observations, indicating that SMCs contribute to the NiV disease by harboring NiV, allowing replication and the production of high amounts of progeny virus, without suffering cytopathic damage.

2. Materials and Methods

2.1. Ethics Statement

Work with NiV and all potentially infectious material was performed in the BSL4 facility at the Rocky Mountain Laboratories (RML), National Institute of Allergy and Infectious Diseases, National Institutes of Health. The Institutional Biosafety Committee (IBC) approved all of the procedures. This study used tissues from Syrian hamsters and African green monkeys that were enrolled in previously published NiV studies [15,25]. These studies were approved by the Institutional Animal Care and Use Committee and performed in accordance with the guidelines of the Association for Assessment and Accreditation of Laboratory Animal Care (AAALAC, Frederick, MD, USA) in an AAALAC-approved facility. Primary human cells for this study were obtained from a commercial source (Lonza, Walkersville, MD, USA) from anonymous donors.

2.2. Cells and Viruses

Vero C1008 cells were obtained from the European Collection of Cell Cultures (Salisbury, UK). Primary human lung microvascular endothelial cells (ECs, CC-2527) and primary smooth muscle cells (SMCs, CC-2581) (Lonza, Walkersville, MD, USA) obtained from human pulmonary arteries were propagated and maintained in specialized media supplemented with growth factors according to the manufacturer's instructions. Experiments are representative of results obtained from two separate human donors for each primary cell type. Although not every experiment was performed using cells from both donors, infection kinetics and the initial fusion experiments were performed with cells from both donors as biological replicates and the results were consistent with no notable donor variation. HeLa cells stably expressing ephrin-B2 or B3 were kindly provided by Dr. Christopher Broder (Uniformed Services University, MD, USA). The Malaysian strain of NiV (NiV-M) was provided by the Special Pathogens Branch of the Centers for Disease Control and Prevention, Atlanta, GA, USA and propagated on Vero E6 cells grown in Dulbecco's Modified Eagle's medium (DMEM) supplemented with 10% fetal calf serum (FCS), 2 mM L-glutamine, 50 IU/mL penicillin, and 50 µg/mL streptomycin (Life Technologies, Carlsbad, CA, USA). Supernatants were clarified by low-speed centrifugation and stocks were stored in liquid nitrogen.

2.3. Histology

Tissue blocks from NiV-infected hamsters [15] and African green monkeys (AGMs) [25] from prior studies at RML that were removed from the BSL4 laboratory according to IBC-approved protocols were used to prepare slides for staining to assess viral tropism. Embedded tissues were processed using a Discovery XT automated processor (Ventana Medical Systems, Oro Valley, AZ, USA) with a DAPMap kit, and stained with hematoxylin and eosin (H&E). Immunohistochemistry (IHC) was performed to detect NiV antigens using a rabbit anti-NiV nucleocapsid (NP) primary antibody at 1:5000 dilution as previously described [26]. Tissues were also stained with a monoclonal mouse anti-smooth muscle actin antibody at 1:100 (Millipore) and a mouse anti-CD31 antibody at 1:700 (LifeSpan BioSciences, Seattle, WA, USA) for identification of SMCs and ECs, respectively.

2.4. In Vitro Infections

Cells were grown to 90–95% confluency in 48-well plates in their respective growth media. The media was then removed and 200 µL of NiV diluted in fresh media was added to the cells at the indicated multiplicity of infections (MOIs). After 1 h, the inoculum was removed, cells were washed in DPBS, and 500 µL of fresh growth media was added to the monolayer. Supernatants were collected by removing and replacing half of the media at the indicated time points until 6 days post-infection (DPI), then a complete media replacement was performed every second day thereafter. In parallel, cells identically treated were stained with a Kwik-Diff kit (Thermo Scientific, Waltham, MA, USA) to visualize the cells and assess their morphology and syncytia formation. Images were captured using a Nikon DS-Fi1 camera. For immunofluorescence assays, cells were grown in 8-well plastic chamber slides (Nunc Lab-Tek) and infected with NiV at a MOI of 5. These cells were fixed in 10% formalin overnight at the indicated time points, prior to removal from the BSL4 following IBC-approved protocols, and then stained.

For lentivirus transductions, monolayers of SMCs or ECs were exposed to lentiviruses encoding red fluorescent protein (RFP) or green fluorescent protein (GFP) according to the manufacturer's instructions (Cellomics Technology, Halethorpe, MD, USA). Briefly, cells were incubated in growth media with 6 µg polybrene and the supplied lentivirus stock for 24 h at 37 °C, 5% CO_2. The virus was then replaced with a growth medium. Following visualization of RFP/GFP expression in 95–100% of cells after 24–48 h, cells were sub-cultured at a 1:1 ratio. The following day, the co-cultures were infected with NiV at a MOI of 5 and visualized by either staining with Kwik-Diff stain and light microscopy or stained for NiV antigen and visualized using confocal microscopy (IFA) as described below.

2.5. Microscopy

Formalin-fixed monolayers of NiV-infected cells grown in 8-well chamber slides were washed in DPBS prior to incubation in 0.2% Triton X-100 for 7 min, followed by blocking in 4% BSA/PBS for 10 min. Slides were then incubated with NiV-specific rabbit antisera, washed twice, then incubated with an anti-rabbit Alexa Fluor 488 antibody (Life Technologies). After washing twice, slides were mounted using ProLong Antifade containing DAPI (Invitrogen, Carlsbad, CA, USA) and visualized by confocal microscopy.

2.6. Virus Quantitation

The tissue culture infectious dose 50% ($TCID_{50}$) method was used to titrate NiV from the supernatants of infected cells as described previously [27]. Briefly, Vero E6 monolayers grown in 96-well plates were inoculated in triplicate with 100 μL of serial dilutions of supernatants in DMEM supplemented with 2% FCS. After 4 to 5 days of incubation at 37 °C, 5% CO_2, wells were examined for cytopathic effect (CPE) and the Spearman-Karber method was used to calculate $TCID_{50}$ values.

2.7. Ephrin-B2/3 Expression and Transfections

Endogenous expression of ephrin-B2 or B3 by cells was quantified by flow cytometry. Cells were collected using 100 mM EDTA and gentle scraping, washed in PBS containing 15 mM EDTA, and incubated with recombinant human EphB4, the ligand for ephrin-B2/3, conjugated to human FC (R&D Biosystems, Minneapolis, MN, USA) for 1 h. After washing in 15 mM EDTA, cells were stained with an anti-human FC antibody conjugated to Alexa Fluor 647 (Life Technologies, Carlsbad, CA, USA) and fixed in 4% PFA. HeLa cells stably expressing ephrin-B2 or B3 were used as controls. Flow cytometry was performed using a LSR II cytometer (BD Biosciences, Franklin Lakes, NJ, USA) and data were analyzed using the FlowJo software (Treestar Inc, Ashland, OR, USA).

Transfection of SMCs with human ephrin-B2 driven by a CMV promotor expression plasmid (Sino Biological Inc., Chesterbrook, PA, USA) or plasmids encoding the fusion protein (F) and glycoprotein (G) of NiV was performed using a Nucleofector kit for primary SMC transfection (Lonza, Walkersville, MD, USA). The NiV F and G expression plasmids were described previously [28]. Cells (5×10^5) in the Nucleofector solution were mixed with plasmid DNA (1 μg of ephrin or 1 μg each of F and G plasmids) and loaded into a cuvette. The A033 program was used, after which the medium was added and cells were plated in 12-well plates. In the case of NiV F and G plasmid transfection, SMCs were co-cultured with ECs the day after transfection, then stained with Kwik-Diff 2 days later, and assessed for syncytia formation using light microscopy. Two days after transfection with plasmids encoding ephrin-B2, SMCs were infected with NiV at a MOI of 5 and then stained with Kwik-Diff and assessed for syncytia formation.

2.8. Statistics

Data obtained from the titration of NiV were plotted as a geometric mean with geometric standard deviation, using the Prism (Graphpad, v9) software (https://www.graphpad.com/support/faq/prism-900-release-notes/).

3. Results

3.1. NiV Infects SMCs and ECs of the Lung of Animal Models of NiV Disease with Distinct Histological Consequences

The disease caused by NiV in Syrian hamsters and African green monkeys (AGM), such as in humans, is characterized by systemic infection and vasculitis [29,30]. In an effort to better understand the cellular targets of NiV across species that are models for

NiV disease, we focused on examining the vasculature of the lung, a major target organ affected, and leading to the respiratory component of NiV disease. The lung tissue from NiV-inoculated hamsters necropsied at 5 DPI, and AGMs at 10 DPI, were stained for NiV NP antigen, and co-stained with markers for ECs (CD31) and SMCs (smooth muscle actin). Multiple lung samples from both hamsters and AGMs showed NiV-positive staining surrounding many small arteries (Figure 1). A closer examination revealed NiV-positive cells within the SMC layer, comprising the tunica media, as well as more abundant NiV-positive cells within the endothelium. Pathologic changes in these tissues were observed in the endothelium, with syncytia formation being a common finding. Conversely, although SMCs of the tunica media contained NiV-positive cells, there were no observable pathologic changes in either hamsters or AGMs, a similar observation to what has been reported in human tissues.

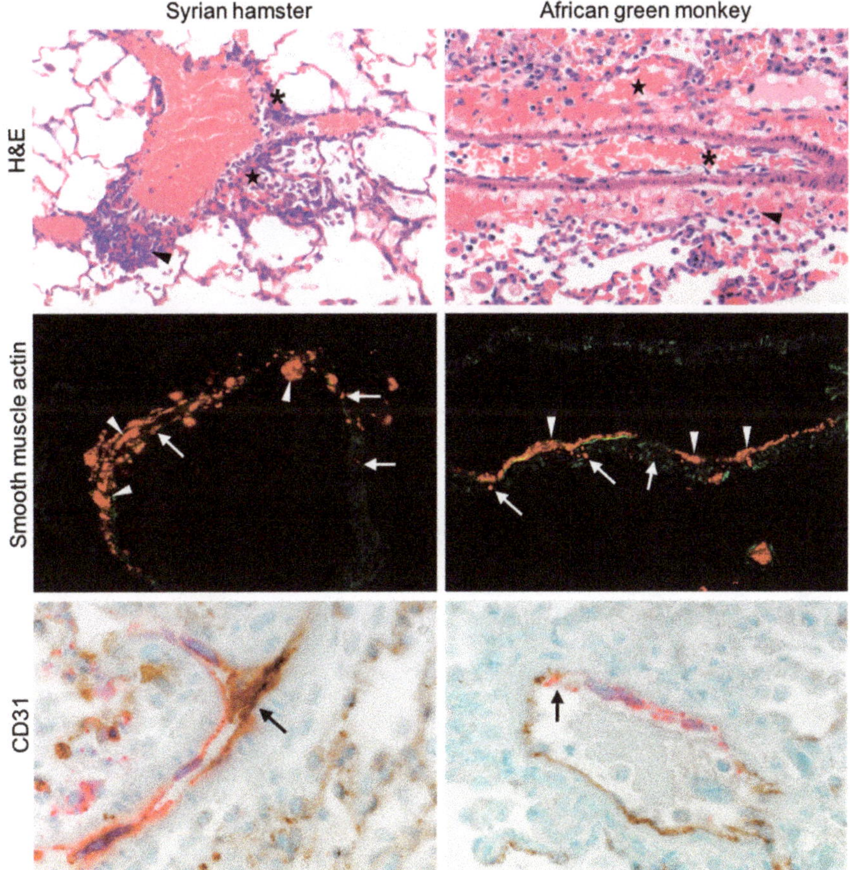

Figure 1. Lung tissue from hamsters (**left column**) and AGMs (**right column**) infected with NiV were sectioned and stained by H&E (**top row**). Asterisks denote degeneration of the endothelium, stars denote hemorrhage, and arrowheads highlight perivascular inflammation. NiV NP (**red**) and smooth muscle actin (**green**) are shown (**middle row**), arrows point to infected SMC, and arrowheads infected ECs, displaying syncytia. NiV NP (**red**) and CD31 (**brown**) are shown (**bottom row**), showing extensive colocalization and infection of ECs and the formation of multinucleated giant cells (**arrows**).

3.2. NiV Productively Infects Human SMCs and ECs, Resulting in Disparate Cytopathogenicity

To determine the susceptibility of human lung primary SMCs and ECs to the NiV infection in vitro, we cultured low-passage (1–2 passages) lung SMCs and ECs with varying amounts of NiV, ranging from a MOI of 0.1–5. The cells were monitored for CPE and fusion events or other morphologic changes throughout the experiment. As early as 1 DPI, ECs began to display cytopathic changes, including fusion and the formation of syncytia, when infected with a MOI of 0.1 (Figure 2A, top). By 3 DPI, all of the EC monolayers had developed extensive CPE and only sparse adherent cells were present. Conversely, SMCs exposed to NiV at the same MOI (0.1) showed no cytopathic changes out to 5 DPI (Figure 2A, bottom) or even through 21 DPI (data not shown).

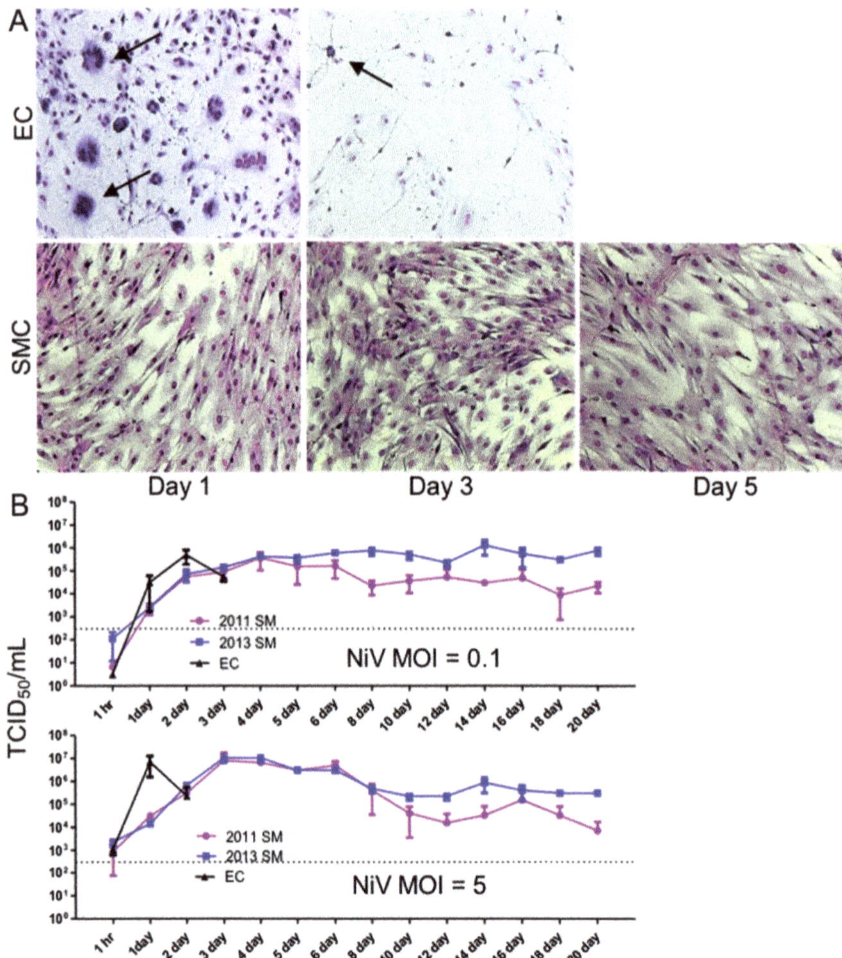

Figure 2. Cytopathic effect and replication of NiV in EC and SMC. (**A**). Monolayers of ECs and SMCs were infected with NiV at a MOI of 0.1 and cells were fixed and stained with Kwik-Diff for visualization. Arrows point to syncytia formation. Images were captured at 10× magnification. (**B**). ECs and SMCs were infected with NiV at a MOI of 0.1 or 5 and supernatants were collected at the indicated time points, and NiV was quantitated by titration using a TCID$_{50}$ assay.

To assess whether SMCs exposed to NiV can be productively infected, as observed in vivo, we compared the production of progeny NiV between SMC and ECs by inoculating monolayers of each cell type with NiV at MOIs of 0.1 or 5. In both cases, peak titers of NiV were reached more quickly in EC cultures than in SMCs (2 days sooner for both MOIs), although ultimately, SMCs produced approximately the same peak titer of progeny virus (Figure 2B). The EC cultures developed extensive CPE, thus the elevated viral titers were only sustained for 2 days at the lower MOI, and 1 day at the higher MOI, and titers dropped the following day, after which the sampling of these cultures was discontinued due to the lack of viable cells. In contrast, the SMCs continuously produced between 1×10^4–5×10^5 TCID$_{50}$/mL out to 21 DPI when inoculated with a MOI of 0.1, and up to 1×10^7 TCID$_{50}$/mL at early time points. After approximately 10 DPI, titers were similar in the SMC cultures, regardless of inoculum, out to 21 DPI. These results were consistent between two independent experiments using two donors of SMCs, designated 2011 SM and 2013 SM. These peak titers are comparable to levels often obtained during NiV propagation on Vero E6 cells (Figure 2B).

3.3. Cell-to-Cell Spread of NiV in SMC Cultures Is Limited, Compared to ECs

Since we did not observe CPE in the NiV-infected SMC cultures, we investigated the initial infection dynamics and the potential of SMCs to fuse, facilitating cell-to-cell spread of NiV. We performed an immunofluorescence time course experiment to directly compare the infection dynamics in SMC and EC cultures. Monolayers of each cell type were exposed to NiV at a MOI of 5 and stained for NiV NP over the course of 2.5 days, at matched time points. The NP of NiV was first detected in ECs 8 h post-infection, and in SMCs 2 h thereafter (Figure 3). By 12 h, ECs showed evidence of the initiation of fusion, followed by large syncytia and infection of 100% of the monolayer by 14 h. A complete destruction of the monolayer was achieved 2–4 h later. In contrast to ECs, infection of SMCs remained focal, with no syncytia formation observed, and no evidence of cell-to-cell spread at any time point. Continuing the experiment until 60 h, many cells remained uninfected, as indicated by the absence of NP staining. This result indicates that not only do SMCs not fuse or form syncytia, only a portion of the cells become infected, despite a high MOI and the continuous presence of infectious progeny virus in the supernatant, as shown in Figure 2B.

3.4. Ephrin-B2/3 Cell Surface Expression Is Undetectable on Primary SMCs

Since the initial infection of NiV in SMCs is inefficient and there is no cell-to-cell fusion following infection, even at a high MOI, we hypothesized that SMCs express little or no ephrin-B2 or B3. Ephrin-B2/3 is the receptor ligand identified for NiV, and is thought to be required for fusion between infected cells expressing the F and G proteins of NiV, and neighboring cells that express ephrin-B2 or B3 on their cell surface. SMCs were incubated with EphB4, the receptor for the ephrin-B2/3 ligand, fused to human FC, as a means to measure the expression of ephrin-B2/3 by flow cytometry. Analysis of both Vero E6 cells and ECs, which are susceptible to NiV-induced fusion, showed measurable amounts of ephrins on their cell surfaces, although not as much as HeLa cells that stably express ephrin-B2 or B3, used as positive controls (Figure 4). In contrast, SMCs were absent for ephrin-B2 or B3.

3.5. Exogenous Expression of Ephrin-B2 in SMCs Permits NiV-Induced Fusion

The lack or low level of ephrin-B2/B3 expression might prevent cell-to-cell fusion and explain the lack of pathology associated with these cells in vivo. To assess whether exogenous expression of ephrin-B2 might render SMCs fusogenic upon NiV infection, we transfected SMC with a plasmid driving the expression of human ephrin-B2 or a control plasmid expressing GFP, prior to the NiV infection. Expression of ephrin-B2 alone did not result in morphological changes in SMCs cultured in a monolayer. However, infection of ephrin-B2-expressing SMCs with NiV resulted in conspicuous syncytia formation (Figure 5 bottom) by 48 h post-infection, similar to what we observe in EC cultures. This demonstrated that NiV-induced cytopathology can occur in SMC when they express ephrin-B2, and that SMCs possess the machinery to fuse, with the exception of a cell surface receptor that interacts with the G protein on neighboring NiV-infected cells, such as ephrin-B2.

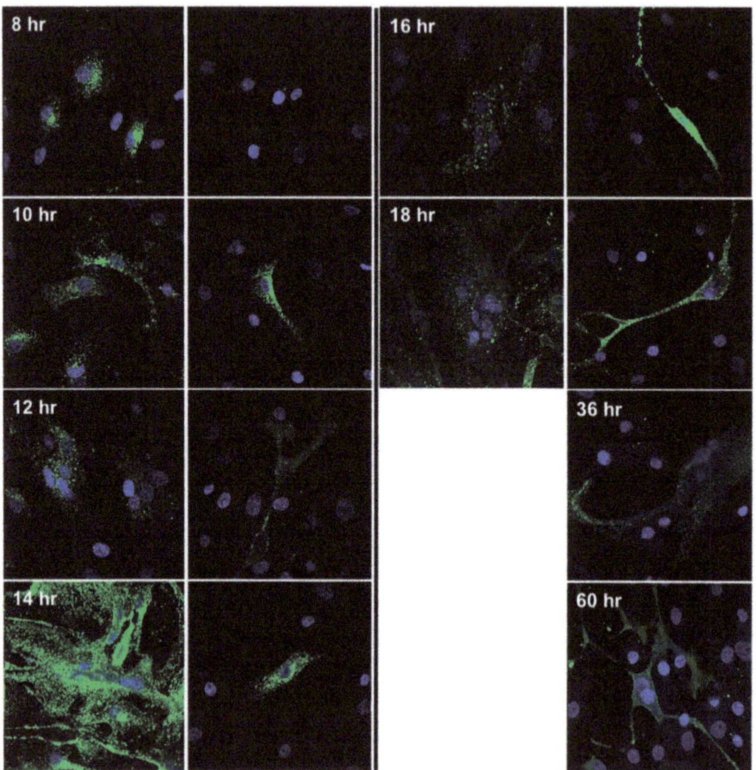

Figure 3. Visualization of cytopathic changes and NiV antigen in ECs and SMCs. Monolayers of the respective cell types were exposed to NiV at a MOI of 5. At the indicated time points, samples were fixed, permeabilized, and stained for NiV NP expression (green), and DAPI for nuclei staining (blue). Images were captured at 20× magnification.

3.6. SMCs Expressing NiV F and G Fuse with Ephrin-B2-Expressing ECs

To further examine whether SMCs are biologically capable of fusing in the context of NiV infection, either with themselves or with ECs, we transfected SMC with plasmids encoding the F and G proteins of NiV to mimic NiV infection, with the notion that SMCs might acquire the ability to fuse with ECs, which endogenously express ephrin-B2 on their surface. SMCs expressing NiV F and G did not self-fuse (Figure 6A), showing again that these cells lack a surface receptor, possibly ephrin-B2, that can interact with F and/or G complexes to initiate fusion. However, when F and G-expressing SMCs were co-cultured with ECs, extensive fusion and syncytia formation was readily observed (Figure 6B), demonstrating that surface expression of F and G on SMCs can interact with an EC-derived surface receptor, most likely ephrin-B2, leading to efficient fusion.

Then, we used an additional approach to confirm the fusogenic capacity of SMC-ECs. We co-cultured SMCs that expressed RFP via lentivirus transduction, with ECs transduced with a GFP-expressing lentivirus, in order to distinguish between the different cell types. Uninfected cultures showed no fusion events, however, following NiV infection, fusion events between SMC and ECs were observed by visualizing syncytia that contained colocalized RFP and GFP, as well as the NiV NP antigen (Figure 6C,D) or sometimes only

GFP and NiV NP, demonstrating EC-EC fusion events. These experiments show that SMCs are capable of fusing with cells that express ephrin-B2 on their surface, when NiV F and G are present on the membrane of SMCs, either by transfection or infection with NiV.

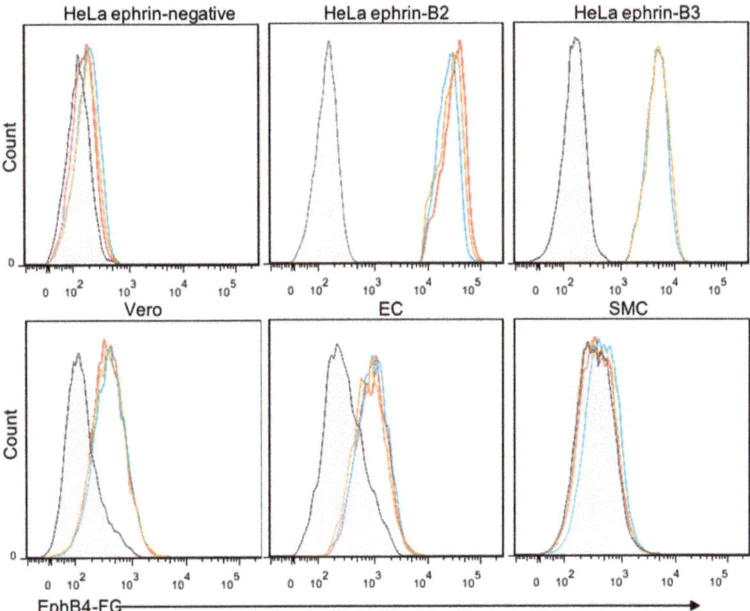

Figure 4. Expression of ephrin-B2/3 on ECs and SMCs. The respective cells were surface stained using recombinant EphB4 (the ligand for ephrin-B2/3) fused to human FC receptor (EphB4-FC), followed by an Alexa 647-conjugated anti-FC secondary antibody. Flow cytometry was performed and grey histograms show negative controls and colored lines represent stained cells from three independent experiments. HeLa cells were used as a negative control, and HeLa cells that stably express ephrin-B2 or B3 were used as positive controls for the assay.

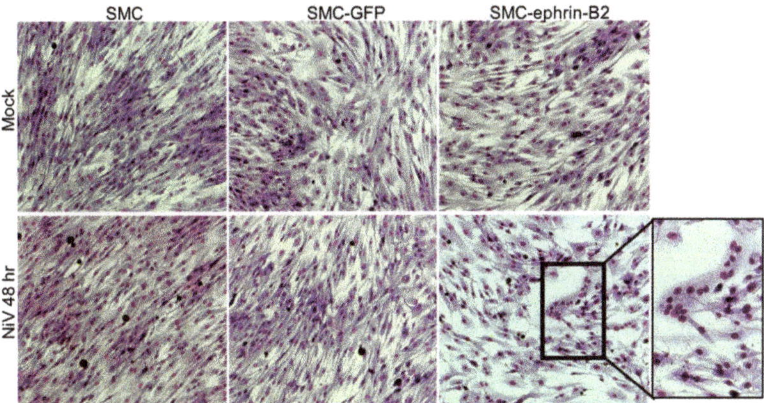

Figure 5. SMC transfected with a plasmid encoding ephrin-B2 can fuse following infection with NiV. SMCs were either mock transfected or transfected with plasmids expression either GFP or ephrin-B2. Following transfection, cells were exposed to NiV at a MOI of 5 or mock-infected (top row). At 48 h, cells were fixed with Kwik-Diff for visualization. Insets highlight morphologic changes. Images were captured at $10\times$ magnification.

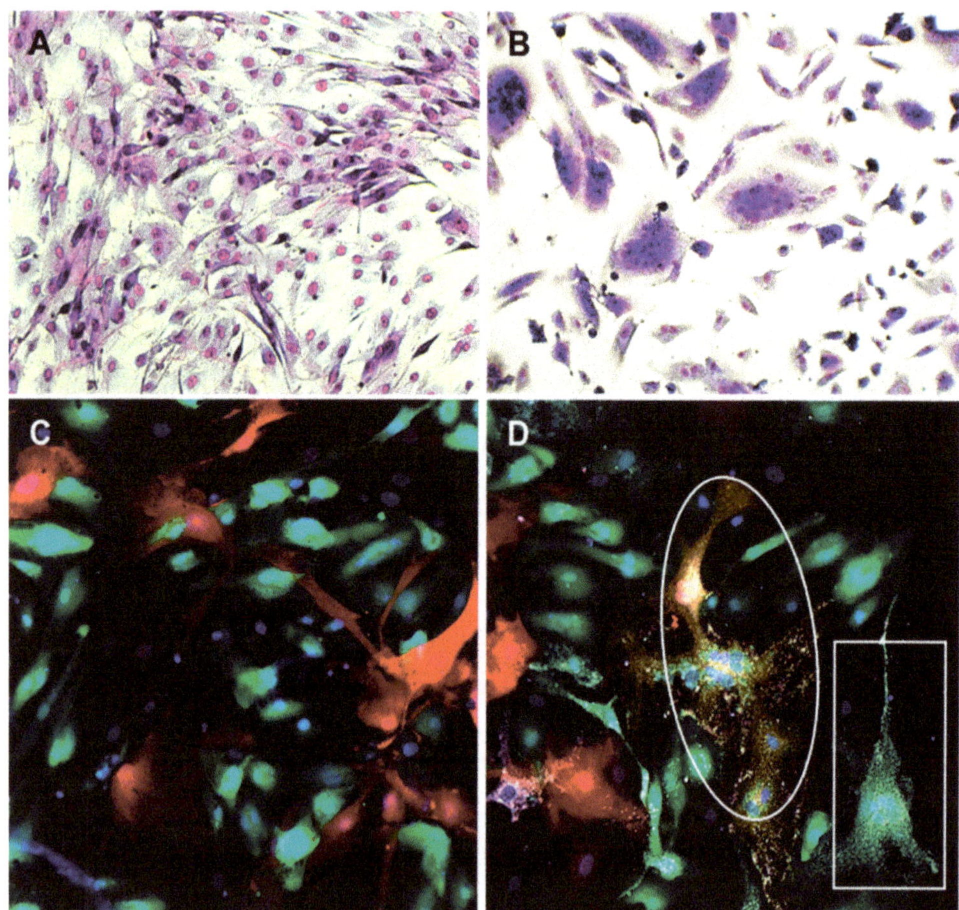

Figure 6. Fusion between SMC expressing NiV F and G and ECs. SMCs were transfected with plasmids encoding NiV F and G proteins and seeded in culture plates alone (**A**) or mixed with primary ECs (1:1) (**B**). Panels C and D show visualization of virus infection and syncytia in mixed cultures of SMCs and ECs. SMC and EC were transduced with lentivirus constructs expressing either RFP or GFP, respectively. After fluorescent protein expression was observed in almost all cells, SMCs and ECs were co-cultured for 24 h (1:1) and then either mock-infected (**C**) or infected with NiV at a MOI of 5 (**D**). Cells were fixed 18 h after infection and stained with anti-NiV NP antibody (white) and DAPI (blue). Syncytia were composed of colocalized RFP and GFP (oval), indicating that both SMCs and ECs were involved in fusion or only GFP (rectangle) showing fusion of ECs alone. All fusions contained NiV NP (white). Images were captured at 10× magnification (**A,B**) or 20× magnification (**C,D**).

4. Discussion

The basic understanding of NiV pathology in humans is solely derived from histological observations from relatively few autopsies, and most of what is known overall has been gleaned from animal studies that model the NiV disease. Two prominent animal models that recapitulate many aspects of human NiV disease are the Syrian hamster, and the African green monkey [29,30]. A hallmark of NiV disease in humans and models, histologically, is the severe vasculitis observed, which is associated with NiV antigen-positive vasculature [6,14]. ECs, which line the lumen of blood vessels, as well as cells of the tunica media are sites of viral replication [6,9,28,30,31]. Infection of ECs results in the formation of multinucleated syncytia and overt disruption of the involved and neighboring cells, which is recapitulated in vitro [9]. However, little is known regarding how cells of the tunica media, SMCs, might contribute to the disease.

In vivo, we observed the NiV antigen systematically throughout the vasculature of infected animal models (hamsters and African green monkeys). Examination of lung tissues from infected animals showed numerous antigen-positive ECs, and although more rare, NiV-positive SMCs. Pathogenic changes, such as fusion and syncytia formation, were readily observed and uniquely associated with the infected endothelium. Identification of SMCs by their expression of smooth muscle actin showed that several of these cells were also NiV antigen positive, often in areas proximal to the infected ECs, and sometimes in areas where the infected ECs were not detected. Moreover, isolated infected SMCs were not associated with syncytia formation or any other overt cytopathic changes [31].

These disparate observations in histologic changes between infected ECs and SMCs were recapitulated in vitro. We used primary ECs and SMCs isolated from human lung tissue to model the susceptibility and consequences of NiV infection in an attempt to expand upon in vivo observations. Both cell types were permissive for NiV infection, however cytopathology was only observed in EC cultures. In contrast, SMC cultures remained unaffected for the duration of the study, up to 3 weeks, even with a continuous production of infectious NiV. ECs showed widespread fusion, resulting in multinucleated cells, as early as 1 DPI, with a complete destruction of the cell monolayer within 2–3 DPI. Other viruses, such as encephalomyocarditis virus, cytomegalovirus, and Epstein-Barr virus infect SMCs [32–36]. Infection of SMC by Epstein-Barr and encephalomyocarditis viruses leads to a lytic infection. In contrast, infection of SMCs by cytomegaloviruses leads to vial latency or persistence similar to what we observe herein [37].

We used several approaches to investigate the consequences and dynamics of NiV infection of SMCs, compared to ECs. Whereas ECs exhibited nearly 100% infection in a monolayer at a high MOI, only approximately 10–20% of SMCs became infected, even with a MOI as high as 5, and even after several days of infection. This lack of susceptibility of all cells might indicate that these cultures of primary cells contain cells in various states in their cell cycle or slight differences in receptor expression. Regardless of the fraction of SMCs that are susceptible to NiV, we were unable to detect ephrin-B2 or B3, the known entry receptor ligand for NiV, on the cell surface of SMCs. It is possible that very low levels of ephrins are expressed in these specific primary cells, accounting for the lack of complete monolayer infection. This is unlikely, however, because even after multiple days of infection, with a constant production of NiV progeny in the supernatant of these cultures, only a limited number of cells became persistently infected and no fusion was observed at any time point, unless ephrin-B2 was exogenously expressed. Following ephrin-B2 transfection, SMCs readily fused upon infection with NiV. These observations suggest that an unidentified entry receptor exists or entry in SMCs can occur by a non-specific mechanism.

Likewise, ECs, which readily fused to each other following the NiV infection, were able to fuse with SMCs only after the SMCs were transfected with plasmids that encode the F and G proteins of NiV or after infection with NiV. Taken together, these data demonstrate that SMCs lack a membrane-bound receptor, such as ephrin-B2/B3, that can interact with the neighboring infected cells, but when F and G is expressed (by transfection or infection) it can interact with the adjacent cells that express a membrane-bound receptor, resulting in fusion. Early studies aimed at identifying NiV entry mechanisms identified micropinocytosis as a possible mechanism, however, ephrin-B2 was still utilized in that instance [23]. It is possible that ephrin-B2/B3 is expressed in these SMCs, but not at the cell surface, and binding and internalization via another attachment receptor or non-specific attachment at the surface might trigger endocytosis and downstream interactions between ephrin-B2/3 in endosomes and the NiV surface proteins. This would prevent fusion between neighboring cells, yet allow for NiV entry. SMCs are known to express ephrins and Eph receptors, but expression is highly spatiotemporal [20,24]. Expression is associated with angiogenesis and development, and it is likely that SMCs of the tunica media express highly variable amounts of ephrins in vivo.

From these in vitro experiments, we can surmise that the CPE observed in ECs is almost entirely a result of fusion, as infected SMCs remained intact for several weeks following the infection. This is not surprising and much of the vasculitis observed in vivo is likely due to the direct effects of the NiV infection, and not entirely dependent on inflammation and immune cell infiltration. An examination of NiV-infected Syrian hamster tissue at very early time points shows infection of the tunica media of larger arterial vessels in the absence of infected endothelium [31]. In this case, it was hypothesized that SMCs play a role in NiV spread from the initially infected epithelial cells to the tunica media, and later to the endothelium, which then leads to syncytia formation within the endothelium and severe disease [31]. This corresponds with the low viremia associated with NiV infections in animal models, in contrast with high viremia observed for many hemorrhagic fever-causing viruses. For NiV infection, it is likely that viral particles gain access to ECs from interstitial tissues, traversing the tunica adventitia and tunica media, as opposed to the direct infection from the vessel lumen, where ECs would be the first cells to become infected. In human tissues assessed post-mortem, there is extensive infection of the endothelium with syncytia formation, as well as adjacent infected SMCs [13]. Together, these observations suggest that SMCs might act as an intermediate between the early-infected parenchyma and terminally infected endothelium. Although infection of SMCs alone is not cytopathic, SMCs likely facilitate pathogenesis by providing an amplifying medium for progeny NiV production and transmission to ECs, without suffering the negative cytopathic consequences seen in the endothelium. Currently, the specific cellular response to the NiV infection in SMCs, and how that response might be involved in the vasculitis-associated pathogenesis independent of cytopathic effects is completely uncharacterized. Future studies should be aimed at determining the entry mechanisms of NiV in SMCs, and elucidating how the response to infection of SMC might contribute to the pathogenesis of the NiV disease.

Author Contributions: Conceptualization, B.L.D., V.W., and J.P.; methodology, B.L.D., D.P.S., R.R. and J.P.; formal analysis, B.L.D. and J.P.; investigation, B.L.D. and J.P., D.P.S. and H.F.; data curation, B.L.D., D.P.S., and J.P.; writing—original draft preparation, B.L.D. and J.P.; writing—review and editing, B.L.D., H.F., V.W., and J.P.; visualization, B.L.D. and J.P.; supervision, H.F. and J.P.; project administration, H.F.; funding acquisition, J.P. and H.F. All authors have read and agreed to the published version of the manuscript.

Funding: This research was funded by the Intramural Research Program of NIAID, NIH. This work was also funded under Agreement No. HSHQDC-15-C-00064 awarded to the Battelle National Biodefense Institute (BNBI) by the Department of Homeland Security (DHS) Science and Technology Directorate (S&T) for the management and operation of the National Biodefense Analysis and Countermeasures Center (NBACC), a Federally Funded Research and Development Center.

Institutional Review Board Statement: Not applicable.

Informed Consent Statement: Not applicable.

Data Availability Statement: Not applicable.

Acknowledgments: We would like to thank Dan Long and Tina Thomas from the Rocky Mountain Veterinary Branch, DIR, NIAID, NIH, for the preparation of histopathologic samples. Shannon Taylor, currently at the Regional Biocontainment Laboratory, UTHSC, contributed ideas and discussions. The views and conclusions contained in this document are those of the authors and should not be interpreted as necessarily representing the official policies, either expressed or implied, of DHS or the US Government. The DHS does not endorse any products or commercial services mentioned in this presentation. In no event shall the DHS, BNBI or NBACC have any responsibility or liability for any use, misuse, inability to use or reliance upon the information contained herein. In addition, no warranty of fitness for a particular purpose, merchantability, accuracy or adequacy is provided regarding the contents of this document. Notice: This manuscript has been authored by Battelle National Biodefense Institute, LLC under Contract No. HSHQDC-15-C-00064 with the US Department of Homeland Security. The United States Government retains and the publisher, by accepting the article for publication, acknowledges that the United States Government retains a non-exclusive, paid up, irrevocable, worldwide license to publish or reproduce the published form of this manuscript or allow others to do so, for the United States Government purposes.

Conflicts of Interest: The authors declare no conflict of interest.

References

1. Chua, K.B. Nipah virus outbreak in Malaysia. *J. Clin. Virol.* **2003**, *26*, 265–275. [CrossRef]
2. Rahman, M.M.A.; Hossain, M.J.; Sultana, S.; Homaira, N.; Khan, S.U.; Rahman, M.M.A.; Gurley, E.S.; Rollin, P.E.; Lo, M.K.; Comer, J.A.; et al. Date palm sap linked to Nipah virus outbreak in Bangladesh, 2008. *Vector Borne Zoonotic Dis.* **2012**, *12*, 65–72. [CrossRef] [PubMed]
3. Chadha, M.S.; Comer, J.A.; Lowe, L.; Rota, P.A.; Rollin, P.E.; Bellini, W.J.; Ksiazek, T.G.; Mishra, A. Nipah virus-associated encephalitis outbreak, Siliguri, India. *Emerg. Infect. Dis.* **2006**, *12*, 235–240. [CrossRef] [PubMed]
4. Hsu, V.P.; Hossain, M.J.; Parashar, U.D.; Ali, M.M.; Ksiazek, T.G.; Kuzmin, I.; Niezgoda, M.; Rupprecht, C.; Bresee, J.; Breiman, R.F. Nipah virus encephalitis reemergence, Bangladesh. *Emerg. Infect. Dis.* **2004**, *10*, 2082–2087. [CrossRef] [PubMed]
5. Lo, M.K.; Rota, P.A. The emergence of Nipah virus, a highly pathogenic paramyxovirus. *J. Clin. Virol.* **2008**, *43*, 396–400. [CrossRef]
6. Wong, K.T.; Shieh, W.J.; Kumar, S.; Norain, K.; Abdullah, W.; Guarner, J.; Goldsmith, C.S.; Chua, K.B.; Lam, S.K.; Tan, C.T.; et al. Nipah virus infection: Pathology and pathogenesis of an emerging paramyxoviral zoonosis. *Am. J. Pathol.* **2002**, *161*, 2153–2167. [CrossRef]
7. Paton, N.I.; Leo, Y.S.; Zaki, S.R.; Auchus, A.P.; Lee, K.E.; Ling, A.E.; Chew, S.K.; Ang, B.; Rollin, P.E.; Umapathi, T.; et al. Outbreak of Nipah-virus infection among abattoir workers in Singapore. *Lancet* **1999**, *354*, 1253–1256. [CrossRef]
8. Hooper, P.; Zaki, S.; Daniels, P.; Middleton, D. Comparative pathology of the diseases caused by Hendra and Nipah viruses. *Microbes Infect.* **2001**, *3*, 315–322. [CrossRef]
9. Maisner, A.; Neufeld, J.; Weingartl, H. Organ- and endotheliotropism of Nipah virus infections in vivo and in vitro. *Thromb. Haemost.* **2009**, *102*, 1014–1023.
10. Ang, B.S.P.; Lim, T.C.C.; Wang, L. Nipah virus infection. *J. Clin. Microbiol.* **2018**, *56*. [CrossRef] [PubMed]
11. Chua, K.B.; Lam, S.K.; Tan, C.T.; Hooi, P.S.; Goh, K.J.; Chew, N.K.; Tan, K.S.; Kamarulzaman, A.; Wong, K.T. High mortality in Nipah encephalitis is associated with presence of virus in cerebrospinal fluid. *Ann. Neurol.* **2000**, *48*, 802–805. [CrossRef]
12. Chua, K.B.; Goh, K.J.; Wong, K.T.; Kamarulzaman, A.; Tan, P.S.; Ksiazek, T.G.; Zaki, S.R.; Paul, G.; Lam, S.K.; Tan, C.T. Fatal encephalitis due to Nipah virus among pig-farmers in Malaysia. *Lancet* **1999**, *354*, 1257–1259. [CrossRef]
13. Wong, K.T.; Shieh, W.J.; Zaki, S.R.; Tan, C.T. Nipah virus infection, an emerging paramyxoviral zoonosis. *Springer Semin. Immunopathol.* **2002**, *24*, 215–228. [CrossRef] [PubMed]
14. Wong, K.T.; Tan, C.T. Clinical and pathological manifestations of human henipavirus infection. *Curr. Top. Microbiol. Immunol.* **2012**, *359*, 95–104. [PubMed]
15. DeBuysscher, B.L.; de Wit, E.; Munster, V.J.; Scott, D.; Feldmann, H.; Prescott, J. Comparison of the pathogenicity of Nipah virus isolates from Bangladesh and Malaysia in the Syrian hamster. *PLoS Negl. Trop. Dis.* **2013**, *7*, e2024. [CrossRef] [PubMed]
16. Bonaparte, M.I.; Dimitrov, A.S.; Bossart, K.N.; Crameri, G.; Mungall, B.A.; Bishop, K.A.; Choudhry, V.; Dimitrov, D.S.; Wang, L.-F.F.; Eaton, B.T.; et al. Ephrin-B2 ligand is a functional receptor for Hendra virus and Nipah virus. *Proc. Natl. Acad. Sci. USA* **2005**, *102*, 10652–10657. [CrossRef]
17. Negrete, O.A.; Levroney, E.L.; Aguilar, H.C.; Bertolotti-Ciarlet, A.; Nazarian, R.; Tajyar, S.; Lee, B. EphrinB2 is the entry receptor for Nipah virus, an emergent deadly paramyxovirus. *Nature* **2005**, *436*, 401–405. [CrossRef]
18. Negrete, O.A.; Wolf, M.C.; Aguilar, H.C.; Enterlein, S.; Wang, W.; Mühlberger, E.; Su, S.V.; Bertolotti-Ciarlet, A.; Flick, R.; Lee, B. Two key residues in EphrinB3 are critical for its use as an alternative receptor for Nipah virus. *PLoS Pathog.* **2006**, *2*, e7. [CrossRef]
19. Flanagan, J.G.; Vanderhaeghen, P. The ephrins and Eph receptors in neural development. *Annu. Rev. Neurosci.* **1998**, *21*, 309–345. [CrossRef] [PubMed]
20. Augustin, H.G.; Reiss, Y. EphB receptors and ephrinB ligands: Regulators of vascular assembly and homeostasis. *Cell Tissue Res.* **2003**, *314*, 25–31. [CrossRef]
21. Pernet, O.; Wang, Y.E.; Lee, B. Henipavirus receptor usage and tropism. *Curr. Top. Microbiol. Immunol.* **2012**, *359*, 59–78. [PubMed]
22. Erbar, S.; Diederich, S.; Maisner, A. Selective receptor expression restricts Nipah virus infection of endothelial cells. *Virol. J.* **2008**, *5*, 142. [CrossRef] [PubMed]
23. Pernet, O.; Pohl, C.; Ainouze, M.; Kweder, H.; Buckland, R. Nipah virus entry can occur by macropinocytosis. *Virology* **2009**, *395*, 298–311. [CrossRef] [PubMed]
24. Foo, S.S.; Turner, C.J.; Adams, S.; Compagni, A.; Aubyn, D.; Kogata, N.; Lindblom, P.; Shani, M.; Zicha, D.; Adams, R.H. Ephrin-B2 controls cell motility and adhesion during blood-vessel-wall assembly. *Cell* **2006**, *124*, 161–173. [CrossRef] [PubMed]
25. Prescott, J.; DeBuysscher, B.L.B.L.; Feldmann, F.; Gardner, D.J.D.J.; Haddock, E.; Martellaro, C.; Scott, D.; Feldmann, H. Single-dose live-attenuated vesicular stomatitis virus-based vaccine protects African green monkeys from Nipah virus disease. *Vaccine* **2015**, *33*, 2823–2829. [CrossRef] [PubMed]
26. Bossart, K.N.; Zhu, Z.; Middleton, D.; Klippel, J.; Crameri, G.; Bingham, J.; McEachern, J.A.; Green, D.; Hancock, T.J.; Chan, Y.P.; et al. A neutralizing human monoclonal antibody protects against lethal disease in a new ferret model of acute nipah virus infection. *PLoS Pathog.* **2009**, *5*, e1000642. [CrossRef]

27. Schountz, T.; Campbell, C.; Wagner, K.; Rovnak, J.; Martellaro, C.; Debuysscher, B.L.; Feldmann, H.; Prescott, J. Differential innate immune responses elicited by nipah virus and cedar virus correlate with disparate in vivo pathogenesis in hamsters. *Viruses* **2019**, *11*, 291. [CrossRef] [PubMed]

28. Debuysscher, B.L.; Scott, D.; Marzi, A.; Prescott, J.; Feldmann, H. Single-dose live-attenuated Nipah virus vaccines confer complete protection by eliciting antibodies directed against surface glycoproteins. *Vaccine* **2014**, *32*, 2637–2644. [CrossRef]

29. Geisbert, T.W.; Daddario-DiCaprio, K.M.; Hickey, A.C.; Smith, M.A.; Chan, Y.P.; Wang, L.F.; Mattapallil, J.J.; Geisbert, J.B.; Bossart, K.N.; Broder, C.C. Development of an acute and highly pathogenic nonhuman primate model of Nipah virus infection. *PLoS ONE* **2010**, *5*, e10690. [CrossRef]

30. Wong, K.T.; Grosjean, I.; Brisson, C.; Blanquier, B.; Fevre-Montange, M.; Bernard, A.; Loth, P.; Georges-Courbot, M.-C.C.; Chevallier, M.; Akaoka, H.; et al. A golden hamster model for human acute Nipah virus infection. *Am. J. Pathol.* **2003**, *163*, 2127–2137. [CrossRef]

31. Baseler, L.; Scott, D.P.; Saturday, G.; Horne, E.; Rosenke, R.; Thomas, T.; Meade-White, K.; Haddock, E.; Feldmann, H.; de Wit, E. Identifying Early Target Cells of Nipah Virus Infection in Syrian Hamsters. *PLoS Negl. Trop. Dis.* **2016**, *10*, e0005120. [CrossRef]

32. Tumilowicz, J.J.; Gawlik, M.E.; Powell, B.B.; Trentin, J.J. Replication of cytomegalovirus in human arterial smooth muscle cells. *J. Virol.* **1985**, *56*, 839–845. [CrossRef]

33. Tumilowicz, J.J. Characteristics of human arterial smooth muscle cell cultures infected with cytomegalovirus. *Vitr. Cell. Dev. Biol. Anim.* **1990**, *26*, 1144–1150. [CrossRef] [PubMed]

34. Lemström, K.B.; Bruning, J.H.; Bruggeman, C.A.; Lautenschlager, I.T.; Häyry, P.J. Cytomegalovirus infection enhances smooth muscle cell proliferation and intimal thickening of rat aortic allografts. *J. Clin. Investig.* **1993**, *92*, 549–558. [CrossRef] [PubMed]

35. Jenson, H.B.; Montalvo, E.A.; McClain, K.L.; Ench, Y.; Heard, P.; Christy, B.A.; Dewalt-Hagan, P.J.; Moyer, M.P. Characterization of natural Epstein-Barr virus infection and replication in smooth muscle cells from a leiomyosarcoma. *J. Med. Virol.* **1999**, *57*, 36–46. [CrossRef]

36. Burch, G.E.; Harb, J.M. Encephalomyocarditis (EMC) virus infection of the mouse aorta. An ultrastructural study. *Am. Heart J.* **1973**, *86*, 669–675. [CrossRef]

37. Melnick, J.L.; Dreesman, G.R.; Mccollum, C.H.; Petrie, B.L.; Burek, J.; Debakey, M.E. Cytomegalovirus antigen within human arterial smooth muscle cells. *Lancet* **1983**, *322*, 644–647. [CrossRef]

cells

Review

How Influenza Virus Uses Host Cell Pathways during Uncoating

Etori Aguiar Moreira [1], Yohei Yamauchi [2] and Patrick Matthias [1,3,*]

1 Friedrich Miescher Institute for Biomedical Research, 4058 Basel, Switzerland; etori.moreira@fmi.ch
2 Faculty of Life Sciences, School of Cellular and Molecular Medicine, University of Bristol, Bristol BS8 1TD, UK; yohei.yamauchi@bristol.ac.uk
3 Faculty of Sciences, University of Basel, 4031 Basel, Switzerland
* Correspondence: Patrick.Matthias@fmi.ch

Abstract: Influenza is a zoonotic respiratory disease of major public health interest due to its pandemic potential, and a threat to animals and the human population. The influenza A virus genome consists of eight single-stranded RNA segments sequestered within a protein capsid and a lipid bilayer envelope. During host cell entry, cellular cues contribute to viral conformational changes that promote critical events such as fusion with late endosomes, capsid uncoating and viral genome release into the cytosol. In this focused review, we concisely describe the virus infection cycle and highlight the recent findings of host cell pathways and cytosolic proteins that assist influenza uncoating during host cell entry.

Keywords: influenza; capsid uncoating; HDAC6; ubiquitin; EPS8; TNPO1; pandemic; M1; virus–host interaction

Citation: Moreira, E.A.; Yamauchi, Y.; Matthias, P. How Influenza Virus Uses Host Cell Pathways during Uncoating. *Cells* **2021**, *10*, 1722. https://doi.org/10.3390/cells10071722

Academic Editors: Thomas Hoenen and Allison Groseth

Received: 7 May 2021
Accepted: 2 July 2021
Published: 8 July 2021

Publisher's Note: MDPI stays neutral with regard to jurisdictional claims in published maps and institutional affiliations.

1. Introduction

Viruses are microscopic parasites that, unable to self-replicate, subvert a host cell for their replication and propagation. Despite their apparent simplicity, they can cause severe diseases and even pose pandemic threats [1–3]. Emerging viral infections, caused by viruses that have not been previously recorded, continue to pose a major threat to global public health [4], as it is the case for the biggest pandemic of the millennium so far, the coronavirus disease 2019 (COVID-19) caused by the severe acute respiratory syndrome coronavirus 2 (SARS-CoV-2) [5].

Entry of an enveloped virus from the extracellular environment into cells proceeds through a number of essential steps [6]. These include binding and attachment of a virus outer protein to its receptor at the cell surface, penetration of the viral particle into the cytoplasm, uncoating of the proteinaceous capsid allowing release of the viral nucleic acids into the cell cytosol, viral genetic material replication, protein synthesis, and finally new viral particle assembly and budding from the infected cell. The dissection of the molecular events and viral–host interactions that take place once the virus binds to the cell surface is essential for understanding how a particular virus infects cells. It also allows the identification of potential new targets for antivirals and therapies for blocking or controlling the infection and onset of diseases. In addition, understanding how viruses adopt or hijack cellular pathways to their advantage often leads to novel insights in the normal functioning of these pathways and is, therefore, of general interest beyond virology.

As viruses recognize target cells by first binding to cell receptors, the discovery of the virus ligands is primordial for understanding the organ tropism, the potential host diversity, and the mechanism of infection [7,8]. Enveloped animal viruses enter their host cells by membrane fusion and two pathways have been described, depending on the characteristics of the virus fusion protein. Fusion can occur at the cell plasma membrane at physiological pH [9–15] or within the endocytic vacuolar system where it is triggered by a low pH [16–22].

Virus capsid opening, the so-called uncoating, enables the virus genetic material to be released in the cytosol and get ready for replication. Our understanding of virus uncoating mechanisms has grown substantially in recent years and defines uncoating as a complex, highly orchestrated, multi-step process that relies on both viral and cellular factors.

In this review, we describe the influenza A virus (IAV) infection cycle and focus on the capsid uncoating process. We explore the latest studies that elucidate IAV capsid uncoating and the host proteins involved.

2. The Infection Cycle of IAV

2.1. Influenza Virus Structure, Proteins and Classification

Influenza viruses are orthomyxoviruses, members of the family Orthomyxoviridae, which comprises the genera Influenzavirus A, B and C, Thogotovirus, Quaranjavirus, and Isavirus [23,24]. IAV is pleiomorphic [25] meaning that viruses with varying morphologies can be produced by an infected cell. The most studied virus shape is spherical, with around 100 nm diameter; the other is filamentous, from 100 nm to 30 μm in length [26–28]. The filamentous morphology is typical of clinical isolates, whereas the spherical shape is common in laboratory-passaged strains [28,29]. Whilst the biological function and consequences of the viral morphology during IAV infection remain unknown, studies have shown that it has implications in transmission, host adaptation and pathogenesis [30–36]. IAV has eight distinct gene segments organized as a single-stranded, negative-sense RNA genome assembled into ribonucleoprotein complexes (vRNPs) that produce at least eleven proteins [37]. The segmented nature of the IAV genome has many implications, the most popular is that it provides an evolutionary benefit by enabling the virus to evolve by reassortment of gene segments between coinfecting viruses (see reference [38] for a review on this topic). Filamentous and spherical particles have their vRNPs arranged in bundle with all segments associated with the M1 from the capsid at the same end of the virus [29,39]. Figure 1 presents a scheme of the virus structure and genome.

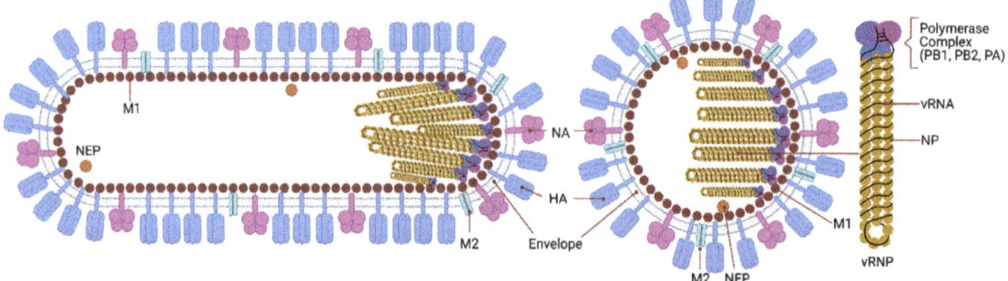

Figure 1. IAV structure and genome. Influenza is an enveloped virus in which structural proteins can be found associated with the virus envelope, a lipid bilayer derived from the plasma membrane of the host cell. The viral envelope contains three of the viral transmembrane proteins: hemagglutinin (HA), neuraminidase (NA), and the matrix ion channel M2. HA and NA proteins are the main proteins at the virus surface and HA is four times more abundant than NA. M2 also penetrates the envelope but represents a minor component of the envelope, with about 20 molecules per virus particle. The matrix protein M1 is found beneath the lipid membrane, and forms a rigid single-helical layer shell, the virus capsid. The nuclear export protein (NEP) is found in the interior of the virus. The IAV genome consists of eight negative-sense RNA segments that form distinct viral ribonucleoproteins (vRNPs). vRNPs are assembled as virus RNA segments where the termini of viral RNAs associate with the viral RNA-dependent RNA polymerase complex, PB1, PB2 and PA, while the rest of the viral RNAs are bound by oligomers of the nucleoprotein, NP. The virus has an asymmetric internal structure, maintained by vRNPs-vRNPs and M1-vRNPs interactions. Not shown in the figure, the interior of IAV bears a substantial number of host proteins (ubiquitin, tubulin, actin, annexin, among others). IAV is known to display a number of shapes. The spherical form of IAVs is typically about 100 nm in diameter. Filamentous forms of IAVs can be over a few μm in length.

The surface glycoproteins hemagglutinin (HA) and neuraminidase (NA) are the most abundant proteins present in the lipid bilayer envelope. Based on the antigenic properties and phylogenetic clustering of HA and NA, IAV can be classified into several subtypes. There are eighteen different HA (H1-H18) and eleven NA (N1-N11) serotypes. The relative abundance of each protein within the virus particle varies among virus subtypes and depends on the genetic background, with the HA/NA ratio being on average 4 to 1 [40]. However, for the IAV/WSN/33 (H1N1) strain it is approximately 10 to 1 [41] and for IAV/Aichi/68 (H3N2) it is 5 to 1 [39]. HA and NA play a role in the recognition and binding to the receptor in target cells and release of the virus during budding from the plasma membrane, respectively [42]. Due to the exposure at the virus surface and their biological functions, these two proteins are the major antigenic targets of neutralizing antibodies. In fact, during a natural infection the majority of antibodies will target HA, with lower amounts targeting NA or even other virus proteins [43].

IAV has two matrix proteins: M1 and M2. They are the main determinants of the spherical or filamentous virus morphology [44]. M1 is the major structural component of the virus, forming a rigid shell, the virus capsid. It acts as an adaptor between the lipid envelope and the vRNPs, besides being the driving force for virus budding [45–47]. A recent study solved the structure of assembled M1 within intact virus particles, gave structural insights on how M1 oligomerizes to form the capsid and how the pH change triggers the capsid disassembly [48]. Five histidine residues contributed by three sequential M1 monomers form a histidine cluster that can serve as the switch for the pH-mediated M1 disassembly [48]. M2 is an ion channel present in low amounts in the virus envelope, with approximately 20 to 60 units on each virus particle [49]. M2 forms tetrameric ion channels that open in response to the endosome low pH, allowing a proton flux into the virus. Lowering the pH of the virus interior is involved in the HA maturation by changing HA conformation from a native (nonfusogenic) structure to a fusion-active (fusogenic) [50–52]. The M2 protein cytoplasmic tail interacts with the M1 protein and influences virus assembly and genome packaging at the site of virus budding [53,54].

Inside the virus, each gene segment is associated with a trimeric RNA-dependent RNA polymerase complex consisting of the PB1, PB2, and PA proteins [55]. Multiple nucleoprotein (NP) molecules bind the viral RNA with high affinity and, together with the polymerase proteins, forms the vRNPs [56]. The nuclear export protein (NEP), also known as non-structural protein 2 (NS2) is found inside virus particles in low amounts where it may interact with M1 [57]. Its main function is the nuclear export of vRNPs.

The non-structural protein 1 (NS1) is abundant in IAV-infected cells but usually not detected in virus particles [58]. Nevertheless, recent studies reported that a low amount of NS1 is present in purified virus particles and suggest that NS1 can be incorporated during assembly [59,60]. Although the relevance of the presence of NS1 in the virus particles is unknown, its incorporation might have to do with its ability to associate with the IAV vRNAs and facilitate the genome packaging at the influenza budding sites [60]. NS1 is a non-essential virulence factor that has multiple functions during the viral life cycle. Its major role is to antagonize type I interferon-mediated antiviral responses [61]. NS1 also controls vRNA splicing and temporal regulation of the RNA synthesis [62,63], induces or suppresses host apoptotic responses [64,65], and has a role in strain-specific pathogenesis [66,67], among others.

2.2. Early Events of Influenza Virus Infection

2.2.1. Receptor Binding and Envelope Fusion with Late Endosome

During the first step of IAV infection of the host cell, attachment, HA binds to the target cell via sialic acid linkages on host glycoproteins [68,69]. The sialic acid binding specificity of HA is one of the major determinants for viral tropism and host specificity; changes in key HA amino acids that control its binding specificity have been identified to contribute to the spillover of avian viruses to humans, leading to new influenza epidemics [70–73]. In general, human IAVs exhibit a strong preference for binding glycans terminating with α2,6-linked sialic acid and replicate in the respiratory tract, whereas avian IAVs have a preference for α2,3-linked sialic acid [74]. In contrast, bat IAV carries the H17 or H18 HA serotypes which cannot bind to sialic acid; rather, they require the host MHC class II proteins to infect cells [75,76]. Figure 2 illustrates the main steps of the virus life cycle.

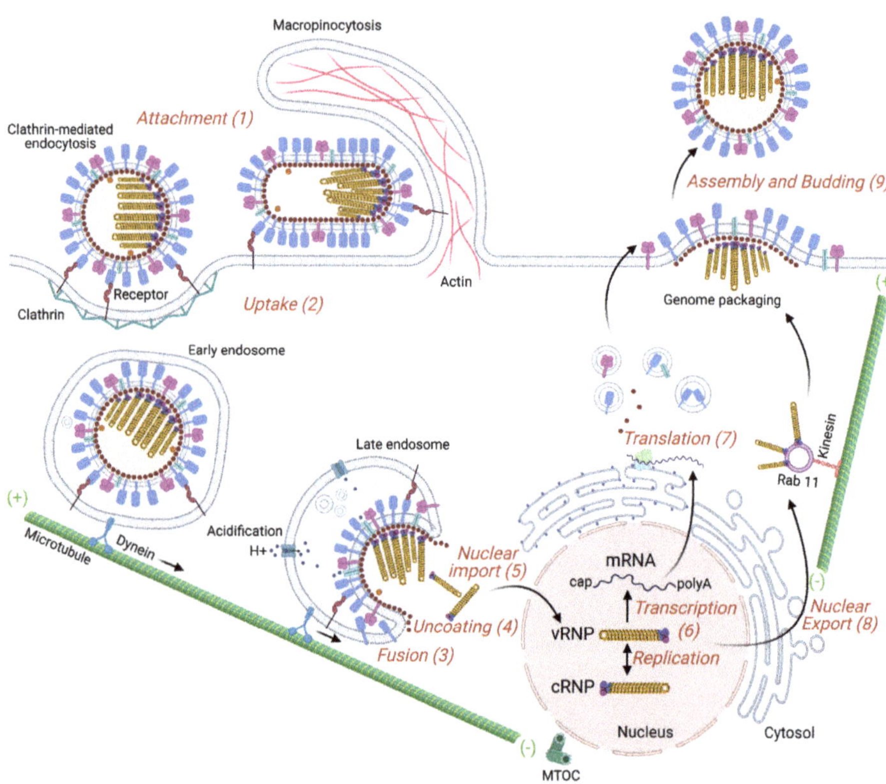

Figure 2. Overview of the IAV replication cycle. The influenza virus life cycle can be divided into several stages: (**1**) Virus binding to the target cell. HA binds to sialic acid found on the surface of the host cell's membrane. (**2**) Entry into the host cell: a clathrin-mediated endocytosis or macropinocytosis takes place. Early endosome containing viruses is transported by dynein along microtubules to the perinuclear region close to the microtubule-organizing center (MTOC). (**3**) Fusion of the virus envelope with the endosomal membrane. Acidification increases progressively from endocytic vesicles to late endosomes and induces a HA conformational change to a fusion-competent state. M2, an acid-activated viral ion channel, is required for efficient viral envelope fusion with the endosomal membrane and nucleocapsid release. (**4**) Uncoating of the virus capsid by disassembly of the M1 proteins and release of the viral ribonucleoproteins (vRNPs) to the cytosol. (**5**) Entry of vRNPs into the nucleus by an active nuclear import pathway. (**6**) Transcription and replication of the viral genome. The IAV genome is composed of negative-sense strand RNAs. The genome is first converted into positive-sense RNAs, forming complementary ribonucleoprotein (cRNP) complexes, that serve as templates to produce viral RNAs. The transcription of the vRNA generates mature viral messenger RNAs (mRNAs) that have a 5′ methylated cap and a poly(A) tail. (**7**) Viral protein translation occurs by free ribosomes or ribosomes on the rough endoplasmic reticulum. Some of these proteins enter the nucleus where they assemble with viral RNAs. (**8**) Export of the vRNPs from the nucleus. vRNPs are exported out of the nucleus via the CRM1 dependent pathway through the nuclear pores. (**9**) Transport of viral components, assembly and budding at the host cell plasma membrane. Viral glycoproteins, HA and NA, associate with lipid rafts, membrane microdomains comprised of densely packed cholesterol and sphingolipids. vRNP complexes are transported as sub-bundles on Rab11 to recycling endosomes close to ER exit sites toward the plasma membrane and are incorporated as a complex of eight different vRNPs into budding viruses. Finally, the plasma membrane containing the viral structural proteins at the assembly site bends releasing infectious virus into the extracellular environment.

Even though the NA is mainly recognized for its role at the virus budding stage, where it removes sialic acid bound to the newly synthesized HA and NA on nascent viruses, it has also been implicated in helping IAV to penetrate the mucus layer and get access to the receptors at the host cell membrane [77,78]. For this, NA locally cleaves sialylated O-linked glycans covering mucins and cell glycocalyx, decreasing the number of sialylated decoys and promoting the motility of IAV towards the receptors on the target cell surface [79]. The spherical IAV bound to the sialic acid-containing receptor proteins at the plasma membrane activates an internalization pathway that is by default clathrin-mediated endocytosis. In addition to this traditional route, IAV may have other entry pathways that could be dependent on the cell type. For instance, filamentous IAV enters host cells by a dynamin-independent route, using macropinocytosis as the primary entry mechanism [80]. The intact filamentous IAVs are trafficked to the acidic late-endosomal compartment within macropinosomes [81]. Similarly, spherical IAV has recently been reported to also use macropinocytosis [80]. Caveolae have already been described as an alternative route to clathrin for mediating the entry of IAV in MDCK cells [82]. By combining inhibitory methods to block both clathrin-mediated endocytosis and uptake by caveolae in HeLa cells, another study demonstrated that a non-clathrin-dependent, non-caveolae-dependent, but dynamin-dependent endocytic pathway also exists [83].

The traffic of viruses within endosomes towards the cell nucleus occurs through the cytoskeleton using actin, myosin and dynein motor protein, and microtubules (MTs) [84,85]. Polarized respiratory epithelium is the target of IAV in vivo, in which it preferentially enters the cells from the apical surface [86–88]. However, most molecular studies on the virus entry have been carried out using non-polarized cell lines. There are significant differences between polarized and non-polarized cells regarding receptor distribution, cytoskeletal structure and the mechanism of endocytosis [85,89]. For instance, IAV seems to depend much more on the actin dynamics in polarized than non-polarized cells [90]. Following infection, the cytoskeleton undergoes structural reorganization and the endosomes harboring viruses travel in a retrograde traffic towards the microtubule organizing center (MTOC), in close proximity to the cellular nucleus [91]. The kinetics of virus-containing endosomes vary according to the cell type and the virus subtypes. It has been reported that IAV/X31 strain can be found in early endosomes, marked by early endosomal autoantigen 1 (EEA1) [92], Rab5 [93,94], and rabenosyn-5 (Rab5 effector) [95,96], around 5 min after adsorption in dendritic cells [97]. While in Chinese hamster ovary CHO cells, the same virus already fuses its envelope with late endosomes, usually marked by lysosomal-associated membrane protein-1 (Lamp1) [98] and Rab7 [93], in the perinuclear region only 8 min after binding [84]. For IAV/WSN/33 in human lung epithelial A549 cells, co-localization of virus proteins and early endosomes peaked at 45 min whereas co-localization with late endosomes only at 120 min [99].

Upon entry through the endocytic pathway, HA only reaches a fusion-competent form when the virus has trafficked beyond early endosomes. This happens because a progressive pH drop by endosomal acidification is needed for HA conformational changes prior to fusion of the envelope and endosome membranes. The acidification of endosomes occurs during their maturation and the M2 proton channel in the virus envelope mediates the flux of protons into the IAV particle upon acid activation (pH $\approx$ 6) [100]. The drop of pH in endosomes has at least two main functions during early IAV stages. First, as mentioned above, the low pH in late endosomes triggers a conformational change in the HA glycoprotein that exposes a fusion peptide. Second, it strips away the M1 matrix protein from the capsid during uncoating.

2.2.2. IAV Capsid Uncoating

The different steps of virus uncoating are regulated by cellular cues which come from cellular receptors, enzymes, and small chemicals including ions [101]. Here, we summarize the IAV uncoating process, and detail it further to discuss the role of host proteins in IAV uncoating (see Section 3). Uncoating refers to the series of events that alter the viral core structure leading to disassembly which is essential for the release of the viral genomic segments into the cytosol. It is a continuous and dynamic event that begins inside acidic endosomes and is completed in the cytosol by multiple host proteins that interact with viral core components. Due to the methodological limitations for identifying and visualizing all the molecules playing a role during uncoating, this process has remained relatively poorly studied. Yet, recently some important cellular proteins have been described to be involved in this process. Histone deacetylase 6 (HDAC6), epidermal growth factor receptor pathway substrate 8 (EPS8), transportin-1 (TRN-1 or TNPO1) have all been shown to be involved in the IAV core uncoating [102–104] and their role for other viruses begins to be examined. Catalyzing the capsid opening for a fast genome release may decrease the probability of the virus genetic material being degraded by cellular RNases as has been shown for iflaviruses, positive-strand RNA viruses from the family Iflaviridae (order Picornavirales) [105]. After release of the viral RNPs into the cytosol in the proximity of the cell nucleus, the vRNPs are imported into the nucleus through nuclear pore complexes using the nucleoprotein (NP) nuclear localization sequence (NLS) motifs and the importin α/β-dependent nuclear import pathway.

2.3. Nuclear Import, IAV Genome Replication, and vRNP Export

As mentioned above, the IAV genome has eight segments of single-stranded negative-sense RNA, each of them is transcribed in the nucleus of the host cell [106]. Three viral RNAs types are synthesized: viral mRNAs of positive sense (mRNA), viral genomic RNAs (vRNA) of negative sense, and complementary RNAs (cRNA) of positive sense. The vRNAs are bound by a RNA-dependent RNA polymerase, forming a viral ribonucleoprotein (vRNP) complex [107]. Although replication is a primer-independent process, during transcription of a viral mRNA the viral RNA polymerase relies on host capped RNAs as cap-donors [108]. As IAV vRNAs are of negative sense, in order for the genome to be transcribed it first must be converted into a positive sense RNA that serves as template for the production of new viral RNAs [59]. The cRNA is the replication intermediate, a full-length complement of the vRNA that works as a template for the synthesis of new copies of vRNA [107]. The formation of new vRNP complexes results from the binding of newly synthesized subunit proteins (PB1, PB2 and PA) and NP proteins to the vRNAs. For further details, the reader is referred to excellent recent reviews [109,110].

The influenza virus infection leads to a slowdown in the synthesis of cellular proteins, this phenomenon is known as cell shutoff [111–114]. The synthesis rates of vRNAs and proteins reach a maximum within the first few hours after infection before dropping [115]. One study reported the production of most viral proteins to peak in the first 8–12 h after infection [114]. The synthesis of the IAV proteins is regulated at the transcriptional level, and the synthesis rate and accumulation level of the mRNAs differ considerably among the eight RNA segments [116,117]. For instance, there is an early production of proteins such as NS1 and NP and a delayed synthesis of M1 [114,118].

Nuclear export of vRNPs is mediated by the cellular protein Crm1, or exportin 1, a member of the importin β family, and putatively by the viral protein NEP/NS2 [119–123]. The nuclear import relies on importin α, which acts as an adaptor between importin β and NLS-cargos [124]. M1 shuttles between the cytoplasm and the nucleus and has important functions in both compartments. In the nucleus, M1 proteins attach to vRNPs forming M1-vRNP complexes and participate in transport of vRNPs to the cytosol [125]. The NEP/NS2 protein contains a highly conserved nuclear export signal (NES) motif in its amino-terminal region and mediates the nuclear export of vRNP-M1-NS2 complexes [126]. The export of vRNP is impaired when cells are infected with a recombinant virus that cannot express NS2 or have mutations in the NS2 NES [127]. Cytoplasmic M1 proteins inhibit the nuclear import of vRNP complexes [128,129], and newly synthesized vRNPs associated with M1 protein are unable to re-enter the cell nucleus [128].

2.4. Late Events of IAV Infection

Trafficking of the vRNPs and IAV Proteins to the Cell Plasma Membrane, Genome Packaging and Virus Budding

Newly synthesized NP and the viral polymerase proteins (PB1, PB2 and PA) form a complex with vRNA, and the formed vRNP is transported to the plasma membrane or to the apical site of polarized epithelial cells for genome packaging and virus budding [130]. Similar to what happens during viral entry, the viral egress pathways depend on the cytoskeleton, transport vesicles, and motor proteins [131]. After nuclear export, vRNPs can be found colocalized with microtubules and concentrated at the MTOC [132]. The small GTPase Rab11 mediates the transport of the vRNPs across the cytoplasm to the viral budding sites at the plasma membrane [133–137]. It mediates the docking of a single vRNP or vRNP sub-bundles to recycling endosomes close to ER exit sites through direct or indirect interaction of its active GTP bound form with the viral polymerase complex proteins, taking the form of liquid viral inclusions [138–142]. Transmission electron tomography of budded IAV virions shows a distinct organization of vRNPs in which a central segment is surrounded by seven different segments of various lengths [143].

In addition to the vRNPs, IAV structral proteins (HA, NA, M1, M2) must be transported to the plasma membrane. Both HA and NA have been shown to possess apical determinants in their transmembrane domain [144,145]. HA, NA, and M2 transmembrane domains contain specific sorting signals that promote their association with sphingoglycolipid rafts at the plasma membrane [146,147]. While the transport of the IAV membrane-associated proteins has been well characterized, the mechanism by which other viral core proteins are transported to the budding sites is vague. M1 is synthesized on free cytosolic polyribosomes and may possess apical determinants or diffuse to the assembly site, or a combination of these pathways. It was shown that the M1 associates with HA and NA at the budding site and only a small fraction of the cytoplasmic M1 associates with cellular membranes in the absence of another viral protein [148]. In contrast, another study found that M1 hast the ability to associate with the membrane independent of the viral glycoproteins [45]. It is also possible that M1 may be able to associate with membranes through electrostatic interactions [149].

Virus assembly is coordinated by M1, which binds to all viral components and the plasma membrane. Interactions of M1 with other M1 proteins, vRNPs, HA and NA facilitate concentration of viral components and exclusion of host proteins from the budding site. M1 also interacts with the cytoplasmic tail and transmembrane domain of the glycoproteins HA and NA and with M2, functioning as a bridge between the viral envelope and the vRNPs [150,151]. Virus bud formation requires membrane bending at the budding site. A combination of factors including the increased concentration of viral proteins and the interaction of M1 with the viral glycoproteins, M1-M1 and M1-vRNPs play an important role for triggering virus budding [151]. Asymmetry of the lipid bilayer in lipid raft is likely to cause a curvature of the plasma membrane at the assembly site leading to bud formation. The matrix M2 transmembrane protein further facilitates virus release from the infected host cell. M2 is able to both contribute to curvature induction and also sense curvature to line up in manifolds where local membrane line tension is high [152]. During viral budding, vRNPs with their polymerase-binding ends at the budding tip are oriented in a parallel or antiparallel fashion [153]. Eventually, fusion of the opposing membranes leading to the closure of the bud will take place and newly formed viruses will be released into the extracellular environment.

3. In Focus: Influenza Virus Capsid Uncoating

3.1. Involvement of Ubiquitin Chains in Influenza Virus Uncoating

Ubiquitin (Ub) is a small 76 amino acids protein with a molecular mass of about 8.6 kDa. It participates in multiple cellular signaling pathways, that are usually involved in the regulation of protein function and homeostasis. The ubiquitin proteasome system (UPS) forms a cellular machinery for the degradation of unwanted proteins. The aggresome processing pathway (APP) is an alternative system, whereby misfolded proteins are accumulated before being degraded by autophagy. Ubiquitination and ubiquitin-like modification is usurped by many viruses to establish infection, and IAV uses ubiquitin-enhanced viral uncoating mechanisms [104,154]. How the APP facilitates efficient IAV uncoating is described below.

Protein ubiquitination is a post-translational modification in which there is an addition of a Ub molecule to one or more sites, most frequently lysine residues of a target protein. Proteins can be monoubiquitinated or poly-monoubiquitinated if they have one or more Ub molecules, respectively. In addition, Ub monomers can be connected to each another forming chains of varying lengths, linkages, and structures.

Protein ubiquitination involves a series of cellular enzymes in cascade. As depicted in Figure 3, ubiquitination starts with the Ub-activating enzyme E1, followed by the Ub-conjugating enzyme E2 and by the Ub ligase E3, which form an isopeptide bond between the carboxyl terminus of Ub and the amino group of a lysine residue on the target protein [155]. E3 ligases determine the substrate specificity of the cascade by the covalent attachment of Ub to substrate proteins, but the E2-conjugating enzyme can also play a role in the substrate selection [156]. Ub is often linked to substrates as polymeric chains that vary in both linkage and length, with important consequences for their function [157]. Ub itself contains seven lysines (K; K6, 11, 27, 29, 33, 48 and 63), all of which can be used by the Ub ligases to generate the different types of chains on the target proteins [158]. The K48-based linkages lead mainly to the proteasome-mediated degradation of the ubiquitinated protein, while K63-based Ub chains control primarily protein endocytosis, as well as trafficking and enzyme activity [159–162]. K63 is also a signal for targeting misfolded proteins to the APP [163]. Free poly-Ub chains, referred to as unanchored Ub chains, arise when deubiquitinases (DUBs) remove a chain from a protein, or they can be generated through E1/E2/E3 cycles [164]. In contrast to Ub chains bound to target proteins, unanchored Ub chains have a free C-terminus which can be bound efficiently by a conserved zinc finger domain found in the DUB isopeptidase T or in HDAC6 [165,166].

Ubiquitination has a vital role in regulating a wide variety of processes in eukaryotes through multiple mechanisms, including protein degradation, protein trafficking, gene expression, DNA repair, control of the cell cycle and signaling [167–171]. The versatility of the Ub system in regulating protein function and cell behavior makes it a particularly attractive target for pathogens such as viruses [172]. Ub was thought to be exclusively a cellular protein until a report described a modified form in baculovirus particles [173]. Similarly, host Ub was reported in purified vaccinia virus and herpes simplex virus particles [174]. Ub was also identified by proteomics in filovirus, such as purified Ebola and Marburg viruses [175]. More recently, liquid chromatography mass spectrometry proteomic analyses revealed a great variety of host proteins in purified extracellular viruses [176]. Among these proteins, Ub (polyubiquitin B and C) was found in Ebola Zaire, Marburg Lake Victoria, HIV-1, moloney murine leukemia virus (MLV), herpes simplex type-1 (HSV-1), vaccinia virus (VACV), human cytomegalovirus (HCMV), vesicular stomatitis virus (VSV), respiratory syncytial virus (RSV) and IAV. Ub has also been reported in HIV-1 cores [177,178]. The importance of these proteins in the different steps of the virus life cycle is unknown but Ub is long speculated to participate in virus uncoating and replication of particles [174].

Unanchored Ub chains within the IAV structural core are exposed following virus envelope and endosome fusion at late endosomes close to the nuclear periphery [104,179]. Unanchored Ub chains can also be produced by DUBs that cleave off ubiquitin chains from substrates targeted to the proteasome. One example of such a DUB is Poh1, a proteasome-associated DUB that generates K63-linked unanchored ubiquitin [180,181]. The interaction of unanchored Ub chains with HDAC6 and the interaction of HDAC6 with motor proteins in microtubules and actin filaments generates physical forces that catalyze the dissociation of the capsid M1 layer.

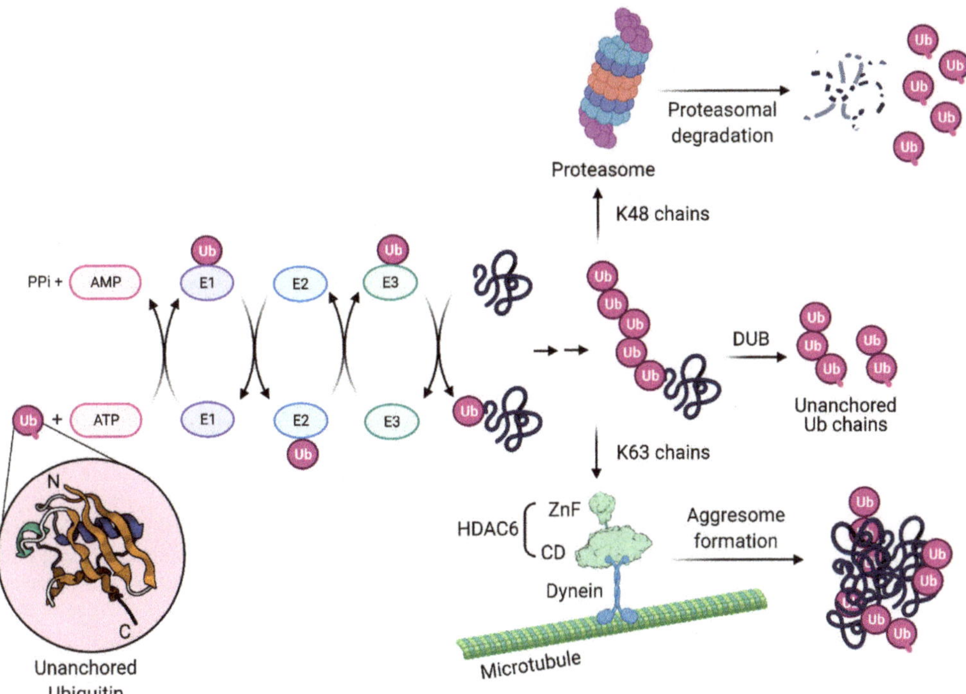

Figure 3. Protein ubiquitination, ubiquitin chains, proteasomal and aggresome-autophagy degradation. Ubiquitin (Ub) is a small, 76 amino acid protein present in all eukaryotic cells that plays a key role in the cellular defense mechanism by functioning as a proteolytic signal for the proteasome. The process of covalent Ub attachment to target proteins is called ubiquitination (also known as ubiquitylation). This post-translational modification forms by an isopeptide bond between the carboxyl terminus of Ub and a lysine residue on the target protein. First, Ub is covalently conjugated to the E1 (Ub-activating enzyme) in an active ATP-dependent reaction and transferred to the E2 (Ub-conjugating enzyme). The E3 (Ub–protein ligase) transfers the Ub from E2 to the target protein and determines the specificity. A monoubiquitinated protein can have the Ub chain elongated by E3 that creates Ub–Ub isopeptide bonds. Chain extension can happen through seven lysine (K) residues on Ub: K6, K11, K27, K29, K33, K48 and K63. Proteasomes recognize K48 chains leading to target protein degradation (upper part). Other lysine chains are involved in different biological functions. K63 chains do not specify degradation but usually facilitate the recruitment of other proteins in the formation of functional complexes involved in cellular signaling such as aggresome formation (lower part). HDAC6 can bind misfolded proteins entangled with Ub K63 chains and bridges to dynein motors, mediating transport to and formation of the aggresome compartment. Free poly-Ub chains, referred to as unanchored Ub chains, have been found in virus particles. Unanchored poly-Ub chains arise when a deubiquitinase (DUB) removes an intact chain from a protein, or they can be generated through E1/E2/E3 cycles. They can be recognized by HDAC6 and activate the aggresome pathway as well IAV capsid uncoating. Ub Protein Data Bank (PDB): 1UBI; NH2 and COOH termini are labeled N and C, respectively. HDAC6 zinc finger (ZnF) and catalytic domain (CD) PDB: 3C5K and 5G0I, respectively.

3.2. The Role of HDAC6 in Influenza Virus Uncoating

Histone deacetylases (HDACs) are enzymes that catalyze the removal of acetyl groups from modified lysine residues of histone and non-histone proteins and several classes of mammalian HDACs exist [182]. Class I HDACs are 400–500 amino acids long and include HDAC1, HDAC2, HDAC3 and HDAC8. Class II HDACs are approximately 1000 amino acids long; class IIa comprises HDAC4, HDAC5, HDAC7 and HDAC9, and class IIb comprises HDAC6 and HDAC10 [183–185]. The class III HDACs, also known as the sirtuins (SIRT1–7), are the silent information regulator 2 (Sir2) family of proteins and have a size ranging from 300 to 750 amino acids [186,187]. Despite the name, HDAC function is not limited to histone deacetylation and the regulation of gene transcription. HDAC6 localizes mainly in the cytosol and targets proteins through one of its deacetylase domains, CD1 or CD2, or its Ub-binding zinc finger domain, ZnF. In humans, HDAC6 contains a Ser Glu-repeat domain (SE14), which acts as a cytoplasmic retention signal and mediates its stable anchorage in the cytoplasm [188] where it deacetylases tubulin [189–191], heat shock protein 90 (Hsp90) [192–194], β-catenin [195,196], cortactin [197], MYH9, Hsc70, DNAJA1 [198] or the DEAD box RNA helicase 3, X-linked (DDX3X) [199]. HDAC6 has been associated with carcinogenesis, neurodegenerative diseases and inflammatory disorders, and has been exploited as a therapeutic target for pharmacological intervention [200–208].

HDAC6 has also been identified to have antiviral effects which have been linked to its enzymatic activity. It was found to inhibit IAV release by downregulating the trafficking of viral components to the plasma membrane via acetylated microtubules [209]. Overexpression of HDAC6 in cells leads to diminished viral budding due to tubulin deacetylation [210]. More recently, it was shown that HDAC6 regulates viral sensing by deacetylating retinoic acid inducible gene I, RIG-I [211], a key cytosolic sensor that detects RNA viruses through its C-terminal region and activates the production of antiviral interferons (IFNs) and proinflammatory cytokines. RIG-I is thought to be the most important sensor of IAV by binding to the virus genomic panhandle promoter region [212,213]. HDAC6 transiently binds to RIG-I and removes lysine 909 acetylation in the presence of viral RNAs, thus promoting RIG-I sensing of viral RNAs. Thus, HDAC6-mediated RIG-I deacetylation is critical for efficient viral RNA detection and IFN production. HDAC6 also acts as a negative regulator of IAV infection by deacetylating lysine 664 of the polymerase complex PA subunit, thereby restricting vRNA transcription and replication.

HDAC6, through its ZnF domain, associates with unanchored Ub chains [165,214,215]. As shown in Figure 3, Ub chains can be generated through E1/E2/E3 cycles and unanchored Ub chains are present in monoubiquitin, or ubiquitin derived from proteasomal degradation or the catalytic action of DUBs on pre-existing Ub chains. HDAC6 ZnF binds Ub with high affinity by recognizing its unanchored C-terminal sequence (-RLRGG-COOH) [216,217] and can recruit misfolded proteins with entangled Ub chains. In addition, HDAC6 interacts with the motor protein dynein and dynactin, the protein complex that links cargo to dynein. In this way, HDAC6 acts as a scaffold that mediates the transport of misfolded protein aggregates along microtubules and promotes formation of the aggresome compartment, which is a crucial pathway to attenuate misfolded protein-induced stress.

Inflammasome complexes are formed in response to pathogen-associated molecules and, for NLR family pyrin domain-containing protein 3 (NLRP3)- and pyrin-mediated inflammasomes, their assembly and downstream functions occur at the MTOC [218]. Similar to the formation of aggresomes, HDAC6 ZnF is required for the interaction with NLRP3 and pyrin inflammasome components and transport of these proteins using the microtubule retrograde transport by dynein for their activation in macrophages [218]. Given the importance of HDAC6 for viral uncoating, one might wonder how formation of novel IAV particles can take place in cells that contain HDAC6. The observation that in IAV-infected cells the C terminus of HDAC6 (encompassing the ZnF domain) gets cleaved off by Caspase-3 at late stages of the infection may help to solve this conundrum [219].

Ablation of class I HDACs in mice is lethal or leads to severe physiological dysfunction [220–222]. In contrast, mice lacking HDAC6 are viable and develop normally, despite having elevated tubulin acetylation in multiple organs [223]. The role of HDAC6 for efficient IAV uncoating was discovered by the observation that in mouse embryonic fibroblast cells lacking HDAC6, the IAV endocytic uptake, acid-induced HA maturation or fusion of virus envelope and late endosome were not affected. In contrast, virus capsid uncoating and the nuclear import of vRNPs were reduced in these cells in comparison to the wild type [104]. During infection, virus proteins and RNAs are detected by pathogen recognition receptors (PRR). This activates protein kinase R (PKR) that mediates the phosphorylation of eukaryotic initiation factor-α (eIF2α) on serine 51 to initiate the assembly of virus-induced stress granules concomitant with repression of cellular proteins translation [224]. By mimicking a misfolded protein aggregate, IAV hijacks the APP to its benefit [104]. The role of HDAC6 and Ub for IAV capsid uncoating can be visualized in Figure 4.

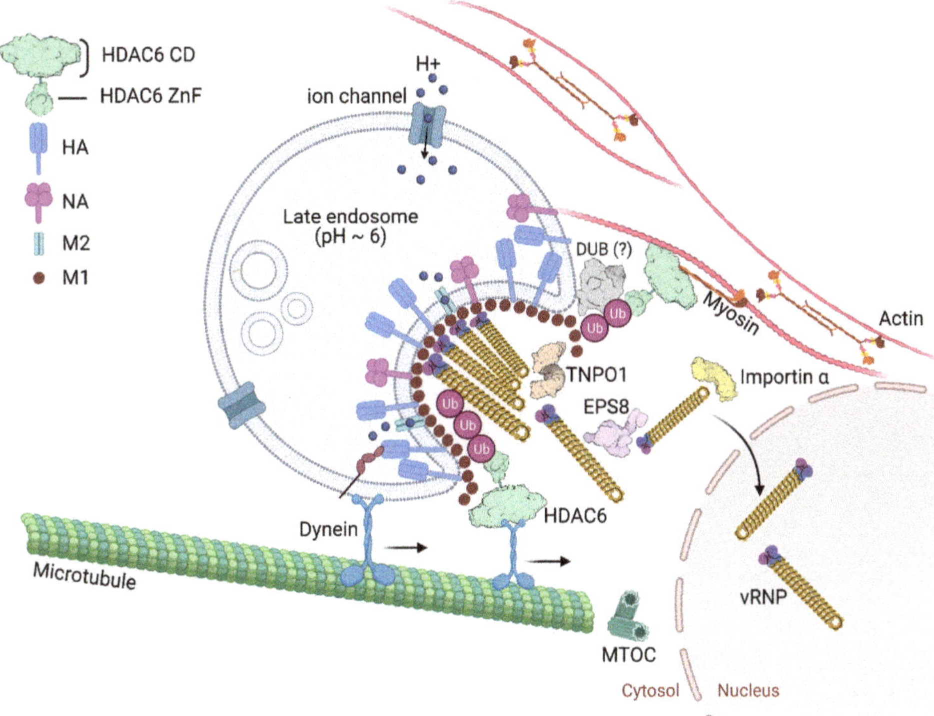

Figure 4. IAV capsid uncoating, genome release and nuclear import. Endosome acidification occurs progressively from the cell periphery toward the microtubule-organizing center (MTOC). Late endosomal acidification (pH~6) triggers change of the homotrimeric glycoprotein HA mediating fusion between the viral envelope and the endosome membrane. Influx of protons and efflux of potassium from the virus core happen through the acid-activated viral ion channel M2. The pH drop triggers the activation of a histidine cluster in the virus capsid, contributed by three sequential M1 monomers, and promotes the capsid disassembly. Further the vRNPs dissociate from the M1 proteins. Free ubiquitin (Ub) chains derived from virus particles activate the aggresome processing pathway (APP) and recruit HDAC6 through its Ub-binding zinc finger domain (HDAC6 ZnF). Deubiquitinases (DUBs) could be involved in unanchored Ub formation. HDAC6 binds to M1 and to NP from vRNPs. HDAC6 by a region between its catalytic domains also binds motor proteins in microtubules and myosin II in actin microfilaments generating physical forces that help dissociate the M1 proteins, disassembling the virus capsid. The epidermal growth factor receptor pathway substrate 8 (EPS8) and transportin-1 (TNPO1) interact with M1 from the capsid and vRNPs, contributing to the disaggregation of the vRNP-associated M1 and vRNP debundling in the cytosol. In this way, vRNPs are transported by importin α/β to the nucleus as individual rod-shaped structures. PDB: TNPO1 (2Z5J), EPS8 (2E8M), importin α (4B18), DUB (6K9P).

Moreover, HDAC6 knockout mice intratracheally infected with IAV showed reduced lung viral titers compared to wild type mice, whereas the antiviral immune responses were comparable ([104] and our unpublished results). This showed that the pro-viral role of HDAC6 ZnF domain during IAV uncoating influenced the infection outcome. In contrast, another recent study in which another strain of HDAC6 knockout mice was infected with IAV showed them to be more susceptible to PR8 H1N1 infection than their wild type counterpart [225]. In this work, it was argued that the absence of HDAC6 leads to a blunted innate response and concomitantly increased susceptibility of mice to IAV infection. The reason for these differences is not known but might possibly reflect the presence of different microbiomes in the two strains. In the future, targeted mutation of the HDAC6 ZnF or CD in mice is desired to fully understand the in vivo pro-viral and antiviral effects of HDAC6, respectively.

3.3. Epidermal Growth Factor Receptor Pathway Substrate 8

The epidermal growth factor receptor pathway substrate 8 (EPS8) is an adaptor protein involved in signaling via the epidermal growth factor receptor (EGFR) [226,227]. EPS8 also directly binds to actin filaments controlling the rate of polymerization and depolymerization by capping the fast-growing ends of actin filaments [228–231]. EPS8 regulates intracellular trafficking of membrane receptors through its direct interaction with the GTPase-activating protein RN-tre, which controls the activity of Rab5, or by interacting with the clathrin-mediated endocytosis machinery.

A screen for host factors involved in IAV infection by correlating WSN H1N1 infectivity with gene expression profiles of 59 distinct cell lines identified EPS8 as the highest confidence pro-viral host gene [102]. Knocking out EPS8 in human A549 lung cells decreased viral titers in the infected-cell supernatant by 10-fold in multicycle replication assays. The loss of EPS8 did not affect virus attachment, uptake or fusion. EPS8 physically associates with incoming virus components possibly through interactions with NP, the viral polymerase, M1, or bridged by other cellular uncoating factors (Figure 4). EPS8 might interact with vRNPs by binding to NP as the viral nucleoprotein specifically co-precipitated with EPS8. Additionally, the import of vRNPs was significantly delayed in cells lacking EPS8 in comparison to WT cells, leading to a reduction in viral gene expression [102]. EGFR signaling, which promotes IAV entry [232], was unaffected by EPS8 depletion [102]. Although mechanistic details are missing, one can speculate that EPS8 regulates actin filaments to enhance IAV uncoating [102].

3.4. SPOPL/Cullin 3 Ubiquitin Ligase Complex and EPS15

The maturation of late endosomes/multivesicular bodies entails the spatial and functional separation of the organelles from early endosomes, preparing them as a feeder pathway to lysosomes [233]. Cellular processes that promote endosome maturation play a critical role in influenza uncoating. Cullin3 (CUL3)-based E3 ubiquitin ligases regulate endocytic trafficking of cargo to lysosomes and endosome maturation. Transfer of cargo from early endosomes to lysosomes depends on an endosomal maturation process regulated by a variety of protein- and lipid-based events. They include a small GTPase Rab5-to-Rab7 switch, a PtdIns(3)P to PtdIns(3,5)P2 conversion, and changes in the luminal ion concentrations, such as decrease in pH and increase in K^+ concentration [233,234]. Using a siRNA screen against 130 human Bric-a-Brac/Tramtrack/Broad (BTB) domain proteins in A549 cells, it was found that the Speckle-type POZ protein-like (SPOPL) was crucial for EPS15 ubiquitination by the Cullin RING E3 ubiquitin ligase 3 (CRL3)^{SPOPL} complex. EPS15, an endocytic adaptor that associates with ESCRT0 proteins HRS and STAM, was necessary for endosome maturation and IAV capsid disassembly [235]. The depletion of SPOP and SPOPL gave a similar phenotype to Cul3 depletion [236], showing retention of viral components in the endocytic system and inhibition of infection. Ubiquitin-modifying enzymes that regulate endosome maturation play a yet incompletely understood but important facilitator role in the successful uncoating of IAVs.

3.5. Transportin 1

In eukaryotic cells, transcription and translation are physically separated by the nuclear membrane; transcription occurs only within the nucleus, and translation occurs only outside the nucleus in the cytoplasm. The nuclear membrane, also known as nuclear envelope, is a phospholipid bilayer that encloses the cell nucleus and is penetrated by nuclear pore complexes. Small molecules (usually less than 60 kDa) diffuse freely through the nuclear pores [237,238]. Alternatively, proteins may shuttle between the cytoplasm and the nucleus in an active way that is mediated by nuclear localization signals (NLSs) or nuclear export signals (NESs). In this way, larger molecules are selected by nuclear transport receptors (also called karyopherins) that carry their cargoes from one compartment to the other by crossing the nuclear envelope at the level of the nuclear pore complexes [239,240]. Importins mediate the nuclear import of cargos and transportin 1 (TNPO1, also known as importin-β2, KPNB2) is one of the best-characterized nuclear import receptors [241].

Many viruses depend on nuclear proteins for replication and their viral genome must enter the nucleus of the host cell. This is the case of most DNA viruses and some RNA viruses, including orthomyxoviruses and retroviruses. Therefore, it is expected that the life cycle of these viruses is dependent on transporters (e.g., importins, exportins, transportins) and regulators (e.g., Ran GTPase). Great effort has been put on deciphering viral nuclear transport mechanisms [242–245]. In the context of IAV infection, a study using RNAi for targeting, among other proteins, nuclear pore proteins identified TNPO1 as an important host factor involved in uncoating. Depletion of TNPO1 in different cells reduced the number of infected cells and the production of new viruses [103]. It was shown that TNPO1 was important not only for the vRNPs nuclear import, but also for the M1 uncoating and vRNP debundling in the cytosol [103]. Moreover, the role of TNPO1 in uncoating was associated with its recognition of a nuclear localization signal as it binds to the exposed M1 N-terminal PY-NLS motif only after capsid acidification. As shown in Figure 4, by recognizing and binding the M1 NLS, TNPO1 promotes the removal of vRNP-associated M1, which leads to dissociation of vRNPs from each other and facilitates further nuclear import by importin α and β via the classical NLS-mediated import pathway. It is noteworthy that as endosomes mature, both decrease in pH and increase in K^+ concentration in the lumen of late endosomes take place, which is important for sufficient priming of the viral core for uncoating [234]. A high K^+ concentration, in particular, promotes dissociation of bundled vRNPs from each other in an in vitro uncoating assay [234]. The segmented nature of IAV vRNPs not only promotes reassortment during co-infection [246] but may also allow the segments to be transported in and out of the nuclear pore individually.

4. Perspectives in the Field of Virus Host Interaction and Capsid Uncoating

Due to their nature, viruses need to constantly interact with their host cells. They are always trying to either counteract or exploit different cellular mechanisms and pathways to their advantage. Better understanding the molecular requirements viruses have on host cells or the immune mechanisms used by the cells to escape infection is important for the development of novel approaches to fight viral infections.

Despite the availability of licensed vaccines, IAV is estimated to be responsible for 290,000 to 650,000 worldwide flu-associated deaths annually [247] and is of major public health interest due to its pandemic potential and constant threat to animals and humans. This review focused on the IAV life cycle highlighting the interactions with the cell host proteins. Capsid uncoating is a dynamic process that has remained relatively poorly studied. However, in recent years, progress in this field has been made with the identification of cellular proteins and pathway involved in IAV uncoating.

Enveloped viruses carry several host proteins in their structural core after budding has taken place. One of these proteins is Ub, which in the form of unanchored Ub chains, can recruit cellular proteins in the infected cells, such as HDAC6. Similarly to what happens to the aggresome and inflammasome pathways, Ub chains recruit HDAC6 that acts as a scaffold protein, interacting with virus proteins from the capsid and virus genome as well as with cytoskeletal motor proteins. These interactions generate physical forces that catalyze the dissociation of the capsid M1 layer underneath the viral envelope. In parallel, EPS8, TNPO1 and possibly other cellular proteins and kinases such as G protein-coupled receptor kinase (GRK2) [248], as well as endosome maturation, together contribute to generate the cellular environment that ultimately leads to uncoating and release of individual vRNPs at the perinuclear area.

Considering that other viruses could use a similar mechanism during their virus cycle, it is important to investigate if the HDAC6-mediated APP is involved in uncoating of other enveloped viruses that also have Ub in their viral mature particles. Similar to IAV, other important viruses including HIV-1, Ebola, rabies, HSV-1, VACV, HCMV, VSV, RSV, also carry Ub in their particles [175–178]. The identification of additional host proteins in viral particles could give hints to the possible pathways used by viruses during their life cycle. Considering that host proteins incorporated by viral particles might play crucial roles, as Ub does, it is important to realize that different host cells may influence the composition of the host protein profile inside virus particles. This was reported for HIV, in which the host protein profile in mature particles was found to be different depending on the cell host from which it originated [177]. This might be even more important for viruses that transition from one host species to another during their life cycle. Arboviruses (Zika, dengue, yellow fever, chikungunya, tick borne encephalitis etc.) are examples of viruses that infect mammals and arthropods during their life cycle. IAV, Ebola and SARS-CoV are other zoonotic viruses that jump from animals to humans and can be at the origin of pandemics.

TNPO1 also plays a role in the uncoating of HIV. Similar to the mechanism of IAV uncoating, TNPO1 binds to HIV capsids, triggers their uncoating and promotes viral nuclear import [249]. Given that some of the host proteins that play a fundamental role in IAV uncoating have been since then shown to participate in the uncoating of other viruses, it is interesting to think about the potential of interfering with this step of the virus life cycle by targeting one or more of these host proteins. Indeed, targeting host processes has a potential advantage of being less likely to give rise to viral resistant variants and to be of broad use for different viruses. Studies in this direction led to the development of the only host-targeting antiviral agent among the 20 approved antiretroviral used to treat HIV patients: maraviroc, a virus entry inhibitor that targets the chemokine receptor CCR5 expressed on the surface of white blood cells [250,251]. However, apart from immunomodulators, almost all antiviral drugs currently approved or under development target viral proteins. For IAV, NA inhibitors, M2 channel blockers, and PA endonuclease inhibitors are the three classes of inhibitors approved for treatment [252,253]. The administration of inhibitors of the M2 ion channel has been discouraged by the CDC due to widespread pre-existing viral resistance among H3N2 and H1N1 strains [253]. This highlights the need for new antiviral strategies with novel mechanisms of action and reduced drug resistance potential. Thus, if one identifies host proteins that are fundamental for the uncoating or replication of multiple viruses, these would have the potential to become novel targets for a broad-spectrum antiviral drug. While the main disadvantage of host-targeted antivirals is the higher risk for host toxicity, an advantage is that the host-targets/proteins can be studied before a new virus emerges. In addition, a host-targeted approach often offers a higher barrier to the appearance of viral drug resistance [254]. The development and approval of such new antivirals could be of great use for viral pandemic preparedness and complement vaccines.

Author Contributions: E.A.M., Y.Y. and P.M. wrote the manuscript; E.A.M. designed the pictures; P.M. and Y.Y. supervised the review. All authors have read and agreed to the published version of the manuscript.

Funding: This work was funded by the Novartis Research Foundation and the European Research Council—ERC Synergy grant number 856581.

Institutional Review Board Statement: Not applicable.

Informed Consent Statement: Not applicable.

Data Availability Statement: Not applicable.

Acknowledgments: We thank lab members for comments and discussion. Figures were created with BioRender.

Conflicts of Interest: The authors declare the absence of any commercial or financial relationships that could be construed as a potential conflict of interest.

References

1. Carrasco-Hernandez, R.; Jácome, R.; López Vidal, Y.; Ponce de León, S. Are RNA Viruses Candidate Agents for the Next Global Pandemic? A Review. *ILAR J.* **2017**, *58*, 343–358. [CrossRef]
2. Burrell, C.J.; Howard, C.R.; Murphy, F.A. Emerging Virus Diseases. *Fenner White's Med. Virol.* **2017**. [CrossRef]
3. Watkins, K. Emerging Infectious Diseases: A Review. *Curr. Emerg. Hosp. Med. Rep.* **2018**, *6*, 86–93. [CrossRef]
4. Ryu, W.-S. Chapter 21—New Emerging Viruses. In *Molecular Virology of Human Pathogenic Viruses*; Ryu, W.-S., Ed.; Academic Press: Boston, MA, USA, 2017; pp. 289–302. [CrossRef]
5. Doherty, P.C. What have we learnt so far from COVID-19? *Nat. Rev. Immunol.* **2021**, *21*, 67–68. [CrossRef]
6. Ryu, W.-S. Virus Life Cycle. *Mol. Virol. Hum. Pathog. Viruses* **2017**. [CrossRef]
7. Ashok, A.; Atwood, W.J. Virus Receptors and Tropism. In *Polyomaviruses and Human Diseases*; Ahsan, N., Ed.; Springer: New York, NY, USA, 2006; pp. 60–72. [CrossRef]
8. Baranowski, E.; Ruiz-Jarabo, C.M.; Domingo, E. Evolution of Cell Recognition by Viruses. *Science* **2001**, *292*, 1102. [CrossRef]
9. Greenstone, H.L.; Santoro, F.; Lusso, P.; Berger, E.A. Human Herpesvirus 6 and Measles Virus Employ Distinct CD46 Domains for Receptor Function. *J. Biol. Chem.* **2002**, *277*, 39112–39118. [CrossRef]
10. Spear, P.G.; Eisenberg, R.J.; Cohen, G.H. Three Classes of Cell Surface Receptors for Alphaherpesvirus Entry. *Virology* **2000**, *275*, 1–8. [CrossRef]
11. Morgan, C.; Rose, H.M.; Mednis, B. Electron Microscopy of Herpes Simplex Virus: I. Entry. *J. Virol.* **1968**, *2*, 507–516. [CrossRef]
12. Fuller, A.O.; Spear, P.G. Anti-glycoprotein D antibodies that permit adsorption but block infection by herpes simplex virus 1 prevent virion-cell fusion at the cell surface. *Proc. Natl. Acad. Sci. USA* **1987**, *84*, 5454–5458. [CrossRef]
13. Wittels, M.; Spear, P.G. Penetration of cells by herpes simplex virus does not require a low pH-dependent endocytic pathway. *Virus Res.* **1991**, *18*, 271–290. [CrossRef]
14. Aguilar, H.C.; Henderson, B.A.; Zamora, J.L.; Johnston, G.P. Paramyxovirus Glycoproteins and the Membrane Fusion Process. *Curr. Clin. Microbiol. Rep.* **2016**, *3*, 142–154. [CrossRef]
15. Chauveau, L.; Donahue, D.A.; Monel, B.; Porrot, F.; Bruel, T.; Richard, L.; Casartelli, N.; Schwartz, O. HIV Fusion in Dendritic Cells Occurs Mainly at the Surface and Is Limited by Low CD4 Levels. *J. Virol.* **2017**, *91*, e01248-17. [CrossRef]
16. Akula, S.M.; Naranatt, P.P.; Walia, N.-S.; Wang, F.-Z.; Fegley, B.; Chandran, B. Kaposi's Sarcoma-Associated Herpesvirus (Human Herpesvirus 8) Infection of Human Fibroblast Cells Occurs through Endocytosis. *J. Virol.* **2003**, *77*, 7978–7990. [CrossRef]
17. Acosta, E.G.; Castilla, V.; Damonte, E.B. Functional entry of dengue virus into Aedes albopictus mosquito cells is dependent on clathrin-mediated endocytosis. *J. Gen. Virol.* **2008**, *89*, 474–484. [CrossRef] [PubMed]
18. Persaud, M.; Martinez-Lopez, A.; Buffone, C.; Porcelli, S.A.; Diaz-Griffero, F. Infection by Zika viruses requires the transmembrane protein AXL, endocytosis and low pH. *Virology* **2018**, *518*, 301–312. [CrossRef]
19. Doms, R.W.; Helenius, A.; White, J. Membrane fusion activity of the influenza virus hemagglutinin. The low pH-induced conformational change. *J. Biol. Chem.* **1985**, *260*, 2973–2981. [CrossRef]
20. Ruigrok, R.W.H.; Aitken, A.; Calder, L.J.; Martin, S.R.; Skehel, J.J.; Wharton, S.A.; Weis, W.; Wiley, D.C. Studies on the Structure of the Influenza Virus Haemagglutinin at the pH of Membrane Fusion. *J. Gen. Virol.* **1988**, *69*, 2785–2795. [CrossRef]
21. Aleksandrowicz, P.; Marzi, A.; Biedenkopf, N.; Beimforde, N.; Becker, S.; Hoenen, T.; Feldmann, H.; Schnittler, H.-J. Ebola Virus Enters Host Cells by Macropinocytosis and Clathrin-Mediated Endocytosis. *J. Infect. Dis.* **2011**, *204*, S957–S967. [CrossRef] [PubMed]
22. Li, L.; Jose, J.; Xiang, Y.; Kuhn, R.J.; Rossmann, M.G. Structural changes of envelope proteins during alphavirus fusion. *Nature* **2010**, *468*, 705–708. [CrossRef]
23. Maclachlin, N.J.; Edward, J.D. (Eds.) *Fenner's Veterinary Virology*, 4th ed.; Academic: Boston, MA, USA, 2011.
24. King, A.M.Q.; Adams, M.J.; Carstens, E.B.; Lefkowitz, E.J. (Eds.) Family—Orthomyxoviridae. In *Virus Taxonomy*; Elsevier: San Diego, CA, USA, 2012; pp. 749–761. [CrossRef]
25. Stevenson, J.P.; Biddle, F. Pleomorphism of influenza virus particles under the electron microscope. *Nature* **1966**, *212*, 619–621. [CrossRef]
26. Stanley, W.M. The size of influenza virus. *J. Exp. Med.* **1944**, *79*, 267–283. [CrossRef]
27. Sieczkarski, S.B.; Whittaker, G.R. Characterization of the host cell entry of filamentous influenza virus. *Arch. Virol.* **2005**, *150*, 1783–1796. [CrossRef]
28. Calder, L.J.; Wasilewski, S.; Berriman, J.A.; Rosenthal, P.B. Structural organization of a filamentous influenza A virus. *Proc. Natl. Acad. Sci. USA* **2010**, *107*, 10685–10690. [CrossRef]
29. Dadonaite, B.; Vijayakrishnan, S.; Fodor, E.; Bhella, D.; Hutchinson, E.C. Filamentous influenza viruses. *J. Gen. Virol.* **2016**, *97*, 1755–1764. [CrossRef]
30. Seladi-Schulman, J.; Steel, J.; Lowen, A.C. Spherical influenza viruses have a fitness advantage in embryonated eggs, while filament-producing strains are selected in vivo. *J. Virol.* **2013**, *87*, 13343–13353. [CrossRef]

31. Basu, A.; Shelke, V.; Chadha, M.; Kadam, D.; Sangle, S.; Gangodkar, S.; Mishra, A. Direct imaging of pH1N1 2009 influenza virus replication in alveolar pneumocytes in fatal cases by transmission electron microscopy. *J. Electron. Microsc.* **2011**, *60*, 89–93. [CrossRef] [PubMed]

32. Kilbourne, E.D. Studies on influenza in the pandemic of 1957–1958. III. Isolation of influenza A (Asian strain) viruses from influenza patients with pulmonary complications; details of virus isolation and characterization of isolates, with quantitative comparison of isolation methods. *J. Clin. Investig.* **1959**, *38*, 266–274. [CrossRef]

33. Badham, M.D.; Rossman, J.S. Filamentous Influenza Viruses. *Curr. Clin. Microbiol. Rep.* **2016**, *3*, 155–161. [CrossRef]

34. Speshock, J.L.; Doyon-Reale, N.; Rabah, R.; Neely, M.N.; Roberts, P.C. Filamentous influenza A virus infection predisposes mice to fatal septicemia following superinfection with Streptococcus pneumoniae serotype 3. *Infect. Immun.* **2007**, *75*, 3102–3111. [CrossRef]

35. Campbell, P.J.; Danzy, S.; Kyriakis, C.S.; Deymier, M.J.; Lowen, A.C.; Steel, J. The M segment of the 2009 pandemic influenza virus confers increased neuraminidase activity, filamentous morphology, and efficient contact transmissibility to A/Puerto Rico/8/1934-based reassortant viruses. *J. Virol.* **2014**, *88*, 3802–3814. [CrossRef] [PubMed]

36. Li, T.; Li, Z.; Deans, E.E.; Mittler, E.; Liu, M.; Chandran, K.; Ivanovic, T. The shape of pleomorphic virions determines resistance to cell-entry pressure. *Nat. Microbiol.* **2021**, *6*, 617–629. [CrossRef]

37. Abbas, M.; Abidin, Z. Proteins of Influenza Virus: A Review. *J. Infect. Mol. Biol.* **2013**, *1*, 1–7.

38. Lowen, A.C. Constraints, Drivers, and Implications of Influenza A Virus Reassortment. *Annu. Rev. Virol.* **2017**, *4*, 105–121. [CrossRef]

39. Harris, A.; Cardone, G.; Winkler, D.C.; Heymann, J.B.; Brecher, M.; White, J.M.; Steven, A.C. Influenza virus pleiomorphy characterized by cryoelectron tomography. *Proc. Natl. Acad. Sci. USA* **2006**, *103*, 19123–19127. [CrossRef]

40. Gaymard, A.; Le Briand, N.; Frobert, E.; Lina, B.; Escuret, V. Functional balance between neuraminidase and haemagglutinin in influenza viruses. *Clin. Microbiol. Infect.* **2016**, *22*, 975–983. [CrossRef]

41. Mitnaul, L.J.; Castrucci, M.R.; Murti, K.G.; Kawaoka, Y. The cytoplasmic tail of influenza A virus neuraminidase (NA) affects NA incorporation into virions, virion morphology, and virulence in mice but is not essential for virus replication. *J. Virol.* **1996**, *70*, 873–879. [CrossRef]

42. Kosik, I.; Yewdell, J.W. Influenza Hemagglutinin and Neuraminidase: Yin-Yang Proteins Coevolving to Thwart Immunity. *Viruses* **2019**, *11*, 346. [CrossRef]

43. Krammer, F. The human antibody response to influenza A virus infection and vaccination. *Nat. Rev. Immunol.* **2019**, *19*, 383–397. [CrossRef]

44. Roberts, P.C.; Lamb, R.A.; Compans, R.W. The M1 and M2 Proteins of Influenza A Virus Are Important Determinants in Filamentous Particle Formation. *Virology* **1998**, *240*, 127–137. [CrossRef]

45. Kretzschmar, E.; Bui, M.; Rose, J.K. Membrane Association of Influenza Virus Matrix Protein Does Not Require Specific Hydrophobic Domains or the Viral Glycoproteins. *Virology* **1996**, *220*, 37–45. [CrossRef] [PubMed]

46. Noton, S.L.; Medcalf, E.; Fisher, D.; Mullin, A.E.; Elton, D.; Digard, P. Identification of the domains of the influenza A virus M1 matrix protein required for NP binding, oligomerization and incorporation into virions. *J. Gen. Virol.* **2007**, *88*, 2280–2290. [CrossRef] [PubMed]

47. Ruigrok, R.; Baudin, F.; Petit, I.; Weissenhorn, W. Role of influenza virus M1 protein in the viral budding process. *Int. Congr. Ser.* **2001**, *1219*, 397–404. [CrossRef]

48. Peukes, J.; Xiong, X.; Erlendsson, S.; Qu, K.; Wan, W.; Calder, L.J.; Schraidt, O.; Kummer, S.; Freund, S.M.V.; Kräusslich, H.-G.; et al. The native structure of the assembled matrix protein 1 of influenza A virus. *Nature* **2020**, *587*, 495–498. [CrossRef] [PubMed]

49. Kang, S.-M.; Kim, M.-C.; Compans, R.W. Virus-like particles as universal influenza vaccines. *Expert Rev. Vaccines* **2012**, *11*, 995–1007. [CrossRef] [PubMed]

50. Betakova, T. M2 protein-a proton channel of influenza A virus. *Curr. Pharm. Des.* **2007**, *13*, 3231–3235. [CrossRef] [PubMed]

51. Alvarado-Facundo, E.; Gao, Y.; Ribas-Aparicio, R.M.; Jiménez-Alberto, A.; Weiss, C.D.; Wang, W. Influenza Virus M2 Protein Ion Channel Activity Helps To Maintain Pandemic 2009 H1N1 Virus Hemagglutinin Fusion Competence during Transport to the Cell Surface. *J. Virol.* **2015**, *89*, 1975–1985. [CrossRef]

52. Grambas, S.; Hay, A.J. Maturation of influenza A virus hemagglutinin–estimates of the pH encountered during transport and its regulation by the M2 protein. *Virology* **1992**, *190*, 11–18. [CrossRef]

53. Iwatsuki-Horimoto, K.; Horimoto, T.; Noda, T.; Kiso, M.; Maeda, J.; Watanabe, S.; Muramoto, Y.; Fujii, K.; Kawaoka, Y. The cytoplasmic tail of the influenza A virus M2 protein plays a role in viral assembly. *J. Virol.* **2006**, *80*, 5233–5240. [CrossRef]

54. McCown, M.F.; Pekosz, A. The influenza A virus M2 cytoplasmic tail is required for infectious virus production and efficient genome packaging. *J. Virol.* **2005**, *79*, 3595–3605. [CrossRef]

55. Pflug, A.; Guilligay, D.; Reich, S.; Cusack, S. Structure of influenza A polymerase bound to the viral RNA promoter. *Nature* **2014**, *516*, 355–360. [CrossRef]

56. Portela, A.; Digard, P. The influenza virus nucleoprotein: A multifunctional RNA-binding protein pivotal to virus replication. *J. Gen. Virol.* **2002**, *83*, 723–734. [CrossRef]

57. Paterson, D.; Fodor, E. Emerging roles for the influenza A virus nuclear export protein (NEP). *PLoS Pathog.* **2012**, *8*, e1003019. [CrossRef] [PubMed]

58. Egorov, A.; Brandt, S.; Sereinig, S.; Romanova, J.; Ferko, B.; Katinger, D.; Grassauer, A.; Alexandrova, G.; Katinger, H.; Muster, T. Transfectant Influenza A Viruses with Long Deletions in the NS1 Protein Grow Efficiently in Vero Cells. *J. Virol.* **1998**, *72*, 6437–6441. [CrossRef] [PubMed]

59. Hutchinson, E.C.; Charles, P.D.; Hester, S.S.; Thomas, B.; Trudgian, D.; Martínez-Alonso, M.; Fodor, E. Conserved and host-specific features of influenza virion architecture. *Nat. Commun.* **2014**, *5*, 4816. [CrossRef] [PubMed]

60. Sha, T.W.; Weber, M.; Kasumba, D.M.; Noda, T.; Nakano, M.; Kato, H.; Fujita, T. Influenza A virus NS1 optimises virus infectivity by enhancing genome packaging in a dsRNA-binding dependent manner. *Virol. J.* **2020**, *17*, 107. [CrossRef] [PubMed]

61. Kochs, G.; García-Sastre, A.; Martínez-Sobrido, L. Multiple Anti-Interferon Actions of the Influenza A Virus NS1 Protein. *J. Virol.* **2007**, *81*, 7011–7021. [CrossRef] [PubMed]

62. Rosário-Ferreira, N.; Preto, A.J.; Melo, R.; Moreira, I.S.; Brito, R.M.M. The Central Role of Non-Structural Protein 1 (NS1) in Influenza Biology and Infection. *Int. J. Mol. Sci.* **2020**, *21*, 1511. [CrossRef] [PubMed]

63. Min, J.Y.; Li, S.; Sen, G.C.; Krug, R.M. A site on the influenza A virus NS1 protein mediates both inhibition of PKR activation and temporal regulation of viral RNA synthesis. *Virology* **2007**, *363*, 236–243. [CrossRef] [PubMed]

64. Schultz-Cherry, S.; Dybdahl-Sissoko, N.; Neumann, G.; Kawaoka, Y.; Hinshaw, V.S. Influenza virus ns1 protein induces apoptosis in cultured cells. *J. Virol.* **2001**, *75*, 7875–7881. [CrossRef]

65. Zhirnov, O.P.; Konakova, T.E.; Wolff, T.; Klenk, H.-D. NS1 Protein of Influenza A Virus Down-Regulates Apoptosis. *J. Virol.* **2002**, *76*, 1617–1625. [CrossRef]

66. Mok, B.W.; Liu, H.; Chen, P.; Liu, S.; Lau, S.Y.; Huang, X.; Liu, Y.C.; Wang, P.; Yuen, K.Y.; Chen, H. The role of nuclear NS1 protein in highly pathogenic H5N1 influenza viruses. *Microbes Infect.* **2017**, *19*, 587–596. [CrossRef]

67. Jackson, D.; Hossain, M.J.; Hickman, D.; Perez, D.R.; Lamb, R.A. A new influenza virus virulence determinant: The NS1 protein four C-terminal residues modulate pathogenicity. *Proc. Natl. Acad. Sci. USA* **2008**, *105*, 4381–4386. [CrossRef]

68. HIRST, G.K. The Agglutination of Red Cells By Allantoic Fluid of Chick Embryos Infected With Influenza Virus. *Science* **1941**, *94*, 22–23. [CrossRef] [PubMed]

69. Ayora-Talavera, G. Sialic acid receptors: Focus on their role in influenza infection. *J. Recept. Ligand Channel Res.* **2018**, *10*, 1–11. [CrossRef]

70. Glaser, L.; Stevens, J.; Zamarin, D.; Wilson, I.A.; García-Sastre, A.; Tumpey, T.M.; Basler, C.F.; Taubenberger, J.K.; Palese, P. A single amino acid substitution in 1918 influenza virus hemagglutinin changes receptor binding specificity. *J. Virol.* **2005**, *79*, 11533–11536. [CrossRef] [PubMed]

71. Vines, A.; Wells, K.; Matrosovich, M.; Castrucci, M.R.; Ito, T.; Kawaoka, Y. The role of influenza A virus hemagglutinin residues 226 and 228 in receptor specificity and host range restriction. *J. Virol.* **1998**, *72*, 7626–7631. [CrossRef] [PubMed]

72. Tumpey, T.M.; Maines, T.R.; Van Hoeven, N.; Glaser, L.; Solórzano, A.; Pappas, C.; Cox, N.J.; Swayne, D.E.; Palese, P.; Katz, J.M.; et al. A two-amino acid change in the hemagglutinin of the 1918 influenza virus abolishes transmission. *Science* **2007**, *315*, 655–659. [CrossRef]

73. Naeve, C.W.; Hinshaw, V.S.; Webster, R.G. Mutations in the hemagglutinin receptor-binding site can change the biological properties of an influenza virus. *J. Virol.* **1984**, *51*, 567–569. [CrossRef] [PubMed]

74. Nicholls, J.M.; Lai, J.; Garcia, J.-M. Investigating the Interaction Between Influenza and Sialic Acid: Making and Breaking the Link. In *Influenza Virus Sialidase—A Drug Discovery Target*; von Itzstein, M., Ed.; Springer: Basel, Switzerland, 2012; pp. 31–45. [CrossRef]

75. Karakus, U.; Thamamongood, T.; Ciminski, K.; Ran, W.; Günther, S.C.; Pohl, M.O.; Eletto, D.; Jeney, C.; Hoffmann, D.; Reiche, S.; et al. MHC class II proteins mediate cross-species entry of bat influenza viruses. *Nature* **2019**, *567*, 109–112. [CrossRef] [PubMed]

76. Giotis, E.S.; Carnell, G.; Young, E.F.; Ghanny, S.; Soteropoulos, P.; Wang, L.F.; Barclay, W.S.; Skinner, M.A.; Temperton, N. Entry of the bat influenza H17N10 virus into mammalian cells is enabled by the MHC class II HLA-DR receptor. *Nat. Microbiol.* **2019**, *4*, 2035–2038. [CrossRef] [PubMed]

77. Matrosovich, M.N.; Matrosovich, T.Y.; Gray, T.; Roberts, N.A.; Klenk, H.D. Neuraminidase is important for the initiation of influenza virus infection in human airway epithelium. *J. Virol.* **2004**, *78*, 12665–12667. [CrossRef]

78. Du, R.; Cui, Q.; Rong, L. Competitive Cooperation of Hemagglutinin and Neuraminidase during Influenza A Virus Entry. *Viruses* **2019**, *11*, 458. [CrossRef]

79. Cohen, M.; Zhang, X.Q.; Senaati, H.P.; Chen, H.W.; Varki, N.M.; Schooley, R.T.; Gagneux, P. Influenza A penetrates host mucus by cleaving sialic acids with neuraminidase. *Virol. J.* **2013**, *10*, 321. [CrossRef] [PubMed]

80. De Vries, E.; Tscherne, D.M.; Wienholts, M.J.; Cobos-Jimenez, V.; Scholte, F.; Garcia-Sastre, A.; Rottier, P.J.; de Haan, C.A. Dissection of the influenza A virus endocytic routes reveals macropinocytosis as an alternative entry pathway. *PLoS Pathog.* **2011**, *7*, e1001329. [CrossRef] [PubMed]

81. Rossman, J.S.; Leser, G.P.; Lamb, R.A. Filamentous Influenza Virus Enters Cells via Macropinocytosis. *J. Virol.* **2012**, *86*, 10950–10960. [CrossRef]

82. Nunes-Correia, I.; Eulálio, A.; Nir, S.; Pedroso de Lima, M.C. Caveolae as an additional route for influenza virus endocytosis in MDCK cells. *Cell. Mol. Biol. Lett.* **2004**, *9*, 47–60. [PubMed]

83. Sieczkarski, S.B.; Whittaker, G.R. Influenza virus can enter and infect cells in the absence of clathrin-mediated endocytosis. *J. Virol.* **2002**, *76*, 10455–10464. [CrossRef] [PubMed]

84. Lakadamyali, M.; Rust, M.J.; Babcock, H.P.; Zhuang, X. Visualizing infection of individual influenza viruses. *Proc. Natl. Acad. Sci. USA* **2003**, *100*, 9280–9285. [CrossRef]

85. Bedi, S.; Ono, A. Friend or Foe: The Role of the Cytoskeleton in Influenza A Virus Assembly. *Viruses* **2019**, *11*, 46. [CrossRef] [PubMed]

86. Wu, N.-H.; Yang, W.; Beineke, A.; Dijkman, R.; Matrosovich, M.; Baumgärtner, W.; Thiel, V.; Valentin-Weigand, P.; Meng, F.; Herrler, G. The differentiated airway epithelium infected by influenza viruses maintains the barrier function despite a dramatic loss of ciliated cells. *Sci. Rep.* **2016**, *6*, 39668. [CrossRef] [PubMed]

87. Zeng, H.; Goldsmith, C.S.; Maines, T.R.; Belser, J.A.; Gustin, K.M.; Pekosz, A.; Zaki, S.R.; Katz, J.M.; Tumpey, T.M. Tropism and infectivity of influenza virus, including highly pathogenic avian H5N1 virus, in ferret tracheal differentiated primary epithelial cell cultures. *J. Virol.* **2013**, *87*, 2597–2607. [CrossRef] [PubMed]

88. Thompson, C.I.; Barclay, W.S.; Zambon, M.C.; Pickles, R.J. Infection of human airway epithelium by human and avian strains of influenza a virus. *J. Virol.* **2006**, *80*, 8060–8068. [CrossRef] [PubMed]

89. Müsch, A. Microtubule Organization and Function in Epithelial Cells. *Traffic* **2004**, *5*, 1–9. [CrossRef] [PubMed]

90. Sun, X.; Whittaker, G.R. Role of the actin cytoskeleton during influenza virus internalization into polarized epithelial cells. *Cell. Microbiol.* **2007**, *9*, 1672–1682. [CrossRef]

91. Simpson, C.; Yamauchi, Y. Microtubules in Influenza Virus Entry and Egress. *Viruses* **2020**, *12*, 117. [CrossRef]

92. Mu, F.-T.; Callaghan, J.M.; Steele-Mortimer, O.; Stenmark, H.; Parton, R.G.; Campbell, P.L.; McCluskey, J.; Yeo, J.-P.; Tock, E.P.C.; Toh, B.-H. EEA1, an Early Endosome-Associated Protein.: EEA1 is a conserved α-helical peripheral membrane protein flanked by cysteine "fingers" and contains a calmodulin-binding IQ motif. *J. Biol. Chem.* **1995**, *270*, 13503–13511. [CrossRef]

93. Chavrier, P.; Parton, R.G.; Hauri, H.P.; Simons, K.; Zerial, M. Localization of low molecular weight GTP binding proteins to exocytic and endocytic compartments. *Cell* **1990**, *62*, 317–329. [CrossRef]

94. Gorvel, J.P.; Chavrier, P.; Zerial, M.; Gruenberg, J. rab5 controls early endosome fusion in vitro. *Cell* **1991**, *64*, 915–925. [CrossRef]

95. Jovic, M.; Sharma, M.; Rahajeng, J.; Caplan, S. The early endosome: A busy sorting station for proteins at the crossroads. *Histol. Histopathol.* **2010**, *25*, 99–112. [CrossRef] [PubMed]

96. Nielsen, E.; Christoforidis, S.; Uttenweiler-Joseph, S.; Miaczynska, M.; Dewitte, F.; Wilm, M.; Hoflack, B.; Zerial, M. Rabenosyn-5, a novel Rab5 effector, is complexed with hVPS45 and recruited to endosomes through a FYVE finger domain. *J. Cell Biol.* **2000**, *151*, 601–612. [CrossRef] [PubMed]

97. Baharom, F.; Thomas, O.S.; Lepzien, R.; Mellman, I.; Chalouni, C.; Smed-Sörensen, A. Visualization of early influenza A virus trafficking in human dendritic cells using STED microscopy. *PLoS ONE* **2017**, *12*, e0177920. [CrossRef] [PubMed]

98. Griffiths, G.; Hoflack, B.; Simons, K.; Mellman, I.; Kornfeld, S. The mannose 6-phosphate receptor and the biogenesis of lysosomes. *Cell* **1988**, *52*, 329–341. [CrossRef]

99. Edinger, T.O.; Pohl, M.O.; Yángüez, E.; Stertz, S. Cathepsin W Is Required for Escape of Influenza A Virus from Late Endosomes. *MBio* **2015**, *6*, e00297-15. [CrossRef] [PubMed]

100. Liang, R.; Swanson, J.M.J.; Madsen, J.J.; Hong, M.; DeGrado, W.F.; Voth, G.A. Acid activation mechanism of the influenza A M2 proton channel. *Proc. Natl. Acad. Sci. USA* **2016**, *113*, E6955–E6964. [CrossRef] [PubMed]

101. Yamauchi, Y.; Greber, U.F. Principles of Virus Uncoating: Cues and the Snooker Ball. *Traffic* **2016**, *17*, 569–592. [CrossRef] [PubMed]

102. Larson, G.P.; Tran, V.; Yú, S.; Caì, Y.; Higgins, C.A.; Smith, D.M.; Baker, S.F.; Radoshitzky, S.R.; Kuhn, J.H.; Mehle, A. EPS8 Facilitates Uncoating of Influenza A Virus. *Cell Rep.* **2019**, *29*, 2175–2183.e2174. [CrossRef] [PubMed]

103. Miyake, Y.; Keusch, J.J.; Decamps, L.; Ho-Xuan, H.; Iketani, S.; Gut, H.; Kutay, U.; Helenius, A.; Yamauchi, Y. Influenza virus uses transportin 1 for vRNP debundling during cell entry. *Nat. Microbiol.* **2019**, *4*, 578–586. [CrossRef] [PubMed]

104. Banerjee, I.; Miyake, Y.; Nobs, S.P.; Schneider, C.; Horvath, P.; Kopf, M.; Matthias, P.; Helenius, A.; Yamauchi, Y. Influenza A virus uses the aggresome processing machinery for host cell entry. *Science* **2014**, *346*, 473–477. [CrossRef] [PubMed]

105. Škubník, K.; Sukeník, L.; Buchta, D.; Füzik, T.; Procházková, M.; Moravcová, J.; Šmerdová, L.; Přidal, A.; Vácha, R.; Plevka, P. Capsid opening enables genome release of iflaviruses. *Sci. Adv.* **2021**, *7*. [CrossRef] [PubMed]

106. Herz, C.; Stavnezer, E.; Krug, R.; Gurney, T., Jr. Influenza virus, an RNA virus, synthesizes its messenger RNA in the nucleus of infected cells. *Cell* **1981**, *26*, 391–400. [CrossRef]

107. Fodor, E. The RNA polymerase of influenza a virus: Mechanisms of viral transcription and replication. *Acta Virol.* **2013**, *57*, 113–122. [CrossRef] [PubMed]

108. Plotch, S.J.; Bouloy, M.; Ulmanen, I.; Krug, R.M. A unique cap(m7GpppXm)-dependent influenza virion endonuclease cleaves capped RNAs to generate the primers that initiate viral RNA transcription. *Cell* **1981**, *23*, 847–858. [CrossRef]

109. Te Velthuis, A.J.; Fodor, E. Influenza virus RNA polymerase: Insights into the mechanisms of viral RNA synthesis. *Nat. Rev. Microbiol.* **2016**, *14*, 479–493. [CrossRef]

110. Fodor, E.; Te Velthuis, A.J.W. Structure and Function of the Influenza Virus Transcription and Replication Machinery. *Cold Spring Harb. Perspect. Med.* **2020**, *10*. [CrossRef]

111. Inglis, S.C. Inhibition of host protein synthesis and degradation of cellular mRNAs during infection by influenza and herpes simplex virus. *Mol. Cell Biol.* **1982**, *2*, 1644–1648. [CrossRef] [PubMed]

112. Katze, M.G.; DeCorato, D.; Krug, R.M. Cellular mRNA translation is blocked at both initiation and elongation after infection by influenza virus or adenovirus. *J. Virol.* **1986**, *60*, 1027–1039. [CrossRef] [PubMed]

113. Rivas, H.G.; Schmaling, S.K.; Gaglia, M.M. Shutoff of Host Gene Expression in Influenza A Virus and Herpesviruses: Similar Mechanisms and Common Themes. *Viruses* **2016**, *8*, 102. [CrossRef] [PubMed]
114. Bogdanow, B.; Wang, X.; Eichelbaum, K.; Sadewasser, A.; Husic, I.; Paki, K.; Budt, M.; Hergeselle, M.; Vetter, B.; Hou, J.; et al. The dynamic proteome of influenza A virus infection identifies M segment splicing as a host range determinant. *Nat. Commun.* **2019**, *10*, 5518. [CrossRef]
115. Etchison, J.; Doyle, M.; Penhoet, E.; Holland, J. Synthesis and cleavage of influenza virus proteins. *J. Virol.* **1971**, *7*, 155–167. [CrossRef] [PubMed]
116. Tekamp, P.A.; Penhoet, E.E. Quantification of influenza virus messenger RNAs. *J. Gen. Virol.* **1980**, *47*, 449–459. [CrossRef] [PubMed]
117. Yamanaka, K.; Ishihama, A.; Nagata, K. Translational regulation of influenza virus mRNAs. *Virus Genes* **1988**, *2*, 19–30. [CrossRef]
118. Shapiro, G.I.; Gurney, T., Jr.; Krug, R.M. Influenza virus gene expression: Control mechanisms at early and late times of infection and nuclear-cytoplasmic transport of virus-specific RNAs. *J. Virol.* **1987**, *61*, 764–773. [CrossRef] [PubMed]
119. Elton, D.; Simpson-Holley, M.; Archer, K.; Medcalf, L.; Hallam, R.; McCauley, J.; Digard, P. Interaction of the influenza virus nucleoprotein with the cellular CRM1-mediated nuclear export pathway. *J. Virol.* **2001**, *75*, 408–419. [CrossRef]
120. Ma, K.; Roy, A.M.; Whittaker, G.R. Nuclear export of influenza virus ribonucleoproteins: Identification of an export intermediate at the nuclear periphery. *Virology* **2001**, *282*, 215–220. [CrossRef] [PubMed]
121. Dong, X.; Biswas, A.; Süel, K.E.; Jackson, L.K.; Martinez, R.; Gu, H.; Chook, Y.M. Structural basis for leucine-rich nuclear export signal recognition by CRM1. *Nature* **2009**, *458*, 1136–1141. [CrossRef] [PubMed]
122. Güttler, T.; Madl, T.; Neumann, P.; Deichsel, D.; Corsini, L.; Monecke, T.; Ficner, R.; Sattler, M.; Görlich, D. NES consensus redefined by structures of PKI-type and Rev-type nuclear export signals bound to CRM1. *Nat. Struct. Mol. Biol.* **2010**, *17*, 1367–1376. [CrossRef] [PubMed]
123. Akarsu, H.; Burmeister, W.P.; Petosa, C.; Petit, I.; Müller, C.W.; Ruigrok, R.W.H.; Baudin, F. Crystal structure of the M1 protein-binding domain of the influenza A virus nuclear export protein (NEP/NS2). *EMBO J.* **2003**, *22*, 4646–4655. [CrossRef] [PubMed]
124. Goldfarb, D.S.; Corbett, A.H.; Mason, D.A.; Harreman, M.T.; Adam, S.A. Importin alpha: A multipurpose nuclear-transport receptor. *Trends Cell Biol.* **2004**, *14*, 505–514. [CrossRef] [PubMed]
125. Bui, M.; Wills, E.G.; Helenius, A.; Whittaker, G.R. Role of the influenza virus M1 protein in nuclear export of viral ribonucleoproteins. *J. Virol.* **2000**, *74*, 1781–1786. [CrossRef]
126. Iwatsuki-Horimoto, K.; Horimoto, T.; Fujii, Y.; Kawaoka, Y. Generation of Influenza A Virus NS2 (NEP) Mutants with an Altered Nuclear Export Signal Sequence. *J. Virol.* **2004**, *78*, 10149–10155. [CrossRef] [PubMed]
127. Neumann, G.; Hughes, M.T.; Kawaoka, Y. Influenza A virus NS2 protein mediates vRNP nuclear export through NES-independent interaction with hCRM1. *EMBO J.* **2000**, *19*, 6751–6758. [CrossRef]
128. Martin, K.; Heleniust, A. Nuclear transport of influenza virus ribonucleoproteins: The viral matrix protein (M1) promotes export and inhibits import. *Cell* **1991**, *67*, 117–130. [CrossRef]
129. Whittaker, G.; Bui, M.; Helenius, A. Nuclear trafficking of influenza virus ribonuleoproteins in heterokaryons. *J. Virol.* **1996**, *70*, 2743–2756. [CrossRef]
130. Rodriguez Boulan, E.; Sabatini, D.D. Asymmetric budding of viruses in epithelial monlayers: A model system for study of epithelial polarity. *Proc. Natl. Acad. Sci. USA* **1978**, *75*, 5071–5075. [CrossRef] [PubMed]
131. Greber, U.F.; Way, M. A Superhighway to Virus Infection. *Cell* **2006**, *124*, 741–754. [CrossRef] [PubMed]
132. Momose, F.; Kikuchi, Y.; Komase, K.; Morikawa, Y. Visualization of microtubule-mediated transport of influenza viral progeny ribonucleoprotein. *Microbes Infect.* **2007**, *9*, 1422–1433. [CrossRef]
133. Chou, Y.Y.; Heaton, N.S.; Gao, Q.; Palese, P.; Singer, R.H.; Lionnet, T. Colocalization of different influenza viral RNA segments in the cytoplasm before viral budding as shown by single-molecule sensitivity FISH analysis. *PLoS Pathog.* **2013**, *9*, e1003358. [CrossRef]
134. Lakdawala, S.S.; Wu, Y.; Wawrzusin, P.; Kabat, J.; Broadbent, A.J.; Lamirande, E.W.; Fodor, E.; Altan-Bonnet, N.; Shroff, H.; Subbarao, K. Influenza a virus assembly intermediates fuse in the cytoplasm. *PLoS Pathog.* **2014**, *10*, e1003971. [CrossRef]
135. Amorim, M.J.; Bruce, E.A.; Read, E.K.C.; Foeglein, Á.; Mahen, R.; Stuart, A.D.; Digard, P. A Rab11- and Microtubule-Dependent Mechanism for Cytoplasmic Transport of Influenza A Virus Viral RNA. *J. Virol.* **2011**, *85*, 4143. [CrossRef] [PubMed]
136. Eisfeld, A.J.; Kawakami, E.; Watanabe, T.; Neumann, G.; Kawaoka, Y. RAB11A is essential for transport of the influenza virus genome to the plasma membrane. *J. Virol.* **2011**, *85*, 6117–6126. [CrossRef] [PubMed]
137. Momose, F.; Sekimoto, T.; Ohkura, T.; Jo, S.; Kawaguchi, A.; Nagata, K.; Morikawa, Y. Apical Transport of Influenza A Virus Ribonucleoprotein Requires Rab11-positive Recycling Endosome. *PLoS ONE* **2011**, *6*, e21123. [CrossRef] [PubMed]
138. Vale-Costa, S.; Amorim, M.J. Clustering of Rab11 vesicles in influenza A virus infected cells creates hotspots containing the 8 viral ribonucleoproteins. *Small GTPases* **2017**, *8*, 71–77. [CrossRef] [PubMed]
139. De Castro Martin, I.F.; Fournier, G.; Sachse, M.; Pizarro-Cerda, J.; Risco, C.; Naffakh, N. Influenza virus genome reaches the plasma membrane via a modified endoplasmic reticulum and Rab11-dependent vesicles. *Nat. Commun.* **2017**, *8*, 1396. [CrossRef]
140. Alenquer, M.; Vale-Costa, S.; Etibor, T.A.; Ferreira, F.; Sousa, A.L.; Amorim, M.J. Influenza A virus ribonucleoproteins form liquid organelles at endoplasmic reticulum exit sites. *Nat. Commun.* **2019**, *10*, 1629. [CrossRef]

141. Bhagwat, A.R.; Le Sage, V.; Nturibi, E.; Kulej, K.; Jones, J.; Guo, M.; Tae Kim, E.; Garcia, B.A.; Weitzman, M.D.; Shroff, H.; et al. Quantitative live cell imaging reveals influenza virus manipulation of Rab11A transport through reduced dynein association. *Nat. Commun.* **2020**, *11*, 23. [CrossRef]

142. Etibor, T.A.; Yamauchi, Y.; Amorim, M.J. Liquid Biomolecular Condensates and Viral Lifecycles: Review and Perspectives. *Viruses* **2021**, *13*, 366. [CrossRef]

143. Noda, T.; Sagara, H.; Yen, A.; Takada, A.; Kida, H.; Cheng, R.H.; Kawaoka, Y. Architecture of ribonucleoprotein complexes in influenza A virus particles. *Nature* **2006**, *439*, 490–492. [CrossRef]

144. Kundu, A.; Avalos, R.T.; Sanderson, C.M.; Nayak, D.P. Transmembrane domain of influenza virus neuraminidase, a type II protein, possesses an apical sorting signal in polarized MDCK cells. *J. Virol.* **1996**, *70*, 6508–6515. [CrossRef]

145. Brewer, C.B.; Roth, M.G. A single amino acid change in the cytoplasmic domain alters the polarized delivery of influenza virus hemagglutinin. *J. Cell Biol.* **1991**, *114*, 413–421. [CrossRef]

146. Barman, S.; Nayak, D.P. Analysis of the Transmembrane Domain of Influenza Virus Neuraminidase, a Type II Transmembrane Glycoprotein, for Apical Sorting and Raft Association. *J. Virol.* **2000**, *74*, 6538–6545. [CrossRef]

147. Hughey, P.G.; Compans, R.W.; Zebedee, S.L.; Lamb, R.A. Expression of the influenza A virus M2 protein is restricted to apical surfaces of polarized epithelial cells. *J. Virol.* **1992**, *66*, 5542–5552. [CrossRef]

148. Enami, M.; Enami, K. Influenza virus hemagglutinin and neuraminidase glycoproteins stimulate the membrane association of the matrix protein. *J. Virol.* **1996**, *70*, 6653–6657. [CrossRef]

149. Ruigrok, R.W.; Barge, A.; Durrer, P.; Brunner, J.; Ma, K.; Whittaker, G.R. Membrane interaction of influenza virus M1 protein. *Virology* **2000**, *267*, 289–298. [CrossRef]

150. Chen, B.J.; Leser, G.P.; Jackson, D.; Lamb, R.A. The influenza virus M2 protein cytoplasmic tail interacts with the M1 protein and influences virus assembly at the site of virus budding. *J. Virol.* **2008**, *82*, 10059–10070. [CrossRef]

151. Nayak, D.P.; Hui, E.K.; Barman, S. Assembly and budding of influenza virus. *Virus Res.* **2004**, *106*, 147–165. [CrossRef] [PubMed]

152. Madsen, J.J.; Grime, J.M.A.; Rossman, J.S.; Voth, G.A. Entropic forces drive clustering and spatial localization of influenza A M2 during viral budding. *Proc. Natl. Acad. Sci. USA* **2018**, *115*, E8595–E8603. [CrossRef]

153. Sugita, Y.; Sagara, H.; Noda, T.; Kawaoka, Y. Configuration of Viral Ribonucleoprotein Complexes within the Influenza A Virion. *J. Virol.* **2013**, *87*, 12879–12884. [CrossRef] [PubMed]

154. Rudnicka, A.; Yamauchi, Y. Ubiquitin in Influenza Virus Entry and Innate Immunity. *Viruses* **2016**, *8*, 293. [CrossRef]

155. Pickart, C.M. Mechanisms underlying ubiquitination. *Annu. Rev. Biochem.* **2001**, *70*, 503–533. [CrossRef]

156. Stewart, M.D.; Ritterhoff, T.; Klevit, R.E.; Brzovic, P.S. E2 enzymes: More than just middle men. *Cell Res.* **2016**, *26*, 423–440. [CrossRef]

157. Dikic, I.; Dötsch, V. Ubiquitin linkages make a difference. *Nat. Struct. Mol. Biol.* **2009**, *16*, 1209–1210. [CrossRef]

158. Wong, J.; Zhang, J.; Si, X.; Gao, G.; Luo, H. Inhibition of the extracellular signal-regulated kinase signaling pathway is correlated with proteasome inhibitor suppression of coxsackievirus replication. *Biochem. Biophys. Res. Commun.* **2007**, *358*, 903–907. [CrossRef]

159. Tanno, H.; Komada, M. The ubiquitin code and its decoding machinery in the endocytic pathway. *J. Biochem.* **2013**, *153*, 497–504. [CrossRef] [PubMed]

160. Kulathu, Y.; Komander, D. Atypical ubiquitylation—the unexplored world of polyubiquitin beyond Lys48 and Lys63 linkages. *Nat. Rev. Mol. Cell Biol.* **2012**, *13*, 508–523. [CrossRef]

161. Komander, D.; Rape, M. The ubiquitin code. *Annu. Rev. Biochem.* **2012**, *81*, 203–229. [CrossRef]

162. Pickart, C.M.; Fushman, D. Polyubiquitin chains: Polymeric protein signals. *Curr. Opin. Chem. Biol.* **2004**, *8*, 610–616. [CrossRef]

163. Olzmann, J.A.; Chin, L.S. Parkin-mediated K63-linked polyubiquitination: A signal for targeting misfolded proteins to the aggresome-autophagy pathway. *Autophagy* **2008**, *4*, 85–87. [CrossRef]

164. Blount, J.R.; Meyer, D.N.; Akemann, C.; Johnson, S.L.; Gurdziel, K.; Baker, T.R.; Todi, S.V. Unanchored ubiquitin chains do not lead to marked alterations in gene expression in *Drosophila melanogaster*. *Biol. Open* **2019**, *8*, bio043372. [CrossRef] [PubMed]

165. Hook, S.S.; Orian, A.; Cowley, S.M.; Eisenman, R.N. Histone deacetylase 6 binds polyubiquitin through its zinc finger (PAZ domain) and copurifies with deubiquitinating enzymes. *Proc. Natl. Acad. Sci. USA* **2002**, *99*, 13425–13430. [CrossRef]

166. Reyes-Turcu, F.E.; Shanks, J.R.; Komander, D.; Wilkinson, K.D. Recognition of polyubiquitin isoforms by the multiple ubiquitin binding modules of isopeptidase T. *J. Biol. Chem.* **2008**, *283*, 19581–19592. [CrossRef]

167. Rape, M. Ubiquitylation at the crossroads of development and disease. *Nat. Rev. Mol. Cell Biol.* **2018**, *19*, 59–70. [CrossRef]

168. Pohl, C.; Dikic, I. Cellular quality control by the ubiquitin-proteasome system and autophagy. *Science* **2019**, *366*, 818–822. [CrossRef] [PubMed]

169. Oh, E.; Akopian, D.; Rape, M. Principles of Ubiquitin-Dependent Signaling. *Annu. Rev. Cell Dev. Biol.* **2018**, *34*, 137–162. [CrossRef]

170. Schwertman, P.; Bekker-Jensen, S.; Mailand, N. Regulation of DNA double-strand break repair by ubiquitin and ubiquitin-like modifiers. *Nat. Rev. Mol. Cell Biol.* **2016**, *17*, 379–394. [CrossRef]

171. Foot, N.; Henshall, T.; Kumar, S. Ubiquitination and the Regulation of Membrane Proteins. *Physiol. Rev.* **2017**, *97*, 253–281. [CrossRef]

172. Randow, F.; Lehner, P.J. Viral avoidance and exploitation of the ubiquitin system. *Nat. Cell Biol.* **2009**, *11*, 527–534. [CrossRef] [PubMed]

173. Guarino, L.A.; Smith, G.; Dong, W. Ubiquitin is attached to membranes of baculovirus particles by a novel type of phospholipid anchor. *Cell* **1995**, *80*, 301–309. [CrossRef]

174. Webb, J.H.; Mayer, R.J.; Dixon, L.K. A lipid modified ubiquitin is packaged into particles of several enveloped viruses. *FEBS Lett.* **1999**, *444*, 136–139. [CrossRef]

175. Spurgers, K.B.; Alefantis, T.; Peyser, B.D.; Ruthel, G.T.; Bergeron, A.A.; Costantino, J.A.; Enterlein, S.; Kota, K.P.; Boltz, R.C.; Aman, M.J.; et al. Identification of essential filovirion-associated host factors by serial proteomic analysis and RNAi screen. *Mol. Cell. Proteom. MCP* **2010**, *9*, 2690–2703. [CrossRef]

176. Gale, T.V.; Horton, T.M.; Hoffmann, A.R.; Branco, L.M.; Garry, R.F. Host Proteins Identified in Extracellular Viral Particles as Targets for Broad-Spectrum Antiviral Inhibitors. *J. Proteome Res.* **2019**, *18*, 7–17. [CrossRef]

177. Santos, S.; Obukhov, Y.; Nekhai, S.; Bukrinsky, M.; Iordanskiy, S. Virus-producing cells determine the host protein profiles of HIV-1 virion cores. *Retrovirology* **2012**, *9*, 65. [CrossRef]

178. Sette, P.; Nagashima, K.; Piper, R.C.; Bouamr, F. Ubiquitin conjugation to Gag is essential for ESCRT-mediated HIV-1 budding. *Retrovirology* **2013**, *10*, 79. [CrossRef]

179. Miyake, Y.; Matthias, P.; Yamauchi, Y. Purification of Unanchored Polyubiquitin Chains from Influenza Virions. In *Influenza Virus: Methods and Protocols*; Yamauchi, Y., Ed.; Springer: New York, NY, USA, 2018; pp. 329–342. [CrossRef]

180. Hao, R.; Nanduri, P.; Rao, Y.; Panichelli, R.S.; Ito, A.; Yoshida, M.; Yao, T.-P. Proteasomes Activate Aggresome Disassembly and Clearance by Producing Unanchored Ubiquitin Chains. *Mol. Cell* **2013**, *51*, 819–828. [CrossRef]

181. Yao, T.; Cohen, R.E. A cryptic protease couples deubiquitination and degradation by the proteasome. *Nature* **2002**, *419*, 403–407. [CrossRef] [PubMed]

182. Seto, E.; Yoshida, M. Erasers of histone acetylation: The histone deacetylase enzymes. *Cold Spring Harb. Perspect. Biol.* **2014**, *6*, a018713. [CrossRef]

183. Schultz, B.E.; Misialek, S.; Wu, J.; Tang, J.; Conn, M.T.; Tahilramani, R.; Wong, L. Kinetics and Comparative Reactivity of Human Class I and Class IIb Histone Deacetylases. *Biochemistry* **2004**, *43*, 11083–11091. [CrossRef]

184. Verdin, E.; Dequiedt, F.; Kasler, H.G. Class II histone deacetylases: Versatile regulators. *Trends Genet.* **2003**, *19*, 286–293. [CrossRef]

185. Yang, X.-J.; Grégoire, S. Class II Histone Deacetylases: From Sequence to Function, Regulation, and Clinical Implication. *Mol. Cell Biol.* **2005**, *25*, 2873–2884. [CrossRef]

186. Marks, P.A.; Miller, T.; Richon, V.M. Histone deacetylases. *Curr. Opin. Pharmacol.* **2003**, *3*, 344–351. [CrossRef]

187. Liu, T.; Liu, P.Y.; Marshall, G.M. The Critical Role of the Class III Histone Deacetylase SIRT1 in Cancer. *Cancer Res.* **2009**, *69*, 1702–1705. [CrossRef]

188. Bertos, N.R.; Gilquin, B.; Chan, G.K.; Yen, T.J.; Khochbin, S.; Yang, X.J. Role of the tetradecapeptide repeat domain of human histone deacetylase 6 in cytoplasmic retention. *J. Biol. Chem.* **2004**, *279*, 48246–48254. [CrossRef]

189. Zhang, Y.; Li, N.; Caron, C.; Matthias, G.; Hess, D.; Khochbin, S.; Matthias, P. HDAC-6 interacts with and deacetylates tubulin and microtubules in vivo. *EMBO J.* **2003**, *22*, 1168–1179. [CrossRef]

190. Matsuyama, A.; Shimazu, T.; Sumida, Y.; Saito, A.; Yoshimatsu, Y.; Seigneurin-Berny, D.; Osada, H.; Komatsu, Y.; Nishino, N.; Khochbin, S.; et al. In vivo destabilization of dynamic microtubules by HDAC6-mediated deacetylation. *EMBO J* **2002**, *21*, 6820–6831. [CrossRef] [PubMed]

191. Hubbert, C.; Guardiola, A.; Shao, R.; Kawaguchi, Y.; Ito, A.; Nixon, A.; Yoshida, M.; Wang, X.F.; Yao, T.P. HDAC6 is a microtubule-associated deacetylase. *Nature* **2002**, *417*, 455–458. [CrossRef]

192. Bali, P.; Pranpat, M.; Bradner, J.; Balasis, M.; Fiskus, W.; Guo, F.; Rocha, K.; Kumaraswamy, S.; Boyapalle, S.; Atadja, P.; et al. Inhibition of histone deacetylase 6 acetylates and disrupts the chaperone function of heat shock protein 90: A novel basis for antileukemia activity of histone deacetylase inhibitors. *J. Biol. Chem.* **2005**, *280*, 26729–26734. [CrossRef] [PubMed]

193. Kovacs, J.J.; Murphy, P.J.; Gaillard, S.; Zhao, X.; Wu, J.T.; Nicchitta, C.V.; Yoshida, M.; Toft, D.O.; Pratt, W.B.; Yao, T.P. HDAC6 regulates Hsp90 acetylation and chaperone-dependent activation of glucocorticoid receptor. *Mol. Cell* **2005**, *18*, 601–607. [CrossRef] [PubMed]

194. Murphy, P.J.; Morishima, Y.; Kovacs, J.J.; Yao, T.P.; Pratt, W.B. Regulation of the dynamics of hsp90 action on the glucocorticoid receptor by acetylation/deacetylation of the chaperone. *J. Biol. Chem.* **2005**, *280*, 33792–33799. [CrossRef]

195. Li, Y.; Zhang, X.; Polakiewicz, R.D.; Yao, T.P.; Comb, M.J. HDAC6 is required for epidermal growth factor-induced beta-catenin nuclear localization. *J. Biol. Chem.* **2008**, *283*, 12686–12690. [CrossRef]

196. Wang, S.; Li, N.; Wei, Y.; Li, Q.; Yu, Z. β-catenin deacetylation is essential for WNT-induced proliferation of breast cancer cells. *Mol. Med. Rep.* **2014**, *9*, 973–978. [CrossRef]

197. Zhang, X.; Yuan, Z.; Zhang, Y.; Yong, S.; Salas-Burgos, A.; Koomen, J.; Olashaw, N.; Parsons, J.T.; Yang, X.J.; Dent, S.R.; et al. HDAC6 modulates cell motility by altering the acetylation level of cortactin. *Mol. Cell* **2007**, *27*, 197–213. [CrossRef]

198. Zhang, L.; Liu, S.; Liu, N.; Zhang, Y.; Liu, M.; Li, D.; Seto, E.; Yao, T.P.; Shui, W.; Zhou, J. Proteomic identification and functional characterization of MYH9, Hsc70, and DNAJA1 as novel substrates of HDAC6 deacetylase activity. *Protein Cell* **2015**, *6*, 42–54. [CrossRef]

199. Saito, M.; Hess, D.; Eglinger, J.; Fritsch, A.W.; Kreysing, M.; Weinert, B.T.; Choudhary, C.; Matthias, P. Acetylation of intrinsically disordered regions regulates phase separation. *Nat. Chem. Biol.* **2019**, *15*, 51–61. [CrossRef]

200. Dallavalle, S.; Pisano, C.; Zunino, F. Development and therapeutic impact of HDAC6-selective inhibitors. *Biochem. Pharmacol.* **2012**, *84*, 756–765. [CrossRef] [PubMed]

201. Krämer, O.H.; Mahboobi, S.; Sellmer, A. Drugging the HDAC6–HSP90 interplay in malignant cells. *Trends Pharmacol. Sci.* **2014**, *35*, 501–509. [CrossRef] [PubMed]
202. Butler, K.V.; Kalin, J.; Brochier, C.; Vistoli, G.; Langley, B.; Kozikowski, A.P. Rational Design and Simple Chemistry Yield a Superior, Neuroprotective HDAC6 Inhibitor, Tubastatin A. *J. Am. Chem. Soc.* **2010**, *132*, 10842–10846. [CrossRef] [PubMed]
203. Ma, J.; Huo, X.; Jarpe, M.B.; Kavelaars, A.; Heijnen, C.J. Pharmacological inhibition of HDAC6 reverses cognitive impairment and tau pathology as a result of cisplatin treatment. *Acta Neuropathol. Commun.* **2018**, *6*, 103. [CrossRef]
204. Gupta, R.; Ambasta, R.K.; Kumar, P. Pharmacological intervention of histone deacetylase enzymes in the neurodegenerative disorders. *Life Sci.* **2020**, *243*, 117278. [CrossRef] [PubMed]
205. LoPresti, P. HDAC6 in Diseases of Cognition and of Neurons. *Cells* **2021**, *10*, 12. [CrossRef]
206. Ganai, S.A. Small-molecule Modulation of HDAC6 Activity: The Propitious Therapeutic Strategy to Vanquish Neurodegenerative Disorders. *Curr. Med. Chem.* **2017**, *24*, 4104–4120. [CrossRef]
207. Pandey, D.; Nomura, Y.; Rossberg, M.; Bhatta, A.; Romer, L.; Berkowitz, D. A5942 Inhibition of Histone Deacetylase 6 activity provides protection against atherogenesis: A role for HDAC6 NEDDylation. *J. Hypertens.* **2018**, *36*, e44–e45. [CrossRef]
208. Prior, R.; Van Helleputte, L.; Klingl, Y.E.; Van Den Bosch, L. HDAC6 as a potential therapeutic target for peripheral nerve disorders. *Expert Opin. Ther. Targets* **2018**, *22*, 993–1007. [CrossRef] [PubMed]
209. Husain, M.; Cheung, C.Y. Histone deacetylase 6 inhibits influenza A virus release by downregulating the trafficking of viral components to the plasma membrane via its substrate, acetylated microtubules. *J. Virol.* **2014**, *88*, 11229–11239. [CrossRef] [PubMed]
210. Husain, M.; Harrod, K.S. Enhanced acetylation of alpha-tubulin in influenza A virus infected epithelial cells. *FEBS Lett.* **2011**, *585*, 128–132. [CrossRef] [PubMed]
211. Choi, S.J.; Lee, H.C.; Kim, J.H.; Park, S.Y.; Kim, T.H.; Lee, W.K.; Jang, D.J.; Yoon, J.E.; Choi, Y.I.; Kim, S.; et al. HDAC6 regulates cellular viral RNA sensing by deacetylation of RIG-I. *EMBO J.* **2016**, *35*, 429–442. [CrossRef]
212. Liu, G.; Park, H.S.; Pyo, H.M.; Liu, Q.; Zhou, Y. Influenza A Virus Panhandle Structure Is Directly Involved in RIG-I Activation and Interferon Induction. *J. Virol.* **2015**, *89*, 6067–6079. [CrossRef] [PubMed]
213. Jiang, Z.; Wei, F.; Zhang, Y.; Wang, T.; Gao, W.; Yu, S.; Sun, H.; Pu, J.; Sun, Y.; Wang, M.; et al. IFI16 directly senses viral RNA and enhances RIG-I transcription and activation to restrict influenza virus infection. *Nat. Microbiol.* **2021**. [CrossRef] [PubMed]
214. Seigneurin-Berny, D.; Verdel, A.; Curtet, S.; Lemercier, C.; Garin, J.; Rousseaux, S.; Khochbin, S. Identification of components of the murine histone deacetylase 6 complex: Link between acetylation and ubiquitination signaling pathways. *Mol. Cell Biol.* **2001**, *21*, 8035–8044. [CrossRef]
215. Kawaguchi, Y.; Kovacs, J.J.; McLaurin, A.; Vance, J.M.; Ito, A.; Yao, T.P. The deacetylase HDAC6 regulates aggresome formation and cell viability in response to misfolded protein stress. *Cell* **2003**, *115*, 727–738. [CrossRef]
216. Dos Santos Passos, C.; Simões-Pires, C.A.; Carrupt, P.A.; Nurisso, A. Molecular dynamics of zinc-finger ubiquitin binding domains: A comparative study of histone deacetylase 6 and ubiquitin-specific protease 5. *J. Biomol. Struct. Dyn.* **2016**, *34*, 2581–2598. [CrossRef]
217. Hard, R.L.; Liu, J.; Shen, J.; Zhou, P.; Pei, D. HDAC6 and Ubp-M BUZ domains recognize specific C-terminal sequences of proteins. *Biochemistry* **2010**, *49*, 10737–10746. [CrossRef] [PubMed]
218. Magupalli, V.G.; Negro, R.; Tian, Y.; Hauenstein, A.V.; Di Caprio, G.; Skillern, W.; Deng, Q.; Orning, P.; Alam, H.B.; Maliga, Z.; et al. HDAC6 mediates an aggresome-like mechanism for NLRP3 and pyrin inflammasome activation. *Science* **2020**, *369*, eaas8995. [CrossRef]
219. Husain, M.; Harrod, K.S. Influenza A virus-induced caspase-3 cleaves the histone deacetylase 6 in infected epithelial cells. *FEBS Lett.* **2009**, *583*, 2517–2520. [CrossRef]
220. Montgomery, R.L.; Davis, C.A.; Potthoff, M.J.; Haberland, M.; Fielitz, J.; Qi, X.; Hill, J.A.; Richardson, J.A.; Olson, E.N. Histone deacetylases 1 and 2 redundantly regulate cardiac morphogenesis, growth, and contractility. *Genes Dev.* **2007**, *21*, 1790–1802. [CrossRef] [PubMed]
221. Lagger, G.; O'Carroll, D.; Rembold, M.; Khier, H.; Tischler, J.; Weitzer, G.; Schuettengruber, B.; Hauser, C.; Brunmeir, R.; Jenuwein, T.; et al. Essential function of histone deacetylase 1 in proliferation control and CDK inhibitor repression. *EMBO J.* **2002**, *21*, 2672–2681. [CrossRef] [PubMed]
222. Haberland, M.; Montgomery, R.L.; Olson, E.N. The many roles of histone deacetylases in development and physiology: Implications for disease and therapy. *Nat. Rev. Genet.* **2009**, *10*, 32–42. [CrossRef] [PubMed]
223. Zhang, Y.; Kwon, S.; Yamaguchi, T.; Cubizolles, F.; Rousseaux, S.; Kneissel, M.; Cao, C.; Li, N.; Cheng, H.-L.; Chua, K.; et al. Mice lacking histone deacetylase 6 have hyperacetylated tubulin but are viable and develop normally. *Mol. Cell Biol.* **2008**, *28*, 1688–1701. [CrossRef]
224. Yoneyama, M.; Jogi, M.; Onomoto, K. Regulation of antiviral innate immune signaling by stress-induced RNA granules. *J. Biochem.* **2016**, *159*, 279–286. [CrossRef]
225. Zanin, M.; DeBeauchamp, J.; Vangala, G.; Webby, R.J.; Husain, M. Histone Deacetylase 6 Knockout Mice Exhibit Higher Susceptibility to Influenza A Virus Infection. *Viruses* **2020**, *12*, 728. [CrossRef] [PubMed]
226. Fazioli, F.; Minichiello, L.; Matoska, V.; Castagnino, P.; Miki, T.; Wong, W.T.; Di Fiore, P.P. Eps8, a substrate for the epidermal growth factor receptor kinase, enhances EGF-dependent mitogenic signals. *EMBO J.* **1993**, *12*, 3799–3808. [CrossRef] [PubMed]

227. Lanzetti, L.; Rybin, V.; Malabarba, M.G.; Christoforidis, S.; Scita, G.; Zerial, M.; Di Fiore, P.P. The Eps8 protein coordinates EGF receptor signalling through Rac and trafficking through Rab5. *Nature* **2000**, *408*, 374–377. [CrossRef]

228. Croce, A.; Cassata, G.; Disanza, A.; Gagliani, M.C.; Tacchetti, C.; Malabarba, M.G.; Carlier, M.F.; Scita, G.; Baumeister, R.; Di Fiore, P.P. A novel actin barbed-end-capping activity in EPS-8 regulates apical morphogenesis in intestinal cells of Caenorhabditis elegans. *Nat. Cell Biol.* **2004**, *6*, 1173–1179. [CrossRef]

229. Disanza, A.; Carlier, M.F.; Stradal, T.E.; Didry, D.; Frittoli, E.; Confalonieri, S.; Croce, A.; Wehland, J.; Di Fiore, P.P.; Scita, G. Eps8 controls actin-based motility by capping the barbed ends of actin filaments. *Nat. Cell Biol.* **2004**, *6*, 1180–1188. [CrossRef]

230. Hertzog, M.; Milanesi, F.; Hazelwood, L.; Disanza, A.; Liu, H.; Perlade, E.; Malabarba, M.G.; Pasqualato, S.; Maiolica, A.; Confalonieri, S.; et al. Molecular basis for the dual function of Eps8 on actin dynamics: Bundling and capping. *PLoS Biol.* **2010**, *8*, e1000387. [CrossRef]

231. Frittoli, E.; Matteoli, G.; Palamidessi, A.; Mazzini, E.; Maddaluno, L.; Disanza, A.; Yang, C.; Svitkina, T.; Rescigno, M.; Scita, G. The signaling adaptor Eps8 is an essential actin capping protein for dendritic cell migration. *Immunity* **2011**, *35*, 388–399. [CrossRef]

232. Eierhoff, T.; Hrincius, E.R.; Rescher, U.; Ludwig, S.; Ehrhardt, C. The epidermal growth factor receptor (EGFR) promotes uptake of influenza A viruses (IAV) into host cells. *PLoS Pathog.* **2010**, *6*, e1001099. [CrossRef]

233. Huotari, J.; Helenius, A. Endosome maturation. *EMBO J.* **2011**, *30*, 3481–3500. [CrossRef]

234. Stauffer, S.; Feng, Y.; Nebioglu, F.; Heilig, R.; Picotti, P.; Helenius, A. Stepwise priming by acidic pH and a high K+ concentration is required for efficient uncoating of influenza A virus cores after penetration. *J. Virol.* **2014**, *88*, 13029–13046. [CrossRef] [PubMed]

235. Gschweitl, M.; Ulbricht, A.; Barnes, C.A.; Enchev, R.I.; Stoffel-Studer, I.; Meyer-Schaller, N.; Huotari, J.; Yamauchi, Y.; Greber, U.F.; Helenius, A.; et al. A SPOPL/Cullin-3 ubiquitin ligase complex regulates endocytic trafficking by targeting EPS15 at endosomes. *eLife* **2016**, *5*, e13841. [CrossRef] [PubMed]

236. Huotari, J.; Meyer-Schaller, N.; Hubner, M.; Stauffer, S.; Katheder, N.; Horvath, P.; Mancini, R.; Helenius, A.; Peter, M. Cullin-3 regulates late endosome maturation. *Proc. Natl. Acad. Sci. USA* **2012**, *109*, 823–828. [CrossRef]

237. Weis, K. Regulating access to the genome: Nucleocytoplasmic transport throughout the cell cycle. *Cell* **2003**, *112*, 441–451. [CrossRef]

238. Nigg, E.A. Nucleocytoplasmic transport: Signals, mechanisms and regulation. *Nature* **1997**, *386*, 779–787. [CrossRef]

239. Tran, E.J.; Wente, S.R. Dynamic nuclear pore complexes: Life on the edge. *Cell* **2006**, *125*, 1041–1053. [CrossRef]

240. Wente, S.R.; Rout, M.P. The nuclear pore complex and nuclear transport. *Cold Spring Harb. Perspect. Biol.* **2010**, *2*, a000562. [CrossRef] [PubMed]

241. Mboukou, A.; Rajendra, V.; Kleinova, R.; Tisné, C.; Jantsch, M.F.; Barraud, P. Transportin-1: A Nuclear Import Receptor with Moonlighting Functions. *Front. Mol. Biosci.* **2021**, *8*, 638149. [CrossRef]

242. Flatt, J.W.; Greber, U.F. Viral mechanisms for docking and delivering at nuclear pore complexes. *Semin. Cell Dev. Biol.* **2017**, *68*, 59–71. [CrossRef]

243. Greber, U.F.; Fornerod, M. Nuclear Import in Viral Infections. In *Membrane Trafficking in Viral Replication*; Marsh, M., Ed.; Springer: Berlin/Heidelberg, Germany, 2005; pp. 109–138. [CrossRef]

244. Yarbrough, M.L.; Mata, M.A.; Sakthivel, R.; Fontoura, B.M.A. Viral Subversion of Nucleocytoplasmic Trafficking. *Traffic* **2014**, *15*, 127–140. [CrossRef]

245. Fontoura, B.M.A.; Faria, P.A.; Nussenzveig, D.R. Viral Interactions with the Nuclear Transport Machinery: Discovering and Disrupting Pathways. *IUBMB Life* **2005**, *57*, 65–72. [CrossRef]

246. McDonald, S.M.; Nelson, M.I.; Turner, P.E.; Patton, J.T. Reassortment in segmented RNA viruses: Mechanisms and outcomes. *Nat. Rev. Microbiol.* **2016**, *14*, 448–460. [CrossRef]

247. Iuliano, A.D.; Roguski, K.M.; Chang, H.H.; Muscatello, D.J.; Palekar, R.; Tempia, S.; Cohen, C.; Gran, J.M.; Schanzer, D.; Cowling, B.J.; et al. Estimates of global seasonal influenza-associated respiratory mortality: A modelling study. *Lancet* **2018**, *391*, 1285–1300. [CrossRef]

248. Yanguez, E.; Hunziker, A.; Dobay, M.P.; Yildiz, S.; Schading, S.; Elshina, E.; Karakus, U.; Gehrig, P.; Grossmann, J.; Dijkman, R.; et al. Phosphoproteomic-based kinase profiling early in influenza virus infection identifies GRK2 as antiviral drug target. *Nat. Commun.* **2018**, *9*, 3679. [CrossRef] [PubMed]

249. Fernandez, J.; Machado, A.K.; Lyonnais, S.; Chamontin, C.; Gärtner, K.; Léger, T.; Henriquet, C.; Garcia, C.; Portilho, D.M.; Pugnière, M.; et al. Transportin-1 binds to the HIV-1 capsid via a nuclear localization signal and triggers uncoating. *Nat. Microbiol.* **2019**, *4*, 1840–1850. [CrossRef] [PubMed]

250. Lingappa, J.R.; Lingappa, V.R.; Reed, J.C. Addressing Antiretroviral Drug Resistance with Host-Targeting Drugs-First Steps towards Developing a Host-Targeting HIV-1 Assembly Inhibitor. *Viruses* **2021**, *13*, 451. [CrossRef] [PubMed]

251. MacArthur, R.D.; Novak, R.M. Maraviroc: The First of a New Class of Antiretroviral Agents. *Clin. Infect. Dis.* **2008**, *47*, 236–241. [CrossRef]

252. Loregian, A.; Mercorelli, B.; Nannetti, G.; Compagnin, C.; Palù, G. Antiviral strategies against influenza virus: Towards new therapeutic approaches. *Cell. Mol. Life Sci. CMLS* **2014**, *71*, 3659–3683. [CrossRef] [PubMed]

253. Toots, M.; Plemper, R.K. Next-generation direct-acting influenza therapeutics. *Transl. Res.* **2020**, *220*, 33–42. [CrossRef] [PubMed]

254. Geraghty, R.J.; Aliota, M.T.; Bonnac, L.F. Broad-Spectrum Antiviral Strategies and Nucleoside Analogues. *Viruses* **2021**, *13*, 667. [CrossRef]

Review

Ultrastructural Features of Membranous Replication Organelles Induced by Positive-Stranded RNA Viruses

Van Nguyen-Dinh and Eva Herker *

Institute of Virology, Philipps-University Marburg, 35043 Marburg, Germany
* Correspondence: eva.herker@uni-marburg.de

Abstract: All intracellular pathogens critically depend on host cell organelles and metabolites for successful infection and replication. One hallmark of positive-strand RNA viruses is to induce alterations of the (endo)membrane system in order to shield their double-stranded RNA replication intermediates from detection by the host cell's surveillance systems. This spatial seclusion also allows for accruing host and viral factors and building blocks required for efficient replication of the genome and prevents access of antiviral effectors. Even though the principle is iterated by almost all positive-strand RNA viruses infecting plants and animals, the specific structure and the organellar source of membranes differs. Here, we discuss the characteristic ultrastructural features of the virus-induced membranous replication organelles in plant and animal cells and the scientific progress gained by advanced microscopy methods.

Keywords: positive-strand RNA viruses; replication organelle; viral replication complex; membrane alterations; electron microscopy

Citation: Nguyen-Dinh, V.; Herker, E. Ultrastructural Features of Membranous Replication Organelles Induced by Positive-Stranded RNA Viruses. *Cells* **2021**, *10*, 2407. https://doi.org/10.3390/cells10092407

Academic Editors: Thomas Hoenen and Allison Groseth

Received: 29 June 2021
Accepted: 2 September 2021
Published: 13 September 2021

1. Introduction

During infection, positive-strand RNA viruses utilize the host's cellular membranes to support every step of their replication cycle, i.e., virus entry, replication of the genome, and assembly and release of virions. These viruses induce (endo)membrane rearrangements in host cells to create a protective microenvironment for replication of their genomes and for subsequent production of new virions [1]. These endomembrane rearrangements form structures termed viral replication organelles (ROs), which are essential for virus replication. They are thought to shield viral replication intermediates from recognition and to protect them from the host cell defense systems, such as RNA silencing or interferon induction [2]. The ROs are confined membranous compartments generated by extensive alteration of (endo)membrane structures. While these membranous structures are essential for viral RNA replication, expression of single viral proteins is often enough to induce them, but size and detailed structural features may be different in the absence of virus replication.

These endomembrane re-arrangements can differ morphologically, from simple dilated membranous structures to very complex structures such as multi-vesicular bodies (Figure 1).

During the infection, viral proteins as well as hijacked host proteins target the (endo)membrane system of the host to remodel it. Through these virus-host interactions different RO structures are created, depending on the virus and the membrane source. The majority of ROs are vesicular structures. The simplest form are single membrane vesicles (SMVs), typically 50–200 nm in diameter with or without small pores that are 10–20 nm diameter which either link vesicles to each other or link the vesicle lumen to the external environment, i.e., the cytosol. These SMVs with pores, also called spherules, are believed to be generated from invaginations or evagination of the host organelle membranes. Multiple SMVs can be packed together in an organelle to form higher order vesicle packets (VPs). Slightly more complicated RO structures are double membrane vesicles (DMVs) normally ranging from 200 to 400 nm in diameter. The biogenesis process of DMVs is not well understood. DMVs can be completely sealed with two membranous layers, the outer membrane can be connected to the organelle membrane the DMV originated from, and sometimes the inner vesicles share the same outer membrane and create a larger complex of DMVs. Some DMVs have open necks linking the internal lumen of DMVs to the external environment. More complex RO organizations such as multi-vesicular bodies (MVBs) also occur. These MVB structures are big vacuoles containing numerous small disordered membranous vesicles. Other, rarer membrane alterations are multi-membrane vesicles (MMVs), which are big multi-layered membranous particles of 300–400 nm in diameter, tubule-like structures of different diameters (20–50 nm) or zippered ER. Finally, massive unstructured membranous aggregates, which are called convoluted membranes (CMs), are frequently observed in virus-infected cells.

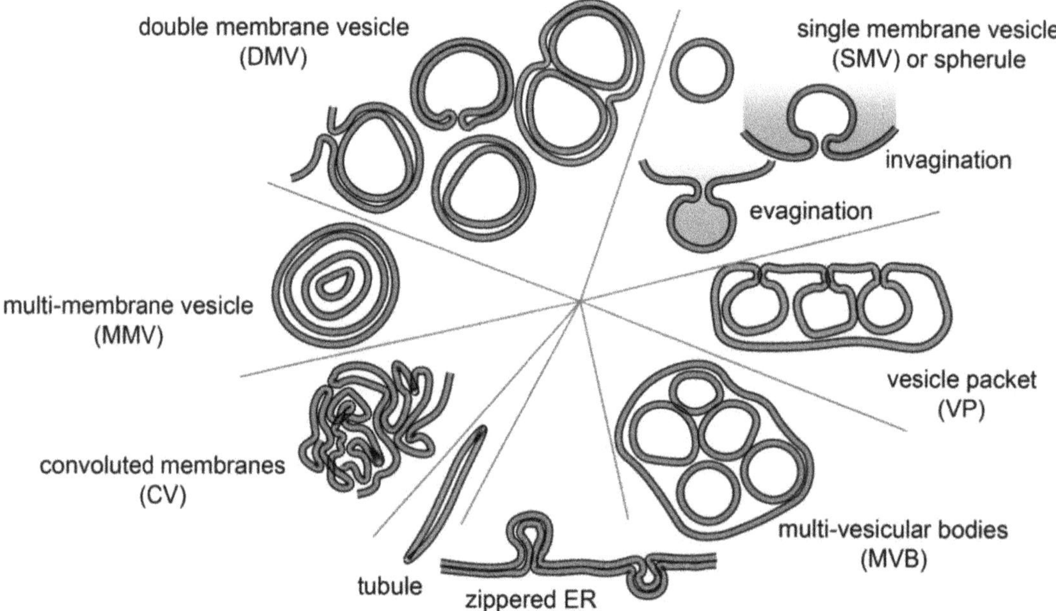

Figure 1. Membranous structures that occur in cell infected with positive-strand RNA viruses. Depicted are the most common membranous structures.

Membrane sources are different membranous organelles such as the endoplasmic reticulum (ER), the Golgi apparatus, peroxisomes, mitochondria, and the plasma membrane, and in plants, chloroplasts, and tonoplasts. In the following sections, we discuss the ultrastructural features and membrane origins of ROs to integrate them into the cell biological context of the infected cell (Table 1).

Table 1. Membrane sources and morphologies of the replication organelles (ROs).

<table>
<tr><th></th><th>Membrane Source</th><th colspan="2">Replication Organelles (RO)</th><th>Virus</th><th>Virus Family</th><th>Ref.</th></tr>
<tr><td rowspan="16">plant viruses</td><td rowspan="5">endoplasmic reticulum (ER)</td><td rowspan="2">vesicle/spherule</td><td>single membrane</td><td>beet black scorch virus (BBSV)</td><td>*Tombusviridae*</td><td>[3]</td></tr>
<tr><td>double membrane</td><td>turnip mosaic virus (TuMV)</td><td>*Potyviridae*</td><td>[4]</td></tr>
<tr><td colspan="2">multi-vesicular body</td><td>peanut clump virus (PCV)</td><td>*Virgaviridae*</td><td>[5]</td></tr>
<tr><td colspan="2">membranous inclusion body</td><td>wheat yellow mosaic virus (WYMV)</td><td>*Potyviridae*</td><td>[6]</td></tr>
<tr><td colspan="2">appressed double-membrane layers</td><td>brome mosaic virus (BMV)</td><td>*Bromoviridae*</td><td>[7]</td></tr>
<tr><td>Golgi</td><td colspan="2">dilated Golgi</td><td>tomato spotted wilt virus (TSWV)</td><td>*Tospoviridae*</td><td>[8]</td></tr>
<tr><td>peroxisomes</td><td colspan="2">multi-vesicular body</td><td>tomato bushy stunt virus (TBSV), cucumber necrosis virus (CNV)</td><td>*Tombusviridae*</td><td>[9,10]</td></tr>
<tr><td>mitochondria</td><td colspan="2">multi-vesicular body</td><td>melon necrotic spot virus (MNSV), Carnation Italian ringspot virus (CIRV)</td><td>*Tombusviridae*</td><td>[11,12]</td></tr>
<tr><td>chloroplast</td><td colspan="2">single membrane vesicle/spherule</td><td>barley stripe mosaic virus (BSMV)</td><td>*Virgaviridae*</td><td>[13]</td></tr>
<tr><td rowspan="2">tonoplast</td><td colspan="2" rowspan="2">single membrane vesicle/spherule</td><td>cucumber mosaic virus (CMV)</td><td>*Bromoviridae*</td><td>[14]</td></tr>
<tr><td>tobacco Necrosis Virus-Serotype A (TNV-A)</td><td>*Tombusviridae*</td><td>[14]</td></tr>
<tr><td rowspan="12">animal viruses</td><td rowspan="6">endoplasmic reticulum (ER)</td><td colspan="2">convoluted membrane</td><td rowspan="2">dengue virus (DENV), West Nil virus (WNV), Zika virus (ZIKV), tick-borne encephalitis virus (TBEV)</td><td rowspan="2">*Flaviviridae*</td><td rowspan="2">[15–18]</td></tr>
<tr><td rowspan="2">vesicle/spherule</td><td>single membrane</td></tr>
<tr><td>tubule-like structure</td><td>tick-borne encephalitis virus (TBEV)</td><td>*Flaviviridae*</td><td>[19]</td></tr>
<tr><td colspan="2">double membrane vesicle</td><td>hepatitis C virus (HCV)</td><td>*Flaviviridae*</td><td>[20–22]</td></tr>
<tr><td colspan="2" rowspan="2">zippered ER</td><td>severe acute respiratory syndrome coronavirus (SARS-CoV), middle east respiratory syndrome coronavirus (MERS-CoV), SARS-CoV2, infectious bronchitis virus (IBV)</td><td>*Coronaviridae*</td><td>[23–26]</td></tr>
<tr><td>Zika virus (ZIKV)</td><td>*Flaviviridae*</td><td>[17]</td></tr>
<tr><td>Golgi</td><td colspan="2">single and double membrane vesicle</td><td>polio virus (PV)</td><td>*Picornaviridae*</td><td>[27,28]</td></tr>
<tr><td>mitochondria</td><td colspan="2">single membrane vesicle/spherule</td><td>flock house virus (FHV)</td><td>*Nodaviridae*</td><td>[29,30]</td></tr>
<tr><td>lysosome</td><td colspan="2">cytopathic vacuole, single membrane vesicle/spherule</td><td>rubella virus (RUBV)</td><td>*Matonaviridae*</td><td>[31]</td></tr>
<tr><td>plasma membrane</td><td colspan="2">evagination, single membrane vesicle/spherule</td><td>sindbis virus (SINV)</td><td>*Togaviridae*</td><td>[32,33]</td></tr>
</table>

2. Structure and Origin of Plant Positive-Strand RNA Virus Replication Organelles

The ROs of positive-strand RNA viruses in plants are derived from different membranous organelles of the host including the ER, Golgi, peroxisomes, chloroplasts, and tonoplasts [34,35] (Figure 2).

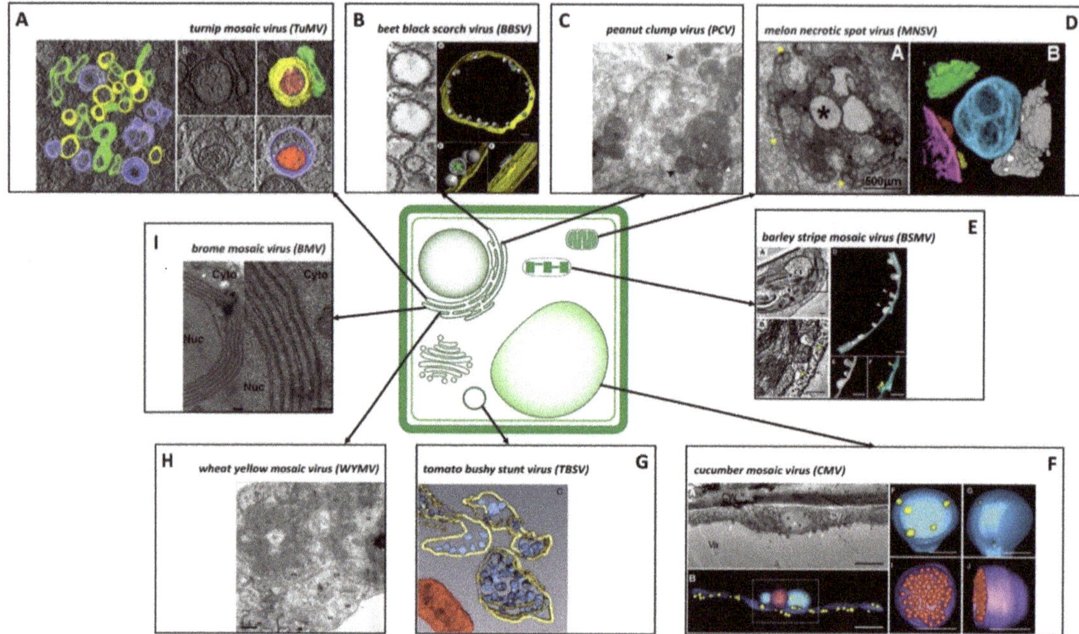

Figure 2. Structure and origin of plant positive-strand RNA virus replication organelles. (A) 3D architecture of TuMV-induced complex membrane structures. Overview of a single slice of a tomogram of a TuMV-infected vascular parenchymal cell. (upper right) The 3D model shows a SMV with fibrillar material inside and with an adjacent intermediate tubular structure. (lower right) 3D model of a DMV with a core of electron-dense materials. Yellow, SMVs; light red, electron-dense materials; green, intermediate tubular structures; light blue, outer membranes of DMVs; dark blue, inner membranes of DMVs; dark red, the electron-dense materials inside DMVs [4]. **(B)** Dilated ER of BBSV-infected cells with SMVs (left) and 3D surface reconstruction of the tomogram corresponding to the intact spherules (right) depicting the outer ER membrane (yellow), BBSV-induced spherules (gray), and fibrillar materials inside the spherules (green). Scale bars 100 nm [3]. **(C)** Electron microscopy of MVB structures in PCV-infected BY-2 protoplasts. White arrows indicate clusters of vesicles. Single arrowheads correspond to MVB; MVB containing disordered membranous vesicles are indicated by black arrowheads, whereas those containing one row of vesicles that are surrounded by a single membrane are indicated by white arrowheads. White asterisks correspond to electron-dense material without detectable vesicles [5]. **(D)** TEM analysis and 3D reconstruction of MNSV-induced altered mitochondria. (left) TEM image of altered mitochondria. Numerous vesicles were observed on the external surface as well as internal large invaginations and internal dilations (star), or both. Yellow arrowheads indicate the pores connecting the lumen of the dilation to the surrounding cytoplasm. (right) 3D model of MNSV-induced altered mitochondria (blue, yellow, red, and purple) with large dilations inside and close interactions with lipid droplets (grey) and chloroplasts (green) [11]. **(E)** BSMV-induced chloroplast membrane rearrangement and 3D model of altered chloroplast membranes. (left) Tomogram slices of altered chloroplast membranes from leaves of BSMV-infected *N. benthamiana*. The arrowheads indicate the same spherules in different slices. (right) 3D model of remodeled chloroplast membranes induced by BSMV indicating the outer chloroplast membrane (cyan), inner chloroplast membrane (gray), and spherules derived from the outer membrane (yellow) [13]. **(F)** 3D visualization of remodeled tonoplasts in CMV-infected cells. (upper left) Tomogram slice of a CMV-infected *N. benthamiana* leaf cell. CMV-induced spherules are observed on a vacuolar membrane and in a MVB (arrowheads). The cell wall (CW), cytosol (Cy), and vacuole (Va) are indicated. Scale bar 500 nm. (lower left) 3D model depicting the vacuolar membrane (dark blue), MVBs (light blue), spherules on the vacuolar membrane and in the MVBs (yellow), and a membrane compartment (purple) with virus particles (red). (upper left) 3D model of the MVB with spherules open to the cytosol. (lower left) 3D model of the membrane compartment with virus particles. Scale bars 200 nm [14]. **(G)** 3D reconstruction of TBSV ROs in wild-type yeast cells characterized by peroxisome-peripheral MVBs depicting the MVB membranes (yellow), vesicle-like spherules (blue) located close to a mitochondrion (red) [9]. **(H)** Electron micrographs of the mesophyll cells of WYMV-infected wheat. The presence of

membranous inclusion body structures in the cytoplasm. The ER, membranous inclusion (MI), mitochondria (Mt), pinwheel inclusion (PW), and virus particles (VP) are labelled [6]. (**I**) A series of 2–7 appressed layers of double-membrane ER in yeast cells expressing both 2a pol and 1a of BMV, double-membrane ER layers are separated by regular, 50–60-nm spaces, the nucleus (Nuc) and cytoplasm (Cyto) are indicated. Scale bars 100 nm [7] Copyright (2004) National Academy of Sciences, U.S.A. The different parts were reproduced with permission.

2.1. The Secretory Pathway Represents a Major Source for Membranes of Replication Organelles

The secretory pathway of plant cells is frequently targeted by viruses as a source of membranes [34]. Like its mammalian counterpart, it is composed of a complex membrane network including the ER, the Golgi apparatus, the trans-Golgi network (TGN), and endosomes. This pathway is involved in the synthesis, modification, and transport of proteins, lipids, and polysaccharides [35]. Among those organelles, the ER is most frequently targeted by viruses for their productive replication. The ER is an extensive membrane network with specialized subdomains that occupies large parts of the cell and is the prime spot for lipid and protein synthesis. Increased protein (and lipid) synthesis occurs during infection with diverse viral species and can trigger ER stress responses.

During infection with positive-strand RNA viruses, the ER structure is often dramatically changed due to the interaction between viral and host membrane proteins to form the ROs. Some viruses, such as brome mosaic virus (BMV), tobacco mosaic virus (TMV), and red clover necrotic mosaic virus (RCNMV), induce massive ER proliferation forming ER aggregates either in the perinuclear region or randomly dispersed in the cytoplasm that are called convoluted membranes or membranous web [36–39]. Other viruses such as beet black scorch virus (BBSV) or tobacco necrosis virus (TNV-W) induce ER membrane dilations and invaginations that are rounded structures of up to 400 nm within the ER cisternae [3,40]. Along with the rearrangement of ER membranes, viruses also form higher order membrane structures called vesicle packets (VPs) containing small vesicle structures which are 50–100 nm in diameter [3]. Most of them are spherules composed of single or double membranes, called single or double membrane vesicles, SMVs or DMVs, respectively. Those vesicular structures are the areas where viruses replicate their genome [3]. The spherule structures in BBSV-infected cells are arranged along the VP membranes and are SMVs. Each spherule vesicle has a narrow neck (5–10 nm in diameter) linked to the VP membrane and thus connecting the spherule interior to the cytoplasm, suggesting that they are formed by invagination of ER membranes [3] (Figure 2B). Those VPs containing spherules with opened necks to the cytoplasm are also reported in other positive-strand RNA viruses that infect animal cells such as viruses in the *Flaviviridae* family [17,41].

In contrast to BBSV, the ROs of peanut clump virus (PCV) in tobacco protoplasts form VPs containing multiple SMVs which are called multivesicular bodies (MVBs). These MVBs contain multiple disordered membranous vesicles of 80–200 nm in diameter often in one row of vesicles and surrounded by a single membrane [5] (Figure 2C). Interestingly, Turnip mosaic viruses (TuMV), do not only induce formation of SMVs but also of DMV–like structures that are found in the perinuclear cytoplasmic region [4] (Figure 2A). The DMVs formed during TuMV infection occur during the late stage of infection concomitantly with massive membrane arrangements leading to altered endomembrane structures such as dilated ER and membranous inclusion bodies [4]. Cytoplasmic or membranous inclusion bodies (MIBs) were observed not only in TuMV infection but also in cells infected with different viruses such as wheat yellow mosaic virus (WYMV). WYMV forms MIBs in infected wheat plants that are large, amorphous, crystalline lattice-like inclusion bodies in the cytoplasm. The periphery of these MIBs appears to be connected to the rough ER [6] (Figure 2H), but high-resolution structural information is not available yet.

However, other membranous structures beside spherular invaginations and vesicles can support RNA virus genome replication. For example, BMV in yeast cells can replicate the RNA at multilayer stacks of appressed double membranes [7] (Figure 2I). In infected cells, the relative expression levels and interactions between viral 1a and 2a-pol proteins can change the structure of perinuclear membrane rearrangements associated with RNA replication from small spherular invaginations to large stacks of 2–7 appressed layers of double-membrane ER. Intriguingly, these membrane stacks are highly ordered with 50–60 nm spaces, which is exactly the same width as the diameter of the spherules. These karmellae-like, multilayer structures are composed of stacks of ER that arise around the nucleus by folding over continuous sheets of ramified, double-membrane ER with its enclosed lumen. The double-membrane layers contain 1a and 2a-pol proteins and support BMV RNA replication but were not observed in yeast cells when only 1a or only 2a-pol proteins were expressed. Individual expression of BMV 1a induces only perinuclear spherules while 2a-pol alone does not cause any membrane alterations [7].

2.2. Peroxisomes and Mitochondria as Membrane Origins

The plant peroxisome is a single membrane-bound organelle that is solely responsible for beta-oxidation of fatty acids and the glyoxylate cycle, reactive oxygen species and reactive nitrogen species metabolism, and is involved pathogen defense. It is also one of the main target organelles for viruses as a membrane source to form ROs, especially for viruses in the *Tombusviridae* family such as tomato bushy stunt virus (TBSV) or cucumber necrosis virus (CNV) [9,10]. TBSV replicates in peroxisome-derived MVBs both in plant and yeast cells that are often found in close proximity to mitochondria (Figure 2G) [9]. Those MVBs are interconnected through membranes and might be nascent peroxisomes whose maturation and detachment from the ER is blocked by viral factors. In *N. tabacum* cell lines, TBSV p33 protein targets to peroxisomes and induces clustering and the formation of peroxisomal ghosts, but not MVBs, when expressed on its own [42]. CNV infection induces peroxisome biogenesis to form ROs [10]. Following infection, the peroxisomal boundary membranes are highly vesiculated, leading to the formation of doughnut- or C-shaped MVBs with the central region containing cytoplasmic material. The interiors of these doughnut-shaped MVBs contain many single-membrane vesicle-like structures with 80–150 nm in diameter. These vesicles appear to be connected to the MVB boundary membrane through a neck, and they provide the sites for CNV genome replication [10]. If peroxules that form in response to oxidative stress, which often occurs during virus infection, are hijacked by viruses as well, is currently unknown.

Interestingly, members of *Tombusviridae* not only target the ER or peroxisomes but also the mitochondria to form ROs to support viral replication as exemplified by melon necrotic spot virus (MNSV) and Carnation Italian ringspot virus (CIRV) [11,12]. In MNSV-infected cells, the mitochondrial structure is dramatically altered, and these abnormal organelles are frequently found close-by lipid droplets and ER membranes [11] (Figure 2D). Ultrastructural changes include dilated cristae and a vesiculated outer membrane. This vesiculated membrane forms multiple single-membrane vesicles with 45–50 nm in diameter which surround the large dilations inside the mitochondria. These vesicles appear to be connected to the cytoplasm or to the internal lumen of the large dilations through neck-like structures. Immuno-EM suggests that MNSV RNA and capsid proteins reside in the large dilations of abnormal mitochondria, suggesting that MNSV performs its genome replication as well as packaging in mitochondria and possibly within the interior of the vesicles [11].

2.3. The Chloroplast and Tonoplast Are Plant-Specific Membrane Sources

One organelle unique in plant cells that is also a target structure for many viruses is the chloroplast. Chloroplasts are membrane-rich organelles that conduct photosynthesis [43]. Barley stripe mosaic virus (BSMV) is a member of family *Virgaviridae* that alters chloroplast morphology during infection. In BSMV-infected plant cells, the membranes of the chloroplasts change dramatically with clusters of outer membrane-derived invaginated spherules (diameter ~50 nm with a neck of 11 nm) within inner membrane-derived packets (average diameter 112 nm) [13] (Figure 2E). The small spherules are linked via neck-like structures to the cytosol and immune-EM analysis revealed the presence of the viral RNA and replication proteins, suggesting that these spherules are the site of BSMV genome replication. In addition, big cytoplasmic invaginations surrounded by double membranes that contained virions were observed inside the chloroplasts [13]. This suggests that in addition to RNA replication, viral assembly takes place within the chloroplast.

The semipermeable membrane surrounding the vacuole is the tonoplast, an organelle that plays an important role in osmotic regulation of turgor pressure and that is targeted by viral infection. Already in the 1980s, cucumber mosaic virus (CMV)-infected leaf cells were shown to harbor tonoplast-associated vesicular structures [44]. The latest findings revealed that vacuole membranes are remodeled and invaginated in cells infected with CMV or tobacco necrosis virus A Chinese isolate (TNV-AC) [14]. Membrane invaginations form spherules at the periphery of the vacuole that are 50–70 nm in diameter (Figure 2F). These spherules contain neck-like structures that connect their interior with the cytosol. Interestingly, in CMV-infected cells, besides the spherules located at the tonoplast membrane, peripheral spherule-containing MVBs were also observed. The spherules inside the MVBs are also open towards the cytoplasm with a neck-like structure and the interior of the MVB seems to be connected to the vacuole. In addition to spherule-containing MVBs, membrane compartments harboring viral particles are found in close proximity to the vacuole and the ROs [14].

3. Structure and Origin of Animal Positive-Strand RNA Virus Replication Organelles

Similar to plant viruses, genome replication of all positive-strand RNA viruses that infect animal cells is intimately associated with membranes. The viral ROs supporting the replication of the viral genomes are generated from different host cellular membranous organelles including the endoplasmic reticulum (ER), the Golgi apparatus, mitochondria, lysosomes, and the plasma membrane [1] (Figure 3).

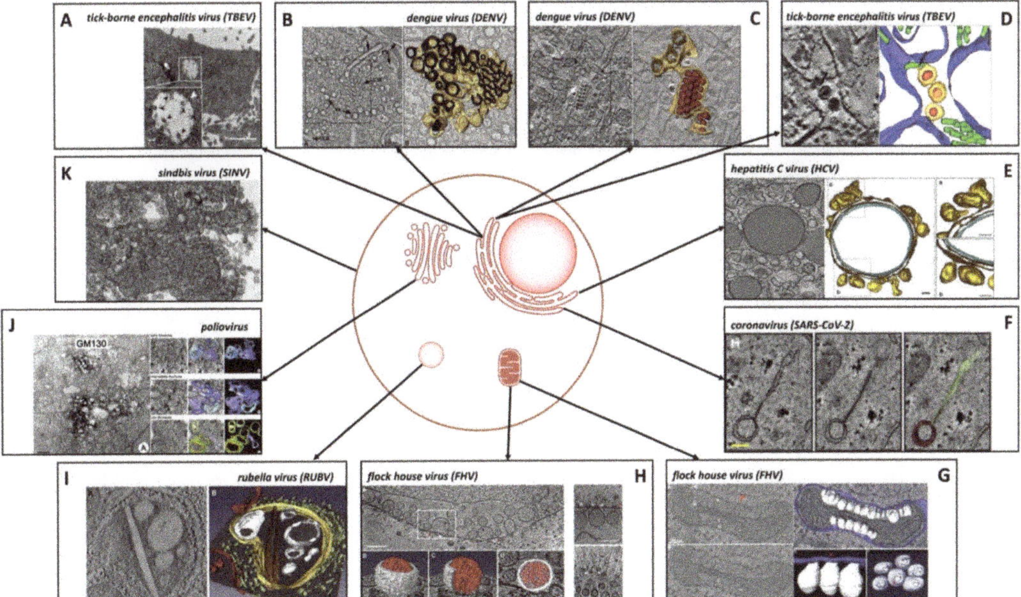

Figure 3. Structure and origin of animal positive-strand RNA virus replication organelles. (**A**) TEM images of HeLa cells transfected with the TBEV DNA replicon. White arrowheads show dilated ER areas; black arrowheads denote replication-vesicle-like structures inside the dilated ER areas. Insets show magnifications of the indicated areas. Scale bars 1 μm [41]. (**B,C**) DENV-infected Huh7 cells. (left) Tomogram slice shows DENV-induced convoluted membranes (CM),

vesicles (Ve), and tubes (T) that form a network of interconnected membranes in continuity with ER membranes. (right) 3D surface model of the membranes in the boxed area. The outer (cytosolic) face of the continuous membrane network is depicted in yellow; the ER lumen is dark [15]. (left) Stacked virus particles are in ER cisternae that are directly connected to virus-induced vesicles (white arrow). (right) 3D surface model of the virus-induced structures in the boxed area showing the continuity of virus-and vesicle-containing ER cisternae. ER membranes are depicted in yellow, inner vesicle membranes in light brown, and virus particles in red [15]. (**D**) Proliferation of the ER in human neuronal cells infected with TBEV. TBEV particles and TBEV-induced vesicles are located inside the proliferated and reorganized cisternae of the rough ER. 3D reconstruction of lamellar whorls, which are surrounded by cisternae arising from the rough ER (blue) and accommodate tubule-like structures (green). Detailed image shows the connection between the envelope (yellow) of a TBEV particle with nucleocapsid (red) and a tubule-like structure (indicated with an arrow) inside the rough ER. Scale bars 50 nm [19]. (**E**) 3D model of the HCV replication organelles surrounding lipid droplets. Electron tomography suggests that DMVs arise from ER membranes that are tightly wrapped around lipid droplets. (Left) Single tomographic slice of an HCV-infected cell with lipid droplets that are tightly wrapped by ER membranes and that stain positive for E2 and NS5A as revealed by fluorescence microscopy (not shown). (right) 3D reconstruction of the membranes surrounding the lipid droplet. ER membranes and DMVs are shown in yellow; the phospholipid monolayer of the lipid droplet monolayer membrane is shown in cyan. Insets illustrate that the DMVs originate from the wrapping ER membrane. Scale bars 100 nm [45]. (**F**) High-resolution analysis of ER-DMV interconnectivity in SARS-CoV-2-infected Calu-3 cells. Tomogram slices depict a membrane connector or zippered ER (light green) in contact with a DMV (red). (right) Superposition of rendered DMV and ER. Scale bars 200 nm [25]. (**G**) Tomogram slices and 3D reconstructions of mitochondria in FHV-infected *Drosophila* cells. (Left) Tomogram slices showing FHV-induced spherule rearrangements of a mitochondrion. Labels denote outer mitochondrial membrane (OM) and inner mitochondrial membrane (IM). White arrowheads indicate the necks that connect spherules to the OM. Asterisks mark two spherules that connect via necks to the OM. A red arrow marks the ~10 nm channel connecting the spherule interior to the cytoplasm. (upper right) 3D tomogram image with blue indicates OM, white indicates FHV spherules. (lower right) A close-up view of the connections between the OM and the spherules and 90° rotation of spherules showing the channels that connect the spherule interiors to the cytoplasm [29]. (**H**) (upper left) Tomogram slice of FHV spherules in a mitochondrion. Mitochondrial outer membrane (red), spherule membrane (blue), interior spherule filaments (black), and spherule openings (white) are indicated with arrowheads. Scale bar 100 nm. (lower left) 3D reconstruction of the spherule outlined in upper panel. Scale bars 50 nm. (right) Filaments are associated with FHV spherule pores. Tomographic slices with arrowheads pointing to the mitochondrial outer membrane (red), the spherule membrane (blue), the spherule opening (white), and the extruding filaments that likely represent viral RNA) (black). Scale bars 100 nm [30]. (**I**) 3D ET volumes of RUBV replication complex in BHK-21 cell. Tomogram slice (left) and the corresponding 3D model (right) of a CPV (yellow) surrounded by the rough ER (light green) and containing a number of vacuoles, vesicles, and a rigid straight sheet (brown) that is connected with the periphery of the CPV; mitochondria (red), vesicles and vacuoles (white) and cytoplasm (grey). Scale bars 200 nm [31]. (**J**) Poliovirus ROs in HeLa cells. (left) Viral replication structures are strongly associated with staining for a Golgi antigen, GM130. Scale bar 500 nm. (right) 3D reconstructions of poliovirus ROs at the early, intermediate, and late stages, 3, 4, and 7 hours post infection, respectively, each depicting central slices in tomographic volumes, central slices with segmented overlays, and segmented volumes, with blue indicating SMVs and yellow and green indicating inner and outer membranes of DMVs, respectively. Scale bars 100 nm [27]. (**K**) Plasma membrane invaginations and vacuole formation in SINV-infected BHK-21 cells. Scale bar 200 nm [32]. The different parts were reproduced with permission.

3.1. The ER Is the Main Hub for Animal Virus RO Fomation

Among the different membrane-bounded organelles, the ER represents the main membrane source for many positive-strand RNA virus ROs in animal cells [46]. The *Flaviviridae* family is one positive-strand RNA virus family that is well-known for ER-based RO formation [16,46,47]. In cells infected with dengue virus (DENV) tick-borne encephalitis virus (TBEV), West Nile virus (WNV), or Zika virus (ZIKV), the ER structure is dramatically altered owing to viral genome translation and replication. These viruses induce the formation of different membranous structures in the cytoplasm: vesicle packets (VPs) inside the ER, convoluted membranes (CMs) (Figure 3B), which are peculiar membranous aggregates with unknown function [18,47,48], and, in some cases, dilated ER, which are enlarged rough ER cisternae filled with granular material e.g. in TBEV infected cells [41,49]. In Hela cells transfected with a TBEV DNA replicon, the dilated ER cisterna grow to big cytoplasmic vacuoles containing small spherule-like structures 80–100 nm in diameter, which have open necks towards the cytoplasm (Figure 3A) [41]. The most prominent membranous structures derived from the ER in flavivirus-infected cells are the VPs that are the sites of viral genome replication and thus represent the ROs [17,41,50]. Early immuno-EM studies in DENV-infected insect cells indicated that VPs (or smooth membrane structures, SMS) are the site of DENV RNA replication [50]. These VPs are ER-derived membranous structures that are dilated ER cisterna containing single-membrane vesicles (SMVs) with a diameter of 80–150 nm [15] (Figure 3C). These SMVs originate from the invagination of the ER membrane into the ER lumen, have a spherule structure with small, 10–15 nm diameter necks opening to the cytoplasm. Necks were also observed linking SMVs inside the VPs in WNV-infected cells [16]. Densely packed viral particles are frequently within the ER in close proximity to VPs [15] (Figure 3C). Interestingly, an electron tomography (ET) study of TBEV-infected human neuronal cells investigated the proliferating ER in infected cells and found additional tubule-like structures of different diameters (20–50 nm) inside ER cisternae [19] (Figure 3D). In some instances, these tubule-like structures have direct contacts with viral particles inside these proliferated ER cisterna [19]. The function of these tubule-like structures is thus far unknown; they may represent membranous structures involved in viral replication, abnormal cellular structures arising due to altered membrane metabolism, or a feature of cellular process to limit the viral infection [51].

Among the members of *Flaviviridae*, hepatitis C virus (HCV) is somewhat unique regarding the prototypical RO structures. In HCV-infected hepatocyte cells, ER membranes are intensively rearranged to form the membranous web (MW). The MW contains vesicles of different morphologies, mainly SMVs or DMVs, embedded in a matrix of membranes which are sometimes close to or wrap tightly around lipid droplets [21,22]. HCV infection as well as expression of single HCV proteins induce different types of membranous vesicles in cells [20–22]. While NS3/4A and NS4B induces only SMVs, NS5A induces MMVs and infrequently DMVs [22]. However, expression of the complete replicase complex (NS3-NS5B) is needed for formation of DMVs that are indistinguishable from the ones observed in infection [22]. In HCV-infected cells there are vesicles in clusters containing SMVs of variable sizes (100–200 nm in diameter), sometimes sticking together and harboring internal invaginations, and SMVs of a homogeneous size (~100 nm in diameter) that are clustered together and sometimes arrayed around lipid droplets. However, the most prominent vesicular structure induced by HCV are DMVs, likely representing the ROs. The DMVs are heterogeneous in size, with an average diameter of 200-400 nm, and are morphologically similar to membrane alterations identified in cells infected with coronaviruses [23] or picornaviruses [27]. These vesicles are characterized by two closely apposed membranes. EM/ET analysis revealed that most of the DMVs are generated from the ER and some of them are still connected to ER sheets via their outer membrane [22]. Although most of DMVs are completely closed structures and it is still unknown why HCV would induce these closed structures, a small percentage of them (8–10%) [22] has an opening neck towards the cytosol. The opened and closed DMVs thus may reflect the different stages of DMV "maturation", early and late, respectively [22]. An immunolabeling study of purified DMVs revealed an enrichment for viral proteins as well as dsRNA suggesting that DMVs indeed play an important role for viral RNA replication [52]. Viral RNA amplification may occur inside DMVs, which would allow the exit of newly synthetized viral genomes as long as the DMV is open, but replication might also occur on the outer surface of DMVs [22,52]. A more recent study using correlative light and electron microscopy (CLEM) indicated that DMVs emerge from ER membranes which are tightly wrapped around lipid droplets [45] (Figure 3E). EM/ET analysis of HCV-infected cell revealed two types of lipid droplets: lipid droplets that are tightly wrapped by the ER and that stain positive for the HCV glycoprotein E2 and nonstructural protein NS5A by immunofluorescence microscopy as well as lipid droplets that are not wrapped by ER and that do not stain positive for E2 and NS5A. These data suggest that HCV proteins trigger wrapping of ER membranes around lipid droplets. This tightly closed contact between DMVs and ER-wrapped lipid droplets may enable short-distance trafficking of viral RNA from replication vesicles to assembly sites at lipid droplet–associated ER membranes [45]. Later during HCV infection, multi-membrane vesicles (MMVs) with an average diameter 350–400 nm are generated, likely originating from DMVs through secondary enwrapping events [22].

DMVs are observed not only in HCV infection but also during infection with other positive-strand RNA viruses, such as members of *Nidovirales*, including coronaviruses and arteriviruses [1]. DMVs are well-known typical ROs of coronaviruses [23,24]. A new study employing 3D reconstructions using FIB-SEM (focused ion beam milling combined with scanning EM) to determine morphological alterations induced in severe acute respiratory syndrome coronavirus 2 (SARS-CoV-2)-infected human lung epithelial cells revealed extensive fragmentation of the Golgi apparatus, alteration of the mitochondrial network, and recruitment of peroxisomes to viral ROs, which are clusters of DMVs [25]. In the SARS-CoV-2-infected cells, the ER network was altered intensively to generate the ROs, which consist predominantly of DMVs with an average diameter of 250–350 nm. Theses DMVs were tightly connected with the ER network linking the outer membrane to ER-derived structures such as ER connectors. Similar to DMVs in HCV-infected cells, the DMVs in SARS-CoV-2-infected cells are mostly closed DMV structures. However, DMV-DMV contacts were observed in SARS-CoV-2-infected cells, either through funnel-like junctions between two DMVs or fused DMVs consisting of multiple vesicles sharing the same outer membrane [25]. As described above, SARS-CoV-2 induces the formation ER connectors between the DMVs and ER tubules [25] (Figure 3F). These membranous structures were also described as zippered ER in gamma- or betacoronaviruses, such as infectious bronchitis virus (IBV) or Middle East respiratory syndrome coronavirus (MERS-CoV) [24,26]. The zippered ER or ER connectors lack luminal space, suggesting that they are formed through zippering or collapsing of ER cisternae. However, in contrast to SARS-CoV-2, electron tomograms showed that IBV-induced spherules are tethered to zippered ER and that there is a small pore connecting the interior of the spherule with the cytoplasm [26]. Of note, in a recent study of ZIKV, zippered ER structures were also observed in infected cells [17]. 3D reconstruction of regions containing zippered ER in ZIKV-infected cells revealed that the collapsed ER was connected to regions containing invaginated replication vesicles [17,47].

3.2. Further Down the Secretory Route, the Golgi Apparatus Supports RO Formation

Many viruses rely on the secretory route through the Golgi apparatus for maturation and release of viral progeny. However, some viruses also employ Golgi membranes to establish their ROs for viral RNA replication, e.g. poliovirus or coxsackieviruses, which are members of *Picornaviridae* family. Membrane alterations in poliovirus-infected cells include the formation of SMVs and DMVs [28]. A recent publication employing immuno-EM with subsequent diaminobenzidine (DAB) labeling suggested that membrane rearrangements in poliovirus-infected cells may occur in a sequential manner [27] (Figure 3J). In the early stage of infection, small clusters of SMVs appear. Later in infection, they are replaced by either round or irregularly shaped DMVs. Interestingly, the small clusters of SMVs of 100–200 nm in diameter strongly stained positive for a Golgi antigen, GM130, a cis-Golgi marker, but not for calnexin, an ER marker. These data suggest that the ROs of polioviruses may originate from the Golgi apparatus. However, it is too early to exclude a role of the ER for biogenesis of these ROs as ER-proteins might be dislocated during RO formation. dsRNA, i.e. viral RNA replication intermediates, as well as metabolically labeled viral RNA were detected in both SMVs and DMVs of poliovirus ROs, suggesting that both structures are relevant sites for poliovirus RNA synthesis [27].

3.3. Mitochondria, Lysosomes, and the Plasma Membrane Are Involved in RO Formation

Interestingly, the flock house virus (FHV), a member of the family *Nodaviridae* targets the mitochondria to form ROs supporting their RNA replication. In FHV-infected *Drosophila* cells, the mitochondrial outer membrane is dramatically altered [29] (Figure 3G). The virus induces the formation of invaginations at the outer mitochondrial membrane into the spherule structures with an average diameter of 50 nm. All spherules are outer mitochondrial membrane invaginations with their lumen connected to the cytoplasm through a small pore of 10 nm in diameter, which is sufficient for ribonucleotide import and product RNA export [29]. A recent cryo-electron tomography study showed the presence of electron-dense structures within the spherules, which likely corresponds to the viral RNA as the volume correlated well with viral RNA length [30] (Figure 3H). This study additionally revealed the structure and symmetry of the proteins that form the pore complex. These pore complexes were frequently associated with long cytoplasmic electron-dense trails, likely representing exported viral RNA [30].

The lysosome is another cellular organelle which is a favorite target for some positive-strand RNA viruses such as rubella virus (RUBV) and members of *Togaviridae*, including Semliki Forest virus (SFV) and sindbis virus (SINV) [53–55]. These viruses alter lysosome and endosome structures to form cytopathic vacuoles (CPVs) that represent the viral ROs [54]. In RUBV-infected cells, the rough ER, mitochondria, and the Golgi are clustered around CPVs, which are linked to the cytosol and enclose vesicular structures [31] (Figure 3I). These organelles contain active ROs from which replicated RNA is transported to virion assembly sites at Golgi membranes. These CPVs have a quite variable diameter of 600–2000 nm. Electron tomography and 3D reconstruction revealed that CPVs enclose a variety of different membrane structures such as stacked membranes, rigid membrane sheets, small vesicles, and larger vacuoles that are connected through membrane contacts with each other and functionally connected to the endocytic pathway. CPVs have additional membrane contact sites to other cellular organelles such as the rough ER and Golgi vesicles, but not to nearby mitochondria. Immunogold labeling confirmed the presence of replicase complex proteins and dsRNA inside CPVs, suggesting that RNA synthesis occurs on or in vesicles within the CPVs [31].

As mentioned above, alphaviruses, such as SFV, SINV and WEEV, are known to induce formation of CPVs in infected cells, which are modified lysosomes and endosomes and the sites of viral RNA replication. Interestingly, in SINV-infected cells spherules containing dsRNA and nonstructural protein (nsP) are initially formed at the plasma membrane [32,33]. Immunofluorescence microscopy and EM revealed that at early times of infection, viral nsPs as well as dsRNA replication intermediates locate to spherules the plasma membrane [33] (Figure 3K). These spherules form as evaginations at the plasma membrane and the presence of plasma membrane-associated dsRNA and ns proteins suggest that they represent ROs. Later in infection, these spherules are internalized by endocytosis; trafficking and maturation to CPVs is dependent on phosphatidylinositol 3-kinase activity and the cytoskeleton [33], highlighting the often complex nature of viral RO formation.

4. Recent Technical Developments and Challenges

For multiplication, viruses need to infect a suitable host cell to be able to replicate their genome, to produce and release new infectious virions, and thus continue the next round of the infectious cycle. The interactions of viruses with their hosts are highly dynamic, diverse and complex, and occur on multiple levels. It is important to elucidate the molecular mechanisms of these virus-host interactions in order to understand virus replication cycles and how viruses affect and alter the cell biology of their host to support viral replication. This knowledge is not only important for better understanding of the biology of viruses but also to support control of viral infections, to predict their effect on ecology and human health, and to design effective antiviral strategies against chronic and emerging viral infectious diseases.

"Seeing is believing", we clearly trust observations that we can visualize. Microscopy, especially high-resolution light/fluorescence and electron microscopy (EM) are important tools for visualizing structures of viral and host cell components and thus for the generation of general concepts governing virus-host interactions. Indeed, EM and virus research developments are deeply intertwined since the invention of EM [56,57]. EM is one of the critical methods to elucidate how viruses replicate in the microstructure environments of the infected cell in order to produce new virions [58]. In general, EM techniques encompass two main applications: transmission EM (TEM) and scanning EM (SEM), which each are different microscopic techniques [59]. The resolution of SEM is lower than that of TEM. In contrast, SEM provides a larger sample scanning ability or a bigger field of view for both surface and volume. Therefore, TEM is the favorite method to study small structures in detail, whereas SEM applications help to expand the sample scales.

The combination of EM with advanced light microscopy techniques termed correlative light and electron microscopy (CLEM) provides even more detailed information as it allows to analyze the dynamics and localization of viral and/or host protein-protein interactions in the context of detailed structural aspects of the intracellular environment. In this method proteins are visualized through fluorescent tags or antibodies using light microscopy in order to find rare biological events or to identify specific structures prior to characterizing the structures and their surroundings at high-resolution using EM. The current full spectrum of state-of-the-art microscopic techniques covers an extensive range of scales, resolutions, and information. Many of the methods mentioned together with the viral RO structures in this review, such as electron (cryo)tomography, CLEM, volume SEM, or 3D TEM have thrived and were further advanced within only two decades, especially since cryo-EM was discovered and developed in the 1980s [60].

The newly advanced electron tomography (ET), including volume SEM and cryotomography, has been a useful method in elucidating the 3D volume architecture of viral ROs. Volume SEMs such as serial block face SEM and focused ion beam milling (FIB)-SEM have been used to explore virus-host interaction with the nanometer resolution in wider and thicker volume samples including tissues. Furthermore, advanced cryo-FIB-SEM techniques are applied on cryo-stage specimens, which can help to avoid the artifacts of conventional EM sample preparation due to chemical fixation and staining processes and can also help to improve the stabilization of native structures in the specimen [61]. Although currently cryotomography of FIB-milled cryo-lamellae is the outstanding method in ET, the area that can be investigated is restricted to a very small and thin cellular region (the cryo-lamella) [62]. Difficulties in sample preparation combined with the need for highly demanding technical skills and high equipment costs are further limitations that are needed to be solved with technology developments in the future [62]. On the contrary to volume SEM, cryotomography methods can yield magnificent structural details with molecular-level resolution of the viral ROs in the cryo-native condition [63]. Cryotomography is currently one of the most powerful methods for investigation and characterization of the biological structures of viral ROs from the macro-structural morphology to the nano-organization of detailed protein structures which were presented in many current studies on viral ROs discussed in this review. Furthermore, current cryo-CLEM application, which combines cryo-light microscopy and cryo-EM opens a new way in investigating the molecular mechanisms of virus-host interactions more specifically and more accurate under cryogenic conditions [63]. However, similar to cryo-FIB-SEM, only a small area of the targeted cellular structure can be processed for investigation and the processing of cryotomography requires highly developed technical skills, limiting the popularity of 3D-cryoEM. For cryo-CLEM, the limited resolution of cryo-light microscopy, mostly based on wide-field light microscope also decreases the accuracy of this technique when it comes to localization of specific structural protein or events [64].

Of course, one main obstacle when investigating virus-infected specimens is the need for inactivation, especially for human pathogenic viruses. Thus, these samples require strong fixation that may cause artifacts. Alternatively, all steps including the image acquisition under cryo conditions have to be performed under biosafety containment, which is difficult to implement. Thus, we may need to rely on non-pathogenic model viruses for some of the advanced microscopy techniques.

5. Conclusions

Positive-stranded RNA viruses dramatically remodel intracellular membranes into distinct RO structures that support the synthesis of viral RNA. ROs provide optimal micro-environments for viral genome replication and shield replication intermediates such as double-stranded RNA (dsRNA) from detection by innate immune sensors. Many questions about the biogenesis process viral ROs remain unanswered, i.e., for many viruses we do not have detailed information on host factors such as proteins and specific lipids that contribute to RO formation. Likewise, the dynamic nature of how and where and when during infection viral proteins required for RO formation interact with host proteins to remodel intracellular membranes into viral ROs and to stabilize the RO morphology remains to be determined. For many viruses pores connecting the RO interieur with the cytosol are observed but how viral proteins interact with host membrane proteins to stabilize these pore structures are still poorly understood [65]. The crown-shaped molecular complexes of some of the pores unveiled in recent studies of positive-stranded RNA viruses have provided us an overview of the protein complex organization of these pores [30,66,67]. However, how flexible the pores are and how the pore proteins regulate the transit of proteins and nucleotides/viral RNA from and to ROs and, possibly, coordinate it with other processes in the viral replication cycle is still poorly investigated. For other viruses, closed ROs have been observed frequently. If they are inactive/old ROs or just open up intermittently is still unclear. Elucidating how the viral replication complexes work on a molecular level and integrating biochemical knowledge with structural information gained by EM analysis are challenging goals for the future.

Author Contributions: Writing—original draft preparation, V.N.-D. and E.H.; writing—review and editing, V.N.-D. and E.H.; visualization, V.N.-D. and E.H; funding acquisition, E.H. All authors have read and agreed to the published version of the manuscript.

Funding: This work was funded by the Deutsche Forschungsgemeinschaft (DFG HE 6889/2 and HE 6889/5) and the LOEWE Center DRUID (Novel Drug Targets against Poverty-related and Neglected Tropical Infectious Diseases).

Institutional Review Board Statement: Not applicable.

Informed Consent Statement: Not applicable.

Data Availability Statement: Not applicable.

Acknowledgments: We would like to apologize to all colleagues whose work we did not cite. The different parts were reproduced with permission or under the Creative Commons CC BY license: Figure 2 (A) [4], (B) [3], (C) [5], (D) [11], (E) [13], (F) [14], (G), [9], (H) [6], (I) [7] Copyright (2004) National Academy of Sciences, U.S.A; Figure 3 (A) [41], (B-C) [15], (D) [19], (E) [45], (F) [25], (G) [29], (H) [30], (I) [31], (J) [27], (K) [32].

Conflicts of Interest: The authors declare no conflict of interest. The funders had no role in the design of the study; in the collection, analyses, or interpretation of data; in the writing of the manuscript, or in the decision to publish the results.

References

1. Romero-Brey, I.; Bartenschlager, R. Membranous replication factories induced by plus-strand RNA viruses. *Viruses* **2014**, *6*, 2826–2857. [CrossRef] [PubMed]
2. Overby, A.K.; Popov, V.L.; Niedrig, M.; Weber, F. Tick-borne encephalitis virus delays interferon induction and hides its double-stranded RNA in intracellular membrane vesicles. *J. Virol.* **2010**, *84*, 8470–8483. [CrossRef] [PubMed]
3. Cao, X.; Jin, X.; Zhang, X.; Li, Y.; Wang, C.; Wang, X.; Hong, J.; Wang, X.; Li, D.; Zhang, Y. Morphogenesis of Endoplasmic Reticulum Membrane-Invaginated Vesicles during Beet Black Scorch Virus Infection: Role of Auxiliary Replication Protein and New Implications of Three-Dimensional Architecture. *J. Virol.* **2015**, *89*, 6184–6195. [CrossRef] [PubMed]
4. Wan, J.; Basu, K.; Mui, J.; Vali, H.; Zheng, H.; Laliberte, J.F. Ultrastructural Characterization of Turnip Mosaic Virus-Induced Cellular Rearrangements Reveals Membrane-Bound Viral Particles Accumulating in Vacuoles. *J. Virol.* **2015**, *89*, 12441–12456. [CrossRef]
5. Dunoyer, P.; Ritzenthaler, C.; Hemmer, O.; Michler, P.; Fritsch, C. Intracellular localization of the peanut clump virus replication complex in tobacco BY-2 protoplasts containing green fluorescent protein-labeled endoplasmic reticulum or Golgi apparatus. *J. Virol.* **2002**, *76*, 865–874. [CrossRef]
6. Sun, L.; Andika, I.B.; Shen, J.; Yang, D.; Chen, J. The P2 of Wheat yellow mosaic virus rearranges the endoplasmic reticulum and recruits other viral proteins into replication-associated inclusion bodies. *Mol. Plant Pathol.* **2014**, *15*, 466–478. [CrossRef]

7.   Schwartz, M.; Chen, J.; Lee, W.M.; Janda, M.; Ahlquist, P. Alternate, virus-induced membrane rearrangements support positive-strand RNA virus genome replication. *Proc. Natl. Acad. Sci. USA* **2004**, *101*, 11263–11268. [CrossRef]

8.   Kikkert, M.; Van Lent, J.; Storms, M.; Bodegom, P.; Kormelink, R.; Goldbach, R. Tomato spotted wilt virus particle morphogenesis in plant cells. *J. Virol.* **1999**, *73*, 2288–2297. [CrossRef]

9.   Fernandez de Castro, I.; Fernandez, J.J.; Barajas, D.; Nagy, P.D.; Risco, C. Three-dimensional imaging of the intracellular assembly of a functional viral RNA replicase complex. *J. Cell Sci.* **2017**, *130*, 260–268. [CrossRef]

10. Rochon, D.; Singh, B.; Reade, R.; Theilmann, J.; Ghoshal, K.; Alam, S.B.; Maghodia, A. The p33 auxiliary replicase protein of Cucumber necrosis virus targets peroxisomes and infection induces de novo peroxisome formation from the endoplasmic reticulum. *Virology* **2014**, *452–453*, 133–142. [CrossRef] [PubMed]

11. Gomez-Aix, C.; Garcia-Garcia, M.; Aranda, M.A.; Sanchez-Pina, M.A. Melon necrotic spot virus Replication Occurs in Association with Altered Mitochondria. *Mol. Plant. Microbe Interact.* **2015**, *28*, 387–397. [CrossRef]

12. Heinze, C.; Wobbe, V.; Lesemann, D.E.; Zhang, D.Y.; Willingmann, P.; Adam, G. Pelargonium necrotic spot virus: A new member of the genus Tombusvirus. *Arch. Virol* **2004**, *149*, 1527–1539. [CrossRef]

13. Jin, X.; Jiang, Z.; Zhang, K.; Wang, P.; Cao, X.; Yue, N.; Wang, X.; Zhang, X.; Li, Y.; Li, D.; et al. Three-Dimensional Analysis of Chloroplast Structures Associated with Virus Infection. *Plant Physiol.* **2018**, *176*, 282–294. [CrossRef]

14. Wang, X.; Ma, J.; Jin, X.; Yue, N.; Gao, P.; Mai, K.K.K.; Wang, X.B.; Li, D.; Kang, B.H.; Zhang, Y. Three-dimensional reconstruction and comparison of vacuolar membranes in response to viral infection. *J. Integr. Plant Biol.* **2021**, *63*, 353–364. [CrossRef]

15. Welsch, S.; Miller, S.; Romero-Brey, I.; Merz, A.; Bleck, C.K.; Walther, P.; Fuller, S.D.; Antony, C.; Krijnse-Locker, J.; Bartenschlager, R. Composition and three-dimensional architecture of the dengue virus replication and assembly sites. *Cell Host Microbe* **2009**, *5*, 365–375. [CrossRef] [PubMed]

16. Gillespie, L.K.; Hoenen, A.; Morgan, G.; Mackenzie, J.M. The endoplasmic reticulum provides the membrane platform for biogenesis of the flavivirus replication complex. *J. Virol.* **2010**, *84*, 10438–10447. [CrossRef] [PubMed]

17. Cortese, M.; Goellner, S.; Acosta, E.G.; Neufeldt, C.J.; Oleksiuk, O.; Lampe, M.; Haselmann, U.; Funaya, C.; Schieber, N.; Ronchi, P.; et al. Ultrastructural Characterization of Zika Virus Replication Factories. *Cell Rep.* **2017**, *18*, 2113–2123. [CrossRef]

18. Miorin, L.; Romero-Brey, I.; Maiuri, P.; Hoppe, S.; Krijnse-Locker, J.; Bartenschlager, R.; Marcello, A. Three-dimensional architecture of tick-borne encephalitis virus replication sites and trafficking of the replicated RNA. *J. Virol.* **2013**, *87*, 6469–6481. [CrossRef] [PubMed]

19. Bily, T.; Palus, M.; Eyer, L.; Elsterova, J.; Vancova, M.; Ruzek, D. Electron Tomography Analysis of Tick-Borne Encephalitis Virus Infection in Human Neurons. *Sci. Rep.* **2015**, *5*, 10745. [CrossRef]

20. Egger, D.; Wolk, B.; Gosert, R.; Bianchi, L.; Blum, H.E.; Moradpour, D.; Bienz, K. Expression of hepatitis C virus proteins induces distinct membrane alterations including a candidate viral replication complex. *J. Virol.* **2002**, *76*, 5974–5984. [CrossRef]

21. Ferraris, P.; Beaumont, E.; Uzbekov, R.; Brand, D.; Gaillard, J.; Blanchard, E.; Roingeard, P. Sequential biogenesis of host cell membrane rearrangements induced by hepatitis C virus infection. *Cell. Mol. Life Sci.* **2013**, *70*, 1297–1306. [CrossRef]

22. Romero-Brey, I.; Merz, A.; Chiramel, A.; Lee, J.Y.; Chlanda, P.; Haselman, U.; Santarella-Mellwig, R.; Habermann, A.; Hoppe, S.; Kallis, S.; et al. Three-dimensional architecture and biogenesis of membrane structures associated with hepatitis C virus replication. *PLoS Pathog.* **2012**, *8*, e1003056. [CrossRef]

23. Knoops, K.; Kikkert, M.; Worm, S.H.; Zevenhoven-Dobbe, J.C.; van der Meer, Y.; Koster, A.J.; Mommaas, A.M.; Snijder, E.J. SARS-coronavirus replication is supported by a reticulovesicular network of modified endoplasmic reticulum. *PLoS Biol.* **2008**, *6*, e226. [CrossRef]

24. Snijder, E.J.; Limpens, R.; de Wilde, A.H.; de Jong, A.W.M.; Zevenhoven-Dobbe, J.C.; Maier, H.J.; Faas, F.; Koster, A.J.; Barcena, M. A unifying structural and functional model of the coronavirus replication organelle: Tracking down RNA synthesis. *PLoS Biol.* **2020**, *18*, e3000715. [CrossRef]

25. Cortese, M.; Lee, J.Y.; Cerikan, B.; Neufeldt, C.J.; Oorschot, V.M.J.; Kohrer, S.; Hennies, J.; Schieber, N.L.; Ronchi, P.; Mizzon, G.; et al. Integrative Imaging Reveals SARS-CoV-2-Induced Reshaping of Subcellular Morphologies. *Cell Host Microbe* **2020**, *28*, 853–866. [CrossRef]

26. Maier, H.J.; Hawes, P.C.; Cottam, E.M.; Mantell, J.; Verkade, P.; Monaghan, P.; Wileman, T.; Britton, P. Infectious bronchitis virus generates spherules from zippered endoplasmic reticulum membranes. *mBio* **2013**, *4*, e00801–e00813. [CrossRef] [PubMed]

27. Belov, G.A.; Nair, V.; Hansen, B.T.; Hoyt, F.H.; Fischer, E.R.; Ehrenfeld, E. Complex dynamic development of poliovirus membranous replication complexes. *J. Virol.* **2012**, *86*, 302–312. [CrossRef]

28. Egger, D.; Teterina, N.; Ehrenfeld, E.; Bienz, K. Formation of the poliovirus replication complex requires coupled viral translation, vesicle production, and viral RNA synthesis. *J. Virol.* **2000**, *74*, 6570–6580. [CrossRef] [PubMed]

29. Kopek, B.G.; Perkins, G.; Miller, D.J.; Ellisman, M.H.; Ahlquist, P. Three-dimensional analysis of a viral RNA replication complex reveals a virus-induced mini-organelle. *PLoS Biol.* **2007**, *5*, e220. [CrossRef] [PubMed]

30. Ertel, K.J.; Benefield, D.; Castano-Diez, D.; Pennington, J.G.; Horswill, M.; den Boon, J.A.; Otegui, M.S.; Ahlquist, P. Cryo-electron tomography reveals novel features of a viral RNA replication compartment. *Elife* **2017**, *6*. [CrossRef]

31. Fontana, J.; Lopez-Iglesias, C.; Tzeng, W.P.; Frey, T.K.; Fernandez, J.J.; Risco, C. Three-dimensional structure of Rubella virus factories. *Virology* **2010**, *405*, 579–591. [CrossRef]

32. Frolova, E.I.; Gorchakov, R.; Pereboeva, L.; Atasheva, S.; Frolov, I. Functional Sindbis virus replicative complexes are formed at the plasma membrane. *J. Virol.* **2010**, *84*, 11679–11695. [CrossRef]

33. Spuul, P.; Balistreri, G.; Kaariainen, L.; Ahola, T. Phosphatidylinositol 3-kinase-, actin-, and microtubule-dependent transport of Semliki Forest Virus replication complexes from the plasma membrane to modified lysosomes. *J. Virol.* **2010**, *84*, 7543–7557. [CrossRef] [PubMed]
34. Patarroyo, C.; Laliberte, J.F.; Zheng, H. Hijack it, change it: How do plant viruses utilize the host secretory pathway for efficient viral replication and spread? *Front Plant Sci* **2012**, *3*, 308. [CrossRef]
35. Bassham, D.C.; Brandizzi, F.; Otegui, M.S.; Sanderfoot, A.A. The secretory system of Arabidopsis. *Arabidopsis Book* **2008**, *6*, e0116. [CrossRef] [PubMed]
36. Restrepo-Hartwig, M.A.; Ahlquist, P. Brome mosaic virus helicase- and polymerase-like proteins colocalize on the endoplasmic reticulum at sites of viral RNA synthesis. *J. Virol.* **1996**, *70*, 8908–8916. [CrossRef] [PubMed]
37. Bamunusinghe, D.; Seo, J.K.; Rao, A.L. Subcellular localization and rearrangement of endoplasmic reticulum by Brome mosaic virus capsid protein. *J. Virol.* **2011**, *85*, 2953–2963. [CrossRef] [PubMed]
38. Turner, K.A.; Sit, T.L.; Callaway, A.S.; Allen, N.S.; Lommel, S.A. Red clover necrotic mosaic virus replication proteins accumulate at the endoplasmic reticulum. *Virology* **2004**, *320*, 276–290. [CrossRef] [PubMed]
39. Reichel, C.; Beachy, R.N. Tobacco mosaic virus infection induces severe morphological changes of the endoplasmic reticulum. *Proc. Natl. Acad. Sci. USA* **1998**, *95*, 11169–11174. [CrossRef] [PubMed]
40. Appiano, A.; Redolfi, P. Ultrastructure and Cytochemistry of Phaseolus Leaf Tissues Infected with an Isolate of Tobacco Necrosis Virus Inducing Localized Wilting. *Protoplasma* **1993**, *174*, 116–127. [CrossRef]
41. Yau, W.L.; Nguyen-Dinh, V.; Larsson, E.; Lindqvist, R.; Overby, A.K.; Lundmark, R. Model System for the Formation of Tick-Borne Encephalitis Virus Replication Compartments without Viral RNA Replication. *J. Virol.* **2019**, *93*. [CrossRef] [PubMed]
42. McCartney, A.W.; Greenwood, J.S.; Fabian, M.R.; White, K.A.; Mullen, R.T. Localization of the tomato bushy stunt virus replication protein p33 reveals a peroxisome-to-endoplasmic reticulum sorting pathway. *Plant Cell* **2005**, *17*, 3513–3531. [CrossRef]
43. Howe, C.J.; Barbrook, A.C.; Nisbet, R.E.; Lockhart, P.J.; Larkum, A.W. The origin of plastids. *Philos. Trans. R. Soc. Lond. B Biol. Sci.* **2008**, *363*, 2675–2685. [CrossRef] [PubMed]
44. Hatta, T.; Francki, R.I.B. Cytopathic Structures Associated with Tonoplasts of Plant-Cells Infected with Cucumber Mosaic and Tomato Aspermy Viruses. *J. Gen. Virol.* **1981**, *53*, 343–346. [CrossRef]
45. Lee, J.Y.; Cortese, M.; Haselmann, U.; Tabata, K.; Romero-Brey, I.; Funaya, C.; Schieber, N.L.; Qiang, Y.; Bartenschlager, M.; Kallis, S.; et al. Spatiotemporal Coupling of the Hepatitis C Virus Replication Cycle by Creating a Lipid Droplet- Proximal Membranous Replication Compartment. *Cell Rep.* **2019**, *27*, 3602–3617. [CrossRef]
46. Romero-Brey, I.; Bartenschlager, R. Endoplasmic Reticulum: The Favorite Intracellular Niche for Viral Replication and Assembly. *Viruses* **2016**, *8*, 160. [CrossRef]
47. Mohd Ropidi, M.I.; Khazali, A.S.; Nor Rashid, N.; Yusof, R. Endoplasmic reticulum: A focal point of Zika virus infection. *J. Biomed. Sci.* **2020**, *27*, 27. [CrossRef]
48. Kaufusi, P.H.; Kelley, J.F.; Yanagihara, R.; Nerurkar, V.R. Induction of endoplasmic reticulum-derived replication-competent membrane structures by West Nile virus non-structural protein 4B. *PLoS ONE* **2014**, *9*, e84040. [CrossRef]
49. Senigl, F.; Grubhoffer, L.; Kopecky, J. Differences in maturation of tick-borne encephalitis virus in mammalian and tick cell line. *Intervirology* **2006**, *49*, 239–248. [CrossRef]
50. Grief, C.; Galler, R.; Cortes, L.M.; Barth, O.M. Intracellular localisation of dengue-2 RNA in mosquito cell culture using electron microscopic in situ hybridisation. *Arch. Virol.* **1997**, *142*, 2347–2357. [CrossRef]
51. Offerdahl, D.K.; Dorward, D.W.; Hansen, B.T.; Bloom, M.E. A three-dimensional comparison of tick-borne flavivirus infection in mammalian and tick cell lines. *PLoS ONE* **2012**, *7*, e47912. [CrossRef]
52. Paul, D.; Hoppe, S.; Saher, G.; Krijnse-Locker, J.; Bartenschlager, R. Morphological and biochemical characterization of the membranous hepatitis C virus replication compartment. *J. Virol.* **2013**, *87*, 10612–10627. [CrossRef]
53. Fontana, J.; Tzeng, W.P.; Calderita, G.; Fraile-Ramos, A.; Frey, T.K.; Risco, C. Novel replication complex architecture in rubella replicon-transfected cells. *Cell. Microbiol.* **2007**, *9*, 875–890. [CrossRef] [PubMed]
54. Froshauer, S.; Kartenbeck, J.; Helenius, A. Alphavirus RNA replicase is located on the cytoplasmic surface of endosomes and lysosomes. *J. Cell Biol.* **1988**, *107*, 2075–2086. [CrossRef]
55. Kujala, P.; Ikaheimonen, A.; Ehsani, N.; Vihinen, H.; Auvinen, P.; Kaariainen, L. Biogenesis of the Semliki Forest virus RNA replication complex. *J. Virol.* **2001**, *75*, 3873–3884. [CrossRef]
56. Haguenau, F.; Hawkes, P.W.; Hutchison, J.L.; Satiat-Jeunemaitre, B.; Simon, G.T.; Williams, D.B. Key events in the history of electron microscopy. *Microsc. Microanal.* **2003**, *9*, 96–138. [CrossRef]
57. Roingeard, P. Viral detection by electron microscopy: Past, present and future. *Biol. Cell.* **2008**, *100*, 491–501. [CrossRef] [PubMed]
58. Richert-Poggeler, K.R.; Franzke, K.; Hipp, K.; Kleespies, R.G. Electron Microscopy Methods for Virus Diagnosis and High Resolution Analysis of Viruses. *Front Microbiol.* **2018**, *9*, 3255. [CrossRef] [PubMed]
59. Peddie, C.J.; Collinson, L.M. Exploring the third dimension: Volume electron microscopy comes of age. *Micron* **2014**, *61*, 9–19. [CrossRef]
60. Nogales, E. The development of cryo-EM into a mainstream structural biology technique. *Nat. Methods* **2016**, *13*, 24–27. [CrossRef] [PubMed]

61. Baena, V.; Conrad, R.; Friday, P.; Fitzgerald, E.; Kim, T.; Bernbaum, J.; Berensmann, H.; Harned, A.; Nagashima, K.; Narayan, K. FIB-SEM as a Volume Electron Microscopy Approach to Study Cellular Architectures in SARS-CoV-2 and Other Viral Infections: A Practical Primer for a Virologist. *Viruses* **2021**, *13*, 611. [CrossRef]

62. Lamers, E.; Walboomers, X.F.; Domanski, M.; McKerr, G.; O'Hagan, B.M.; Barnes, C.A.; Peto, L.; Luttge, R.; Winnubst, L.A.; Gardeniers, H.J.; et al. Cryo DualBeam Focused Ion Beam-Scanning Electron Microscopy to Evaluate the Interface Between Cells and Nanopatterned Scaffolds. *Tissue Eng. Part C Methods* **2011**, *17*, 1–7. [CrossRef] [PubMed]

63. Quemin, E.R.J.; Machala, E.A.; Vollmer, B.; Prazak, V.; Vasishtan, D.; Rosch, R.; Grange, M.; Franken, L.E.; Baker, L.A.; Grunewald, K. Cellular Electron Cryo-Tomography to Study Virus-Host Interactions. *Annu. Rev. Virol.* **2020**, *7*, 239–262. [CrossRef] [PubMed]

64. Cortese, K.; Diaspro, A.; Tacchetti, C. Advanced correlative light/electron microscopy: Current methods and new developments using Tokuyasu cryosections. *J. Histochem. Cytochem.* **2009**, *57*, 1103–1112. [CrossRef] [PubMed]

65. Kumar, M.; Altan-Bonnet, N. Viral pores are everywhere. *Mol. Cell* **2021**, *81*, 2061–2063. [CrossRef]

66. Unchwaniwala, N.; Zhan, H.; Pennington, J.; Horswill, M.; den Boon, J.A.; Ahlquist, P. Subdomain cryo-EM structure of nodaviral replication protein A crown complex provides mechanistic insights into RNA genome replication. *Proc. Natl. Acad. Sci. USA* **2020**, *117*, 18680–18691. [CrossRef]

67. Wolff, G.; Limpens, R.; Zevenhoven-Dobbe, J.C.; Laugks, U.; Zheng, S.; de Jong, A.W.M.; Koning, R.I.; Agard, D.A.; Grunewald, K.; Koster, A.J.; et al. A molecular pore spans the double membrane of the coronavirus replication organelle. *Science* **2020**, *369*, 1395–1398. [CrossRef]

Review

New Perspectives on the Biogenesis of Viral Inclusion Bodies in Negative-Sense RNA Virus Infections

Olga Dolnik , Gesche K. Gerresheim and Nadine Biedenkopf *

Institute for Virology, Philipps-University Marburg, 35043 Marburg, Germany;
dolnik@staff.uni-marburg.de (O.D.); gesche.gerresheim@staff.uni-marburg.de (G.K.G.)
* Correspondence: nadine.biedenkopf@staff.uni-marburg.de; +49-(0)-64212825307

Abstract: Infections by negative strand RNA viruses (NSVs) induce the formation of viral inclusion bodies (IBs) in the host cell that segregate viral as well as cellular proteins to enable efficient viral replication. The induction of those membrane-less viral compartments leads inevitably to structural remodeling of the cellular architecture. Recent studies suggested that viral IBs have properties of biomolecular condensates (or liquid organelles), as have previously been shown for other membrane-less cellular compartments like stress granules or P-bodies. Biomolecular condensates are highly dynamic structures formed by liquid-liquid phase separation (LLPS). Key drivers for LLPS in cells are multivalent protein:protein and protein:RNA interactions leading to specialized areas in the cell that recruit molecules with similar properties, while other non-similar molecules are excluded. These typical features of cellular biomolecular condensates are also a common characteristic in the biogenesis of viral inclusion bodies. Viral IBs are predominantly induced by the expression of the viral nucleoprotein (N, NP) and phosphoprotein (P); both are characterized by a special protein architecture containing multiple disordered regions and RNA-binding domains that contribute to different protein functions. P keeps N soluble after expression to allow a concerted binding of N to the viral RNA. This results in the encapsidation of the viral genome by N, while P acts additionally as a cofactor for the viral polymerase, enabling viral transcription and replication. Here, we will review the formation and function of those viral inclusion bodies upon infection with NSVs with respect to their nature as biomolecular condensates.

Keywords: negative strand RNA viruses (NSV); viral inclusion bodies; biomolecular condensates; liquid-liquid phase separation (LLPS); viral replication; nucleoprotein; phosphoprotein

Citation: Dolnik, O.; Gerresheim, G.K.; Biedenkopf, N. New Perspectives on the Biogenesis of Viral Inclusion Bodies in Negative-Sense RNA Virus Infections. *Cells* **2021**, *10*, 1460. https://doi.org/10.3390/cells10061460

Academic Editors: Thomas Hoenen and Allison Groseth

Received: 19 May 2021
Accepted: 8 June 2021
Published: 10 June 2021

Publisher's Note: MDPI stays neutral with regard to jurisdictional claims in published maps and institutional affiliations.

1. Introduction

As viruses are obligatory intracellular parasites, their replication cycle relies on essential processes in the infected host cell. Viruses thereby exploit and remodel the cellular architecture by inducing structural, functional, or biochemical changes to enable efficient viral replication.

During infection, many viruses induce the formation of distinct and specialized intracellular compartments that facilitate viral replication. Those specialized intracellular compartments are very heterogenous and designated as viral inclusions, inclusion bodies (IBs), viroplasms, virosomes, or viral factories and present a hallmark of viral infection [1,2]. Some of those compartments are connected directly to membranes, such as the endoplasmatic reticulum (ER) in Hepatitis C virus [3,4], dengue virus [5] or severe acute respiratory syndrome coronavirus (SARS CoV) 2 [6,7] infections, lysosomes in Semliki forest virus infection [8] or mitochondria (Flock House virus) [9]. These single- and double-membrane vesicles, convoluted membranes or tubular structures are a typical feature of infection by positive strand RNA viruses [10–15]. In contrast, viral inclusions during infections with many negative-sense RNA viruses are membrane-less but still localize in special cytoplasmic areas (summarized in Figure 1).

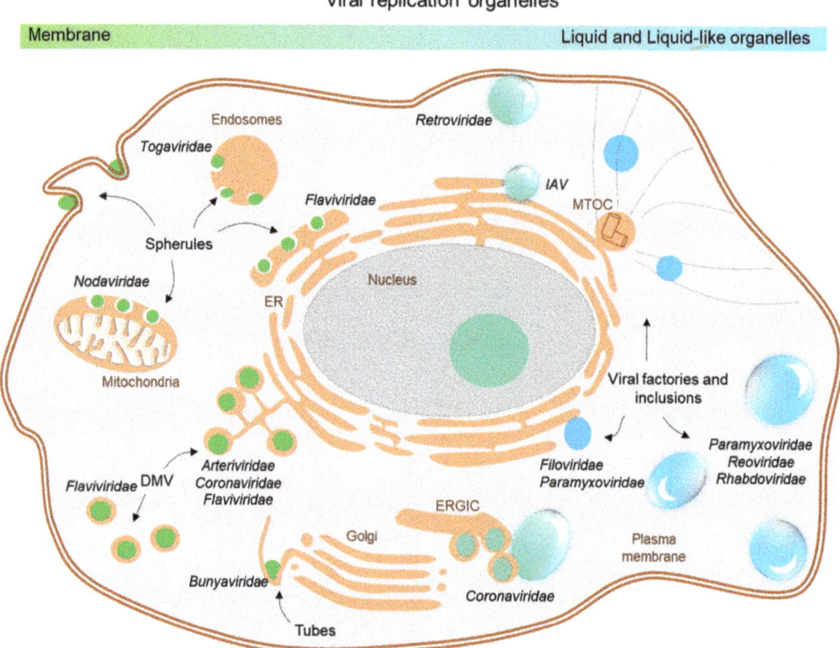

Figure 1. Overview on viral replication organelles in cells upon infection with different RNA viruses. Viral replication compartments associated with membranes are depicted in green, membrane-less compartments are indicated in blue. Those with liquid phase properties are depicted as droplets. DMV, double membrane vesicles. ER, endoplasmatic reticulum. IAV, Influenza A virus. MTOC, microtubule organizing center, ERGIC, endoplasmtic-reticulum–Golgi intermediate compartment.

Recent investigations could demonstrate that many of those IBs share common properties with liquid organelles or biomolecular condensates. Those active, biochemically functional and membrane-less cellular compartments have become an emerging interest during the last decade. Biomolecular condensates display the properties of liquids and are highly dynamic and regulated structures in the cell involved in many different biological processes [16,17]. The underlying biophysical mechanism is, in most cases, regulated by liquid-liquid phase separation (LLPS), a mechanism similar to a water-in-oil-mixture [18,19]. The emerging investigations of these highly dynamic structures lead also to a paradigm change with respect to viral IB formation and function. Typical features of biomolecular condensates like dynamics, fusion activity, and reversibility are also characteristic for viral IB formation [19–22].

Here, we will review the current state of viral IB formation and function in infections with negative-sense RNA viruses, especially with respect to the emerging field of viral inclusions with properties of biomolecular condensates.

1.1. Biomolecular Condensates

In the last years, the physical properties of cellular molecules as a key factor for cellular organization gained more and more interest [16,17,22–24]. The mechanism and localization of biochemical processes have been for long years attributed solely to membrane-surrounded organelles. However, studies over the last decade have proven evidence of a cellular compartmentalization lacking a lipid boundary. Those membrane-less cellular structures are very heterogeneous in size and composition. Although they share similarities with the surrounding cytoplasm, they present separated, sometimes impenetrable cytoplasmic organelles that show a high dynamic plasticity and assemble/disassemble rapidly [16,25–27]. Owing to their biophysical properties that they share with liquids (like droplet fusion, surface tension, etc.), these biochemical functional compartments have been referred to as liquid (droplet) organelles or, more common in cellular biology, as biomolecular condensates [16–19,27–30]. Biomolecular condensates have been observed in different cells among eukaryots, bacteria, yeast, and archae [23,31–34]. They are ubiquitous observed across cellular compartments. In the cytoplasm, they are represented by stress granules, P-bodies, G-bodies [23,35–38] or in the nucleus, by nucleolus or Cajal bodies [39–41], for example. The main biophysical mechanism underlying the formation of biomolecular condensates is the LLPS, a mechanism similar to a water-in-oil mixture that leads to the separation of two (or even more) phases when left unperturbed [42,43]. The transition from soluble molecules to condensates (saturation concentration), liquid crystals, or aggregates is strongly regulated by thermodynamical factors like temperature, concentration, valency, and interaction strength between molecules [28,44–47]. The interphase of the different phases results intracellularly in the membrane-less boundary of biomolecular condensates that allow penetration by molecules with similar properties, while it excludes molecules with dissimilar features [20,22,43]. In cell biology, LLPS originates from protein:protein, protein:RNA, or RNA:RNA interactions that lead to the remodeling of a soluble phase into a condensated, dense phase. A key factor here is the multivalency of the molecules itself: multiple inter- and intramolecular connections that can lead to the formation of condensates with multiple interaction partners [28,29]. Certain properties of protein:protein interfaces have already been shown to drive protein phase separation: arginine-glycine-glycine/arginine-glycine (RGG/RG) motifs [48], charge-charge interactions and intrinsically disordered regions (IDRs) [49–51]. Interestingly, many IDR-containing proteins also have RNA-interaction interfaces. Under high concentration and molecular crowding, structured protein domains have also been described to drive LLPS [29,52]. Accordingly, posttranslational modifications such as phosphorylation or methylation also have a big impact on the formation of biomolecular condensates [53–56]. Expression of RNA and proteins, changes in their ratio, as well as RNA-scaffolded assembly of proteins all contribute to condensation and dynamics of LLPS in cells [35,57–59].

1.2. Viral Replication Cycle of Negative Strand RNA Viruses (NSV)

The members of negative strand RNA viruses (NSV) comprise viruses that have a single-stranded, negative-sense RNA genome. They can be divided into virus families that have segmented genomes such as Orthomyxoviruses, Arenaviruses, and Bunyaviruses, or into non-segmented negative-sense RNA viruses (nsNSV, also termed *Mononegavirales*). The latter comprises several virus families (for example, *Paramyxoviridae, Bornaviridae, Rhabdoviridae, Pneumoviridae, Filoviridae*) with high relevance of individual representatives as human pathogens such as Measles virus (MeV), Nipah virus (NiV), Rabies virus (RABV), Respiratory Syncytial Virus (RSV), or Marburg and Ebola virus (MARV, EBOV) [60]. These viruses share a common architecture of their genomes. The RNA genome length varies between 12 (RSV, VSV) and 19 (filoviruses) kb in length and contains essential untranslated regions (UTRs) at their 3′- (*leader*) and 5′- (*trailer*) terminal ends important for viral transcription, replication, and encapsidation [61–65]. While the number of genes encoded by nsNSV varies among its families (from 5 to 10), the organization and relative position of the structural genes is highly conserved: 3′- *leader*- Nucleoprotein N- Phosphoprotein P- Matrix protein (M)- Glycoprotein (G)- RNA-dependent RNA polymerase (L for large protein)- *trailer*-5′ (Figure 2A,B).

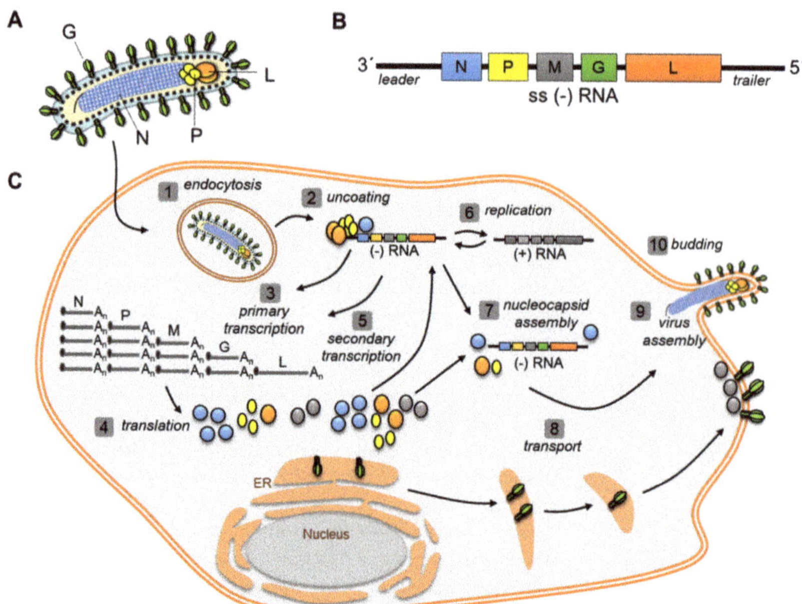

Figure 2. (**A**) Schematic diagram of a NSV particle. (**B**) General genome organization of a NSV. (**C**) Replication cycle of an NSV (based on a filoviral replication cycle). After entry into the cell (**1**) and release of the nucleocapsid into the cytoplasm (**2**), primary viral transcription (**3**) is initiated by the integrated viral polymerase complex. Viral mRNAs are translated by the host translation machinery (**4**). Synthesized viral proteins support new rounds of viral transcription (**5**), replication (**6**) and nucleocapsid assembly (**7**). Nucleocapsids are transported (**8**) to the cell periphery where they assemble to virions (**9**) and bud from the plasma membrane (**10**).

The RNA is tightly encapsidated in a non-covalently manner by the nucleoprotein (N, NP) that forms together with the other viral nucleocapsid or accessory proteins a helical Ribonucleoprotein complex (RNP) [61,66–71]. NSV are enveloped viruses that integrate their surface protein(s) (G, GP, Hemagglutinin H, or fusion protein F) into the host-derived membrane (Figure 2A). A layer of viral matrix protein(s) (M, VP40) represents the matrix that connects the membrane with the nucleocapsid. The replication cycle of nsNSV takes place in the cytoplasm of the host cell, with the exception of Bornaviruses that have a nuclear phase during their replication [72]. Entry of the virus is mediated by the attachment and binding of the surface protein to its receptor and fusion of the viral with the cellular membrane [73–75] (Figure 2C). Subsequently, the viral RNP is released into the cytoplasm of the cell. The RNP serves as template for viral RNA synthesis that starts (owing to the negative-sense genomic RNA) with primary viral transcription accomplished by the incorporated viral polymerase complex [76,77]. The RNA-dependent RNA polymerase (RdRp) L forms together with the phosphoprotein (P, VP35) the viral polymerase complex that enables mRNA synthesis of the viral genes [78,79]. Some representatives of the nsNSV encode additional viral nucleocapsid or accessory proteins that are essential viral transcription factors often regulated by phosphorylation (for example, VP30 for filoviruses or M2-1 for RSV) [77,80,81]. mRNA synthesis starts at the 3′-end of the genomic RNA and results in short, uncapped leader RNAs and 5′-capped, 3′-polyadenylated mRNAs [82,83]. Transcription of the monocistronic mRNAs is assumed to follow a start-stop mechanism regulated by highly conserved gene start and gene end sequences located in UTRs [78,84]. Polyadenylation of the viral mRNAs by the viral polymerase slows down transcription at the gene ends that may result in dissociation of the RdRp from the template. The result is a descending gradient of viral mRNAs from the first (N) to the last (L) gene, suggesting that the RdRp initiates transcription predominantly at the $3'$-end of the viral genome and not from internal genes [76,82,85–89]. Following transcription, the cellular translation machinery translates the mRNAs into new viral proteins. Replication is carried out via the synthesis of a full-length antigenome in positive orientation that serves as a template for replication of the negative-sense genomic RNA. The switch from viral transcription to viral replication, when the RdRp ignores the transcription start and stop signals to synthesize the full-length antigenome, is not completely understood. It is suggested that the amount of newly synthesized N plays an important role to enable encapsidation of the nascent full-length antigenomic RNA during viral replication. N is synthesized as a monomer but starts to oligomerize quite rapidly and forms nucleocapsid-like structures, also with cellular RNA [90–92]. To prevent encapsidation of cellular RNA by N, N is kept soluble by the interaction with P [93]. The N^0P complex, allows a concerted and regulated encapsidation of the viral RNA template [94]. However, different pools of polymerase complexes complemented by cellular and/or viral co-factors are also discussed to define either transcriptase or replicase activity of the RdRp [95]. Simultaneously with viral replication, genomic RNA serves again as template for further rounds of viral transcription accomplished by the newly synthesized polymerase complex components (secondary transcription).

Encapsidated genomic full-length RNA assembles together with the other nucleocapsid proteins to mature nucleocapsids that are condensed and transported along the cytoskeleton to the sites of viral budding at the cell periphery. The surface protein G co-localizes at the budding sites with the matrix protein M that drives the incorporation of the nucleocapsids into virions that are subsequently released from the plasma membrane [96–98].

1.3. Characteristics of Viral Inclusion Bodies (IBs)

New insights into the attributes of biomolecular condensate formation have also led to a reconsideration of viral IB formation in the virology field. High similarities of viral IB formation with biomolecular condensates driven by LLPS are obvious. Many viral IBs upon infection with nsNSVs have a high dynamic plasticity, they assemble/dissemble rapidly during infection, grow in size and appearance, and allow transport of exclusive molecules from in- or outwards. A major driving force in NSV IB formation is the expression of N and P proteins that are suggested as the basic scaffold in IB formation during infection. Two types of N-P interactions involving different interaction domains have been described: A monomeric N^0P complex preventing association of N with cellular RNA, and nucleocapsid-associated P upon N oligomerization following its binding to genomic and antigenomic RNA [99]. All these steps involve multiple protein:protein and protein:RNA interactions that are mediated by highly conserved IDRs in both oligomeric proteins, P and N. All these attributes would contribute to the multivalent interactions underlying LLPS. Furthermore, several studies have demonstrated that de novo RNA synthesis occurs in viral IBs [100–105]. While the cellular protein synthesis itself is often shut down due to the viral infection, viral protein synthesis starts on a large scale. Simultaneously with an excess of viral protein expression, the viral RNA is subsequently replicated, encapsidated by the nucleoprotein and packaged with the nucleocapsid proteins.

All these different steps lead inevitably to strong changes in viral protein:protein interactions or protein:RNA interfaces, that might also contribute to LLPS in the viral IB and its surrounding. Understanding the biophysical mechanisms of viral IB biogenesis and regulation will also contribute to understanding the role and function of IBs for viral multiplication.

2. Viral Inclusions Formed upon Infection with Non-Segmented Negative Strand RNA Viruses (nsNSV)

IB formation and changes in the phase separation due to viral infections might lead to the induction of essential subsequent steps of the viral life cycle like, viral RNA synthesis, encapsidation, assembly of nucleocapsid, and their transport to the cellular periphery. In the last years, there is significant new information about the replication of individual nsNSV in correlation with LLPS, which we will review in greater detail (summarized in Figure 3).

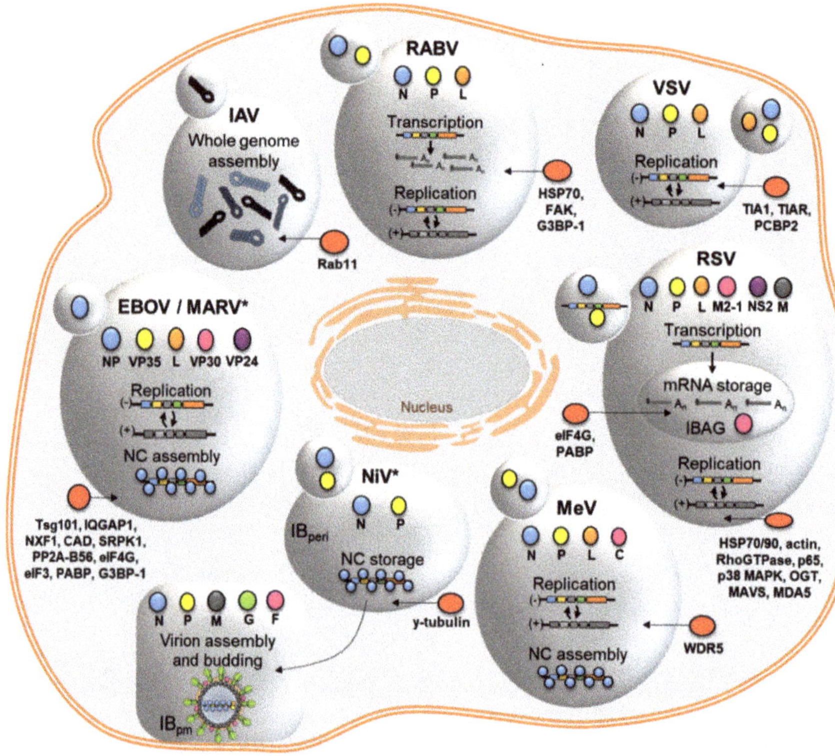

Figure 3. Summary of inclusion body (IB) formation upon infection with NSV. Small IBs indicate minimal required viral proteins for IB formation, while larger IBs represent mature IBs as biomolecular condensates formed by LLPS. * IB formation by LLPS suspected. Different steps of the viral life cycle taking place in IBs as indicated. In red cellular proteins that localize to IBs. RABV, Rabies virus [106]. VSV, vesicular stomatitis virus [103,107]. RSV, Respiratory syncytial virus [101,108]. MeV, Measles virus [109,110]. NiV, Nipah virus [111]. EBOV, Ebola virus [104,105]. MARV, Marburg virus [112,113]. IAV, Influenza A virus [114,115]. IBAG, IB associated granules. IBperi, perinuclear IB. IBpm, IB plasma membrane. NC, nucleocapsid. HSP70/90, heat shock protein 70/90. FAK, focal adhesion kinase. G3BP, Ras GTPase-activating protein-binding protein 1. TIA1, T-cell restricted intracellular antigen 1. TIAR, TIA1-related protein. PCBP2, Poly(RC) Binding Protein 2. p65, NF-κB subunit p65. p65 MAPK, p38 mitogen-activated protein kinase. OGT, O-linked N-acetylglucosamine transferase. MAVS, mitochondrial antiviral-signaling protein. MDA5, melanoma differentiation-associated protein 5. WDR5, WD repeat protein 5. Tsg101, tumor susceptibility gene 101. IQGAP1, Ras GTPase-activating-like protein 1. NXF1, Nuclear RNA export factor 1. CAD, carbamoyl-phosphate synthetase 2, aspar-tate transcarbamylase, and dihydroorotase. SRPK1, Serine-arginine protein kinase. PP2A-B56, protein phosphatase 2 B56 subunit. eIF4G, Eukaryotic translation initiation factor 4 G. eIF3, Eukaryotic initiation factor 3. PABP, Poly(A)-binding protein. Rab11, Ras-related protein Rab-11.

2.1. Rhabdoviridae: IBs of RABV and VSV

A prototype of IB formation upon infection with nsNSV are the Negri bodies that are formed in neurons upon infection with RABV [116]. These cytoplasmic IBs were named after their discoverer Aldechi Negri in 1903 and present a hallmark of rabies diagnosis in the central nervous system (CNS). Negri bodies have been described as places of viral transcription and replication [102]. Components of the viral replication machinery are hence localized in Negri bodies as well as the matrix protein M. Apart from that, cellular proteins like HSP70 and focal adhesion kinase (FAK) are recruited to those IBs [117–119]. Negri bodies were the first viral IBs that have been demonstrated to present organelles with liquid properties [106]. Using fluorescently labelled RABV together with live-cell imaging and FRAP (fluorescence recovery after photobleaching) technologies, the nature of Negri bodies as biomolecular condensates formed by LLPS was demonstrated. Negri bodies are spherical structures with fusion capacity; they show transit with vesicles and can be in reversible form once they encounter a physical barrier [106,116]. The highly dynamic formation of Negri bodies was shown by applying a hypertonic shock to the RABV-infected cells that resulted in the dis- and reappearance of Negri bodies in only 15 min. Interestingly, at later time points of infection, the shape of Negri bodies was changed and they were associated with membranes, most likely derived from the ER [106,116]. The minimal requirement for Negri body formation as a biomolecular condensate was the recombinant expression of N and P alone. However, the typical pinching off events seen from Negri bodies (most likely RNPs) were missing upon N-P expression, suggesting further viral or cellular factors that contribute to the final nature of Negri bodies. The key domains of P that mediate Negri body appearance in complex with N were narrowed down by mutational approaches to the dimerization domain, the amino-terminal part of its second intrinsically disordered domain (IDD2) as well as the C-terminus. IDDs in general have no stable three-dimensional structure, but instead show a high degree in flexibility that can result in binding to other proteins or RNA, as well as in post-translational modifications like phosphorylation [120]. In this regard, IDD1 and IDD2 of P are flanking a dimerization domain (DD) and, like the C-terminus, are phosphorylated. However, phosphorylation of P did not impact Negri body formation [106]. While it was previously shown that stress granules, also liquid organelles, are formed in close proximity to Negri bodies, fusion events or exchange of proteins between both could not be demonstrated, suggesting that both cellular compartments present separate phases within the cytoplasm [106,121].

VSV IBs appear first around 4 h post infection and are also the major site of VSV RNA synthesis. Primary viral transcription, however, is suggested to take place in the cytoplasm prior to IB formation [103]. VSV IBs were recently shown by live-cell imaging to present liquid organelles, whose formation is dependent on LLPS [107]. Disrupting the microtubule cytoskeleton with nocodazol resulted in round inclusions containing eGFP-P labeled VSV. Those IBs were able to fuse by random motion supporting the hypothesis of intrinsic surface tension of VSV IBs, a characteristic feature of LLPS. In contrast to other members of the nsNSV, besides the expression of P and N, IB formation additionally requires the expression of the VSV polymerase L [103,107]. This was tested by recombinant expression of the proteins and complementary by depleting viral protein expression in VSV-infected cells using puromycin, as global protein synthesis inhibitor, or protein-specific PPMOs (peptide-conjugated morpholino oligomers). In addition, the inhibition of M protein expression using a specific PPMO had no effect on the formation or properties of the IBs [103,107]. Using an inactive mutant of L, L G174A, revealed that IB formation is independent of viral RNA synthesis, suggesting that the nature of the protein:protein interaction is the driving force of VSV IB formation via LLPS.

2.2. Pneumoviridae: IBs of RSV

RSV IBs have been described as spherical cytoplasmic structures where viral transcription and replication occurs and to which all viral proteins of polymerase complex, N, P, L, M2-1 are recruited to enable viral RNA synthesis [101,122–124]. Besides the components of the viral polymerase complex, the nonstructural protein NS2 and the matrix protein M are recruited to RSV IBs [125,126]. RSV IBs also recruit cellular proteins involved in translation initiation, like the poly A binding protein PABP, translation initiation factor eIF4G [101], protein phosphatase 1 (for regulating RSV transcription mediated by M2-1 phosphorylation) [81] or heat shock proteins HSP90 and HSP70 [127,128]. Additionally, cellular proteins involved in nucleocapsid assembly and -transport like actin, actin-associated proteins and rhoGTPases like rac1, rhoA and cdc42 colocalize in IBs [129–131].

While genomic RNA could be detected in RSV IBs [124,132], a recent study confirmed additionally viral mRNA synthesis to be present in RSV IBs, independent of their size [101]. Live-cell imaging and pulse chase analyses with a fluorescently labelled recombinant RSV (M2-1 GFP fusion protein) underlined the dynamics of IB formation during the RSV replication cycle. A very intriguing finding of this study was the identification of a subcompartment inside the IBs by super-resolution microscopy, called IBAGs (for IB-associated granules), where newly synthesized viral mRNA accumulated together with the viral transcription activator M2-1, while N, P, L, and genomic RNA were excluded [101]. Formation of IBAGs was strongly dependent on viral RNA synthesis as their number increased during the viral replication cycle from 12 h p.i. on. Interestingly, while nascent viral mRNA and the cellular proteins PABP or eIF4G involved in translation initiation co-localized in IBAGs, other components of the cellular translation machinery, like the ribosomal subunit proteins S6 or L4, did not concentrate on IBs at all. As pulse-chase experiments could demonstrate that newly synthesized viral mRNA only transits through IBAGs, it is suggested that they might present rather transient mRNA storage sites but not sites of viral mRNA translation that most likely occurs in the cytoplasm. IBAGs share similarities with cellular stress granules that are formed by LLPS [133], although IBAGs do not contain typical stress granule proteins like G3BP or TIA-1. The minimal requirement of RSV IB formation is, like for RABV, expression of N and P alone [108,134]. The assembly of IBs was shown to be dependent on the RNA binding- and oligomerization capability of N and P, as N mutation towards a N^0P complex was not sufficient to induce IB assembly in transfected cells [108]. With respect to P, it was demonstrated that the oligomerization domain as well as its C-terminus were essential for IB formation. FRAP experiments on expressed mCherry-tagged N and P proteins could demonstrate *in cellula* as well as in vitro that the formation of RSV IBs occurs by LLPS mediated by N- and P interactions [108].

2.3. Paramyxoviridae: IBs of MeV and NiV

In MeV-infected cells, all the components of the MeV polymerase complex N, P, and L, as well as C colocalize in cytoplasmic IBs where also viral RNA synthesis takes place [135–139]. However, IB formation is initiated by the recombinant expression of N and P alone, even in the absence of viral replication [109]. Extensive studies during the last years have been made to identify N and P domains that contribute to their interaction. Both proteins show a high plasticity with structured and disordered domains. N has been described to consist of a folded domain (N_{CORE}) responsible for RNA binding, with two terminal arms followed by a highly flexible region called N_{TAIL} [140–143]. The tetrameric P contains a long intrinsically disordered (P_{TAIL}) and a shorter disordered domain (P_{LOOP}) [140,144,145]. P_{LOOP} is terminated by a small C-terminal three-helix bundle (XD) that has been shown to interact with RNA-associated N_{TAIL} and, for parainfluenza-virus 5 also with L [144,146]. In contrast, interaction in the N^0P complex is mediated via the C-terminal domain of P [147].

Recent studies could demonstrate that MeV IBs represent biomolecular condensates formed by LLPS [109]. Characteristics of biomolecular condensates and LLPS, like a highly dynamic exchange between materials inside the IB with its surrounding, were observed by live-cell imaging upon MeV infection. IB formation was highly dynamic from small spherical structures to large inclusions. Interestingly, while smaller IBs were ubiquitously distributed in the cytoplasm, larger IBs appeared at the perinuclear region. By inhibition of dynein, a motor protein, the formation of perinuclear larger IBs was reduced suggesting that small cytoplasmic IBs transported dynein-dependent along microtubules to the cell nucleus to fuse towards larger IBs. Two further important assets of biomolecular condensates could be detected for MeV IB: the recruitment of cellular proteins (for example, eGFP- or mCherry-tagged WD-repeat protein (WDR) 5) [138], and recovery from photobleaching [109]. LLPS was initiated by the interaction of the C-terminal disordered region of N and P. N mutants that were unable to bind RNA could still form N- and P-mediated IBs. This suggests that RNP complexes, an often described driving force for LLPS [35,57–59], do not contribute to

MeV IB formation [109], in contrast to RSV [108]. In vitro experiments using co-expressed N^0P complexes and different mutants thereof confirmed that phase separation in vitro is also mediated by P and N interactions [110]. Interestingly interaction of P_{XD} and N_{TAIL} has been previously described to mediate the transport of the polymerase complex to the nucleocapsid prior to RNA synthesis [144,145,148–150]. Preventing the interaction between P and N by an N S491L mutation, a mutation that reduced viral transcription in cells [148,151], resulted in a complete abrogation of LLPS *in vitro*. The same was true upon mutation of P S86A and S151A indicating that P phosphorylation also contributes to phase separation. Interestingly, adding RNA to N-P droplets in vitro leads to the recruitment of RNA to droplets and triggered the encapsidation of RNA by N to nucleocapsid-like structures. The rate of encapsidation in those droplets measured by real-time NMR was enhanced when compared to the dilute phase [110]. These data confirmed the formation of nucleocapsid-like structures in these droplets and suggested a role of LLPS for the maturation of MeV nucleocapsids.

IB formation is also found upon infection by Mumps virus [152], Parainfluenzavirus 3 [153,154] and 5 [155] and NiV [111] and also initiated mainly by N and P expression. However, whether those IBs have properties of biomolecular condensates that contribute to efficient viral replication is so far not clear.

NiV, a highly pathogenic member of the *Paramyxoviridae*, differs from the other nsNSV by the induction of the formation of two distinct types of IB during infection. While one type is localized as spherical structures in the perinuclear region (IB_{peri}), the second type characterized by a square shape is found at the plasma membrane (IB_{pm}) [111]. Both types show not only different localization in the cell but also differ in their kinetics of formation and their content of proteins. While IB_{peri} are rapidly formed by N and P proteins upon transfection or early in infection, the matrix protein M is only found inside IB_{pm}, suggesting that they present places of virion assembly and budding. However, fusion events could not be observed between IB_{pm}, neither in transfected nor in infected cells suggesting a transport of nucleocapsids through the cell from one IB to another. Another very interesting finding was that IB_{peri} did not contain positive-sense RNA (mRNA or antigenomic RNA), suggesting that they represent no places of viral RNA synthesis, which is in contrast to many other nsNSV inclusions bodies. From this study, it is suggested that viral RNA synthesis takes place in a network of membrane-like reticular structures close to the ER [156], which is supported by the detection of nucleocapsids outside the IB_{peri}. Whether LLPS and phase separation plays a role in the formation of IB_{peri} and IB_{pm} during NiV replication is so far not clear.

2.4. Filoviridae: IBs of MARV and EBOV

For Filoviruses, it is not clear whether IBs represent virus induced liquid-like compartments characterized by LLPS, as published so far for other NSV [106,107,110]. However, the current literature provides some evidence that this mechanism of compartmentalization, resulting in high functional dynamic and flexibility during the replication cycle, might be applied by filoviruses as well.

Live-cell imaging and time course studies showed that first small IBs appear in the perinuclear region of filovirus-infected cells [104,105] (and own unpublished data for MARV). The small IBs grow with time, they can fuse with each other, and in addition undergo fission events generating smaller IBs from bigger ones [104] (and own unpublished data for MARV). These observations were made using recombinant viruses expressing fluorophore-tagged nucleocapsid proteins like L-mCherry or VP30-GFP and might suggest LLPS processes during IB formation [104,112]. Earlier studies using single protein expression showed that the nucleoprotein NP alone induces the formation of IBs in transfected cells. All other nucleocapsid-associated proteins VP35 (a P analogue), viral nucleocapsid proteins VP30 and VP24, as well as L are diffusely distributed in the cytosol upon single expression and become IBs localized when co-expressed with NP [157–159]. The nucleocapsid proteins are important for the formation and structure of infectious nucleocapsids and possess in addition a wide range of functions in the filoviral replication cycle. VP35 is the co-factor of the viral polymerase L and inhibits IFN-signaling [160,161]. VP30 is a phosphorylation-dependent viral transcription factor necessary to initiate the formation of viral mRNAs [77]. VP24 is important for the formation and condensation of nucleocapsids and inhibits viral transcription and replication as well as innate immune response by interfering with interferon-mediated signaling [162–165].

It was recently shown that the C-terminal domain of EBOV NP is necessary for IB formation and that co-expression of VP35 can rescue IB formation upon expression of a C-terminal deleted NP [166]. This experiment suggests that IB formation and other functions of NP and VP35 involved in transcription and replication of viral RNA are separated processes, since RNA synthesis could not be rescued in this setting. Functional separation of different protein forms, for example, due to modifications like phosphorylation or different protein:protein complexes can occur by LLPS. Here, MARV and EBOV VP30 phosphorylation represent an example of how this modification changes protein:protein and protein:RNA interactions and influences its functions and localization in IB [77,167–170].

Viral RNA is the second important component detected in filovirus induced IBs [104,105,113]. Since early in infection, when primary transcription of viral mRNA takes place, IBs are not detectable, at later time points when protein translation starts and secondary mRNA transcription is initiated, IB formation colocalizes with *de novo* RNA synthesis and large IBs coincident with RNA replication [104,105]. Interestingly, IBs with different compositions of viral nucleocapsids proteins like L and VP35 were detected, suggesting the existence of different subsets of IBs with different functional properties [104]. The regulation of transcription and replication in filovirus IBs is still not understood and it has to be worked out if and how LLPS might favor one or the other process by formation of subcompartments, as shown for RSV IBAGs [101].

Ultrastructural analysis using electron microscopy identified IBs in filovirus-infected cells that contained nucleocapsids with different electron-densities [91,162,163,171,172]. Ectopic expression of nucleocapsid proteins revealed that thin-walled helices are formed in the presence of NP [171]. Thick-walled helices with high electron density can only be observed in the presence of NP, VP35, and VP24 [171]. The thick-walled helices are mainly located in the periphery of IBs, at the plasma membrane during viral budding and in extracellular virus particles [171]. It is presumed that the thin-walled helices represent RNPs, which serve as templates for the viral polymerase, and the thick-walled electron dense helices represent mature and transport-competent nucleocapsids in infected cells or nucleocapsid-like structures in transfected cells [162,171,173]. Therefore, a proper ratio of NP and VP35 in IBs seems to control the morphogenesis of nucleocapsids in EBOV-infected cells [174].

The formation of transport-competent nucleocapsids that have to be transported from the IBs to the budding sites seems to be highly dependent on VP24 functions in RNP condensation, which in turn blocks EBOV genome replication [162,165,175–177]. Ejection of transport-competent nucleocapsids from MARV and EBOV IBs correlates with high dynamics and the nature of described biomolecular condensates or liquid-like viral factories, which exchange material with the surrounding cytosol, as reviewed by Su and colleagues [112,178,179]. In addition, the transport of EBOV and MARV nucleocapsids from IBs to budding sites depends on actin polymerization, and the dynamic of IB assembly and disassembly is dependent on microtubules, representing a further characteristic described for liquid-like viral factories of other NSVs [104,112,178,180].

Filovirus IBs are not membrane-enclosed, as shown in many ultrastructural images; however, often located in close proximity to different cellular membrane compartments like ER, endosomal vesicles, and mitochondria [181–183]. Which and how host cell factors contribute to filovirus IB formation is not known. Several cellular proteins, like Tsg101, IQ-GAP1, NXF1, CAD and SRPK1, and others have been identified inside IBs, being important for different steps of the filovirus replication cycle [172,177,184–187]. Interestingly, it was published recently that ER contact sites regulate the dynamics of membrane-less organelles like P-bodies [188]. It is therefore also likely that filoviral IBs contact the different cellular compartments to enable material exchange, for example, viral and cellular proteins, and RNA, to favor different viral replication steps (transcription and translation, replication, assembly, condensation and transport of nucleocapsids). It remains to be analyzed if the required ATP provided by mitochondria and the necessary translation of viral and cellular proteins in close proximity to IBs might be covered and orchestrated by the mechanisms of liquid-to-solid transitions [22,189,190].

3. Viral Inclusions Formed upon Infection with Segmented Negative Strand RNA Viruses (sNSV)

IAV, a member of the *Orthomyxoviridae*, belongs to the segmented NSVs containing eight segments of RNPs inside the virion. The fact that most virions contain precisely eight segments of each type indicates that genome packaging in IAV infection is a highly regulated process [171,191,192]. It is suggested that the whole genome assembly of the eight segments takes place before transport to the plasma membrane, where the final assembly of the virion takes place [171,193,194]. In contrast to most other members of the NSVs, IAV replicate their genome in the nucleus. The eight viral RNP segments exit the nucleus and accumulate in IBs in a perinuclear region that enlarge in the course of infection [114,115,195]. Since IAV replication takes place in the nucleus, IBs are no sites of viral RNA synthesis. However, a recent study could demonstrate that IAV IB formation displays characteristics of liquid organelles or biomolecular condensates. Their formation in close proximity of the ER exit sites is spatially regulated, dependent on Rab11-GTPase and shows continuous cycling events of vesicles between the ER and the Golgi apparatus [115,196]. As expression of a single viral RNP could already initiate the formation of viral inclusions, viral IBs obviously occur before the assembly of whole genome RNP complexes. Sharing properties of biomolecular condensates, it is supposed that IAV IBs segregate viral RNPs from the cytosol to increase their concentration at hotspots that, in turn, facilitate the recruitment of other viral RNPs to allow assembly of whole IAV genome complexes [115]. Given the special feature of IAV genome reassortment, it is likely that IAV IB formation plays an important role in the assembly of newly reassorted IAV genomes.

The genus *Bunyavirales* contains viruses with either bi- or tripartite genomes containing the L, (M,) and S segments. In contrast to other bunyaviruses, the nonstructural proteins (NSs) of Severe fever with thrombocytopenia syndrome (SFTS) virus were able to form viral IBs upon transfection and infection, whichwas dependent on NSs self-interaction. It could be further demonstrated that those NSs-induced IBs contain the nucleoprotein and are places of viral RNA synthesis [197]. Interestingly, a colocalization of the SFTS IBs with lipid droplets was observed, and inhibition of lipid metabolism negatively affected SFTS replication.

For Bunyamwera virus, viroplasms have been described as tubular structures associated with the Golgi apparatus and the rough ER that are places of viral RNA synthesis and assembly [198]. For Junín virus (an Arenavirus), the nucleoprotein N was shown to induce the formation of discrete cytosolic IBs that may present viral transcription and replication centers. In contrast to most other nsNSV, those structures were associated with membranes and contained lipid metabolites [199].

However, whether these structures share biochemical properties with biomolecular condensates is so far not clear.

4. Role of NSV IBs in Antiviral Response

Given the spatial segregation of viral IB from the surrounding cytoplasm, it is also conceivable that IB formation may function as an additional viral escape strategy to avoid recognition by intracellular components of the antiviral defense machinery. Activation of pattern recognition receptors (PPRs) like RIG-1 and MDA5 recognizing cytosolic dsRNA leads to the activation of type 1 interferon and inflammatory responses combating viral infection [200–202]. A key determinant of antiviral activity are the viral phosphoproteins that are also key regulators for viral IB formation. The P proteins and their analogues have been described to block, for example, phosphorylation of the interferon regulatory factor 3 (IRF3) or IRF 7 [203–208], bind to dsRNA, and prevent RIG-I signaling or PKR activation [207,209,210].

Preventing activation of cell-intrinsic defense by IB formation could either be enabled by sterical exclusion or by concentrated sequestration of antiviral sensors avoiding activation of downstream pathways [211].

RSV antagonizes the innate immune response by sequestering cellular proteins involved in antiviral response activities into the IBs, such as NF-κB subunit p65, p38 mitogen-activated protein kinase (MAPK), O-linked N-acetylglucosamine transferase (OGT), mitochondrial antiviral-signaling protein (MAVS), and MDA5 [132,212,213]. Sequestration of MAPK p38 and OGT was suggested to suppress MK2 activity and formation of stress granules [213]. The cellular proteins were recruited to the IBs most likely via their interaction with N or P, suggesting an immune evasion strategy independent of the immunomodulatory RSV proteins NS1, NS2, or SH [212]. However, although the NF-κB subunit p65 was recruited to IBs, there was no co-localization with N and P suggesting that p65 localization might be regulated by other multivalent interactions within IBs [212].

Stress granules (SG), also liquid organelles with a role in antiviral activity [133,214] have been found in close proximity to RABV as well as VSV IBs [106,121,215]. While active fusion events between both biomolecular condensates could not be observed for RABV, the SG marker protein G3BP was found in some of the RABV IBs [106]. The function of G3BP localization in RABV IBs is unknown but may point towards the direction that LLPS may exclude antiviral proteins inside viral IBs to block antiviral downstream effectors [106]. For VSV IBs, in contrast, some SG proteins such as T-cell restricted intracellular antigen 1 (TIA1), TIA1-related protein (TIAR) or Poly(RC) Binding Protein 2 (PCBP2) co-localized to IBs [215]. The same is true for the EBOV IFN antagonist VP35 that can disrupt SG formation by sequestration of SG proteins into EBOV IBs (eIF4G, eIF3, PABP, and G3BP-1, but no TIA-1) to block innate immune responses [186,187].

For SFTS virus, a bunyavirus, it was demonstrated that sequestering of antiviral factors like IRF7, RIG-I, or STAT2 into viral IBs via the interaction with NSs leads to the suppression of IFN-alpha and -beta signaling pathways [216–219].

5. Conclusions

Over the last decade, the understanding of the intracellular architecture has changed tremendously by the discovery that intracellular membrane-less compartments represent liquid organelles or biomolecular condensates formed by LLPS. This also led to a paradigm change in the field of virology, especially with respect to the underlying mechanism of viral IB formation and maturation. For many NSVs, the liquid properties of IBs could be already demonstrated, with strong evidence that expression of N and P proteins are mostly the minimal requirement for IB formation (Figure 3). This could be attributed to their special protein architecture that includes multiple disordered regions and RNA-binding domains, hence multivalent interaction interfaces that contribute to LLPS. While RNA synthesis does take place in some of the NSV IBs, the structural role of RNA synthesis for LLPS formation and contribution to IB maturation is not fully understood, as well as the assembly of viral nucleocapsids in or from matured IBs. One may speculate that the molecular crowding of viral (and also cellular) proteins upon viral infection initiates the formation of IBs above a certain threshold, laying the foundation for the induction of further steps of the viral life cycle, possibly also driven by LLPS. In that regard, nucleocapsid assembly may be triggered as a result of the environmental changes induced by N and P expression and RNA synthesis.

Different cellular proteins interacting and co-localizing with viral proteins inside IBs have been identified so far. How they contribute to IB formation and LLPS is until now elusive. It is also feasible that many more cellular proteins might be recruited towards IBs due to their similar physicochemical properties, and maybe not all of them by their direct interaction with a viral protein. These interactions might be transient and require more live cell imaging and time laps studies and the use of super-resolution techniques. Research on the composition of IBs in cells will be an exciting field in the future, although challenging, since the liquid properties will make IBs purification difficult. The role of IBs in innate immunity, and how sequestration of cellular antiviral proteins into viral IBs may contribute actively to counteract antiviral activity will be also of great interest in the next years.

Future research on the biogenesis of viral IB formation and the underlying biophysical mechanism will help to understand how IBs promote viral replication, and may lay the foundation of the development of future antivirals, leading to the disassembly of viral IBs or that that may block viral RNA synthesis in place.

Author Contributions: All listed authors (O.D., G.K.G., N.B.) have written, edited, and reviewed the manuscript. G.K.G. provided the visualization. All authors have read and agreed to the published version of the manuscript.

Funding: This work was supported by the Philipps-University Marburg: O.D. (BSL4 facility core staff) and via a junior research group fund from the FCMH (Research Campus of Central Hessen) to N.B. (junior group leader), G.K.G. (post-doctoral position).

Institutional Review Board Statement: Not applicable.

Informed Consent Statement: Not applicable.

Data Availability Statement: Not applicable.

Acknowledgments: In this section, you can acknowledge any support given which is not covered by the author contribution or funding sections. This may include administrative and technical support, or donations in kind (e.g., materials used for experiments).

Conflicts of Interest: The authors declare no conflict of interest.

Abbreviations

NSV	Negative strand virus
RNA	Ribonucleic acid
nsNSV	Non-segemented NSV
LLPS	Liquid-liquid phase separation
IBs	Inclusion bodies
ER	Endoplasmatic reticulum
IDR	Intrinsically disordered region
MeV	Measles virus
NiV	Nipah virus
RABV	Rabies virus
RSV	Respiratory Syncytial Virus
MARV	Marburg virus
EBOV	Ebola virus
IAV	Influenza A Virus
SFTS	Severe fever with thrombocytopenia syndrome Virus
N/NP	Nucleoprotein
P	Phosphoprotein
NS	Nonstructural protein
RNP	Ribonucleoprotein complex
PPMO	Peptide-conjugated morpholino oligomers
GFP	Green fluorescent protein
SG	Stress granules

References

1. Novoa, R.R.; Calderita, G.; Arranz, R.; Fontana, J.; Granzow, H.; Risco, C. Virus factories: Associations of cell organelles for viral replication and morphogenesis. *Biol. Cell* **2005**, *97*, 147–172. [CrossRef] [PubMed]
2. Netherton, C.L.; Wileman, T. Virus factories, double membrane vesicles and viroplasm generated in animal cells. *Curr. Opin. Virol.* **2011**, *1*, 381–387. [CrossRef] [PubMed]

3.   Romero-Brey, I.; Merz, A.; Chiramel, A.; Lee, J.-Y.; Chlanda, P.; Haselman, U.; Santarella-Mellwig, R.; Habermann, A.; Hoppe, S.; Kallis, S.; et al. Three-dimensional architecture and biogenesis of membrane structures associated with hepatitis C virus replication. *PLoS Pathog.* **2012**, *8*, e1003056. [CrossRef] [PubMed]

4.   Romero-Brey, I.; Berger, C.; Kallis, S.; Kolovou, A.; Paul, D.; Lohmann, V.; Bartenschlager, R. NS5A Domain 1 and Polyprotein Cleavage Kinetics Are Critical for Induction of Double-Membrane Vesicles Associated with Hepatitis C Virus Replication. *mBio* **2015**, *6*, e00759. [CrossRef] [PubMed]

5.   Welsch, S.; Miller, S.; Romero-Brey, I.; Merz, A.; Bleck, C.K.E.; Walther, P.; Fuller, S.D.; Antony, C.; Krijnse-Locker, J.; Bartenschlager, R. Composition and three-dimensional architecture of the dengue virus replication and assembly sites. *Cell Host Microbe* **2009**, *5*, 365–375. [CrossRef]

6.   Klein, S.; Cortese, M.; Winter, S.L.; Wachsmuth-Melm, M.; Neufeldt, C.J.; Cerikan, B.; Stanifer, M.L.; Boulant, S.; Bartenschlager, R.; Chlanda, P. SARS-CoV-2 structure and replication characterized by in situ cryo-electron tomography. *Nat. Commun.* **2020**, *11*, 5885. [CrossRef]

7.   Snijder, E.J.; Limpens, R.W.A.L.; de Wilde, A.H.; de Jong, A.W.M.; Zevenhoven-Dobbe, J.C.; Maier, H.J.; Faas, F.F.G.A.; Koster, A.J.; Bárcena, M. A unifying structural and functional model of the coronavirus replication organelle: Tracking down RNA synthesis. *PLoS Biol.* **2020**, *18*, e3000715. [CrossRef]

8.   Spuul, P.; Balistreri, G.; Kääriäinen, L.; Ahola, T. Phosphatidylinositol 3-kinase-, actin-, and microtubule-dependent transport of Semliki Forest Virus replication complexes from the plasma membrane to modified lysosomes. *J. Virol.* **2010**, *84*, 7543–7557. [CrossRef]

9.   Short, J.R.; Speir, J.A.; Gopal, R.; Pankratz, L.M.; Lanman, J.; Schneemann, A. Role of Mitochondrial Membrane Spherules in Flock House Virus Replication. *J. Virol.* **2016**, *90*, 3676–3683. [CrossRef] [PubMed]

10.   Wolff, G.; Limpens, R.W.A.L.; Zevenhoven-Dobbe, J.C.; Laugks, U.; Zheng, S.; de Jong, A.W.M.; Koning, R.I.; Agard, D.A.; Grünewald, K.; Koster, A.J.; et al. A molecular pore spans the double membrane of the coronavirus replication organelle. *Science* **2020**, *369*, 1395–1398. [CrossRef] [PubMed]

11.   Wolff, G.; Melia, C.E.; Snijder, E.J.; Bárcena, M. Double-Membrane Vesicles as Platforms for Viral Replication. *Trends Microbiol.* **2020**, *28*, 1022–1033. [CrossRef] [PubMed]

12.   Knoops, K.; Kikkert, M.; van den Worm, S.H.E.; Zevenhoven-Dobbe, J.C.; van der Meer, Y.; Koster, A.J.; Mommaas, A.M.; Snijder, E.J. SARS-coronavirus replication is supported by a reticulovesicular network of modified endoplasmic reticulum. *PLoS Biol.* **2008**, *6*, e226. [CrossRef] [PubMed]

13.   Oudshoorn, D.; Rijs, K.; Limpens, R.W.A.L.; Groen, K.; Koster, A.J.; Snijder, E.J.; Kikkert, M.; Bárcena, M. Expression and Cleavage of Middle East Respiratory Syndrome Coronavirus nsp3-4 Polyprotein Induce the Formation of Double-Membrane Vesicles That Mimic Those Associated with Coronaviral RNA Replication. *mBio* **2017**, *8*. [CrossRef]

14.   Wong, L.H.; Edgar, J.R.; Martello, A.; Ferguson, B.J.; Eden, E.R. Exploiting Connections for Viral Replication. *Front. Cell Dev. Biol.* **2021**, *9*, 640456. [CrossRef]

15.   Hernandez-Gonzalez, M.; Larocque, G.; Way, M. Viral use and subversion of membrane organization and trafficking. *J. Cell Sci.* **2021**, *134*. [CrossRef] [PubMed]

16.   Banani, S.F.; Lee, H.O.; Hyman, A.A.; Rosen, M.K. Biomolecular condensates: Organizers of cellular biochemistry. *Nat. Rev. Mol. Cell Biol.* **2017**, *18*, 285–298. [CrossRef]

17.   Uversky, V.N. Intrinsically disordered proteins in overcrowded milieu: Membrane-less organelles, phase separation, and intrinsic disorder. *Curr. Opin. Struct. Biol.* **2017**, *44*, 18–30. [CrossRef] [PubMed]

18.   Hyman, A.A.; Weber, C.A.; Jülicher, F. Liquid-liquid phase separation in biology. *Annu. Rev. Cell Dev. Biol.* **2014**, *30*, 39–58. [CrossRef]

19.   Alberti, S. Phase separation in biology. *Curr. Biol.* **2017**, *27*, R1097–R1102. [CrossRef]

20.   Feng, Z.; Chen, X.; Wu, X.; Zhang, M. Formation of biological condensates via phase separation: Characteristics, analytical methods, and physiological implications. *J. Biol. Chem.* **2019**, *294*, 14823–14835. [CrossRef] [PubMed]

21.   Alberti, S.; Gladfelter, A.; Mittag, T. Considerations and Challenges in Studying Liquid-Liquid Phase Separation and Biomolecular Condensates. *Cell* **2019**, *176*, 419–434. [CrossRef]

22.   Etibor, T.A.; Yamauchi, Y.; Amorim, M.J. Liquid Biomolecular Condensates and Viral Lifecycles: Review and Perspectives. *Viruses* **2021**, *13*. [CrossRef]

23.   Brangwynne, C.P.; Eckmann, C.R.; Courson, D.S.; Rybarska, A.; Hoege, C.; Gharakhani, J.; Jülicher, F.; Hyman, A.A. Germline P granules are liquid droplets that localize by controlled dissolution/condensation. *Science* **2009**, *324*, 1729–1732. [CrossRef] [PubMed]

24.   Boeynaems, S.; Alberti, S.; Fawzi, N.L.; Mittag, T.; Polymenidou, M.; Rousseau, F.; Schymkowitz, J.; Shorter, J.; Wolozin, B.; van den Bosch, L.; et al. Protein Phase Separation: A New Phase in Cell Biology. *Trends Cell Biol.* **2018**, *28*, 420–435. [CrossRef]

25.   Courchaine, E.M.; Lu, A.; Neugebauer, K.M. Droplet organelles? *EMBO J.* **2016**, *35*, 1603–1612. [CrossRef]

26.   Tena-Solsona, M.; Wanzke, C.; Riess, B.; Bausch, A.R.; Boekhoven, J. Self-selection of dissipative assemblies driven by primitive chemical reaction networks. *Nat. Commun.* **2018**, *9*, 2044. [CrossRef]

27.   Gomes, E.; Shorter, J. The molecular language of membraneless organelles. *J. Biol. Chem.* **2019**, *294*, 7115–7127. [CrossRef] [PubMed]

28. Jalihal, A.P.; Pitchiaya, S.; Xiao, L.; Bawa, P.; Jiang, X.; Bedi, K.; Parolia, A.; Cieslik, M.; Ljungman, M.; Chinnaiyan, A.M.; et al. Multivalent Proteins Rapidly and Reversibly Phase-Separate upon Osmotic Cell Volume Change. *Mol. Cell* **2020**, *79*, 978–990.e5. [CrossRef] [PubMed]

29. Jalihal, A.P.; Schmidt, A.; Gao, G.; Little, S.; Pitchiaya, S.; Walter, N.G. Hyperosmotic phase separation: Condensates beyond inclusions, granules and organelles. *J. Biol. Chem.* **2020**. [CrossRef] [PubMed]

30. Mitrea, D.M.; Kriwacki, R.W. Phase separation in biology; functional organization of a higher order. *Cell Commun. Signal.* **2016**, *14*, 1. [CrossRef] [PubMed]

31. Franzmann, T.M.; Jahnel, M.; Pozniakovsky, A.; Mahamid, J.; Holehouse, A.S.; Nüske, E.; Richter, D.; Baumeister, W.; Grill, S.W.; Pappu, R.V.; et al. Phase separation of a yeast prion protein promotes cellular fitness. *Science* **2018**, *359*. [CrossRef]

32. Abbondanzieri, E.A.; Meyer, A.S. More than just a phase: The search for membraneless organelles in the bacterial cytoplasm. *Curr. Genet.* **2019**, *65*, 691–694. [CrossRef]

33. Belott, C.; Janis, B.; Menze, M.A. Liquid-liquid phase separation promotes animal desiccation tolerance. *Proc. Natl. Acad. Sci. USA* **2020**, *117*, 27676–27684. [CrossRef]

34. Emenecker, R.J.; Holehouse, A.S.; Strader, L.C. Biological Phase Separation and Biomolecular Condensates in Plants. *Annu. Rev. Plant Biol.* **2021**. [CrossRef]

35. Sanders, D.W.; Kedersha, N.; Lee, D.S.W.; Strom, A.R.; Drake, V.; Riback, J.A.; Bracha, D.; Eeftens, J.M.; Iwanicki, A.; Wang, A.; et al. Competing Protein-RNA Interaction Networks Control Multiphase Intracellular Organization. *Cell* **2020**, *181*, 306–324.e28. [CrossRef]

36. Anderson, P.; Kedersha, N. RNA granules. *J. Cell Biol.* **2006**, *172*, 803–808. [CrossRef] [PubMed]

37. Lin, Y.; Protter, D.S.W.; Rosen, M.K.; Parker, R. Formation and Maturation of Phase-Separated Liquid Droplets by RNA-Binding Proteins. *Mol. Cell* **2015**, *60*, 208–219. [CrossRef]

38. Jin, M.; Fuller, G.G.; Han, T.; Yao, Y.; Alessi, A.F.; Freeberg, M.A.; Roach, N.P.; Moresco, J.J.; Karnovsky, A.; Baba, M.; et al. Glycolytic Enzymes Coalesce in G Bodies under Hypoxic Stress. *Cell Rep.* **2017**, *20*, 895–908. [CrossRef] [PubMed]

39. Brangwynne, C.P.; Mitchison, T.J.; Hyman, A.A. Active liquid-like behavior of nucleoli determines their size and shape in Xenopus laevis oocytes. *Proc. Natl. Acad. Sci. USA* **2011**, *108*, 4334–4339. [CrossRef] [PubMed]

40. Handwerger, K.E.; Gall, J.G. Subnuclear organelles: New insights into form and function. *Trends Cell Biol.* **2006**, *16*, 19–26. [CrossRef] [PubMed]

41. Feric, M.; Vaidya, N.; Harmon, T.S.; Mitrea, D.M.; Zhu, L.; Richardson, T.M.; Kriwacki, R.W.; Pappu, R.V.; Brangwynne, C.P. Coexisting Liquid Phases Underlie Nucleolar Subcompartments. *Cell* **2016**, *165*, 1686–1697. [CrossRef]

42. Sawyer, I.A.; Sturgill, D.; Dundr, M. Membraneless nuclear organelles and the search for phases within phases. *Wiley Interdiscip. Rev. RNA* **2019**, *10*, e1514. [CrossRef]

43. Hyman, A.A.; Simons, K. Cell biology. Beyond oil and water–phase transitions in cells. *Science* **2012**, *337*, 1047–1049. [CrossRef]

44. Dignon, G.L.; Best, R.B.; Mittal, J. Biomolecular Phase Separation: From Molecular Driving Forces to Macroscopic Properties. *Annu. Rev. Phys. Chem.* **2020**, *71*, 53–75. [CrossRef]

45. Dignon, G.L.; Zheng, W.; Kim, Y.C.; Mittal, J. Temperature-Controlled Liquid-Liquid Phase Separation of Disordered Proteins. *ACS Cent. Sci.* **2019**, *5*, 821–830. [CrossRef] [PubMed]

46. Dignon, G.L.; Zheng, W.; Mittal, J. Simulation methods for liquid-liquid phase separation of disordered proteins. *Curr. Opin. Chem. Eng.* **2019**, *23*, 92–98. [CrossRef]

47. Wang, J.; Choi, J.-M.; Holehouse, A.S.; Lee, H.O.; Zhang, X.; Jahnel, M.; Maharana, S.; Lemaitre, R.; Pozniakovsky, A.; Drechsel, D.; et al. A Molecular Grammar Governing the Driving Forces for Phase Separation of Prion-like RNA Binding Proteins. *Cell* **2018**, *174*, 688–699.e16. [CrossRef]

48. Thandapani, P.; O'Connor, T.R.; Bailey, T.L.; Richard, S. Defining the RGG/RG motif. *Mol. Cell* **2013**, *50*, 613–623. [CrossRef] [PubMed]

49. Lin, Y.-H.; Forman-Kay, J.D.; Chan, H.S. Theories for Sequence-Dependent Phase Behaviors of Biomolecular Condensates. *Biochemistry* **2018**, *57*, 2499–2508. [CrossRef]

50. Posey, A.E.; Holehouse, A.S.; Pappu, R.V. Phase Separation of Intrinsically Disordered Proteins. *Methods Enzymol.* **2018**, *611*, 1–30. [CrossRef] [PubMed]

51. Protter, D.S.W.; Rao, B.S.; van Treeck, B.; Lin, Y.; Mizoue, L.; Rosen, M.K.; Parker, R. Intrinsically Disordered Regions Can Contribute Promiscuous Interactions to RNP Granule Assembly. *Cell Rep.* **2018**, *22*, 1401–1412. [CrossRef]

52. Zhou, H.-X.; Nguemaha, V.; Mazarakos, K.; Qin, S. Why Do Disordered and Structured Proteins Behave Differently in Phase Separation? *Trends Biochem. Sci.* **2018**, *43*, 499–516. [CrossRef]

53. Rai, A.K.; Chen, J.-X.; Selbach, M.; Pelkmans, L. Kinase-controlled phase transition of membraneless organelles in mitosis. *Nature* **2018**, *559*, 211–216. [CrossRef]

54. Bah, A.; Forman-Kay, J.D. Modulation of Intrinsically Disordered Protein Function by Post-translational Modifications. *J. Biol. Chem.* **2016**, *291*, 6696–6705. [CrossRef] [PubMed]

55. Owen, I.; Shewmaker, F. The Role of Post-Translational Modifications in the Phase Transitions of Intrinsically Disordered Proteins. *Int. J. Mol. Sci.* **2019**, *20*, 5501. [CrossRef] [PubMed]

56. Hofweber, M.; Dormann, D. Friend or foe-Post-translational modifications as regulators of phase separation and RNP granule dynamics. *J. Biol. Chem.* **2019**, *294*, 7137–7150. [CrossRef] [PubMed]

57. Banerjee, P.R.; Milin, A.N.; Moosa, M.M.; Onuchic, P.L.; Deniz, A.A. Reentrant Phase Transition Drives Dynamic Substructure Formation in Ribonucleoprotein Droplets. *Angew. Chem. Int. Ed Engl.* **2017**, *56*, 11354–11359. [CrossRef] [PubMed]

58. Maharana, S.; Wang, J.; Papadopoulos, D.K.; Richter, D.; Pozniakovsky, A.; Poser, I.; Bickle, M.; Rizk, S.; Guillén-Boixet, J.; Franzmann, T.M.; et al. RNA buffers the phase separation behavior of prion-like RNA binding proteins. *Science* **2018**, *360*, 918–921. [CrossRef]

59. Berry, J.; Weber, S.C.; Vaidya, N.; Haataja, M.; Brangwynne, C.P. RNA transcription modulates phase transition-driven nuclear body assembly. *Proc. Natl. Acad. Sci. USA* **2015**, *112*, E5237–E5245. [CrossRef]

60. Amarasinghe, G.K.; Ayllón, M.A.; Bào, Y.; Basler, C.F.; Bavari, S.; Blasdell, K.R.; Briese, T.; Brown, P.A.; Bukreyev, A.; Balkema-Buschmann, A.; et al. Taxonomy of the order Mononegavirales: Update 2019. *Arch. Virol.* **2019**, *164*, 1967–1980. [CrossRef]

61. Lamb, R.A. Mononegavirales. *Fields Virol.* **2013**, *1*, 880–884.

62. Le Mercier, P.; Kolakofsky, D. Bipartite promoters and RNA editing of paramyxoviruses and filoviruses. *RNA* **2019**, *25*, 279–285. [CrossRef] [PubMed]

63. Whelan, S.P.; Wertz, G.W. Regulation of RNA synthesis by the genomic termini of vesicular stomatitis virus: Identification of distinct sequences essential for transcription but not replication. *J. Virol.* **1999**, *73*, 297–306. [CrossRef] [PubMed]

64. Pong, L.Y.; Rabu, A.; Ibrahim, N. The critical region for viral RNA encapsidation in leader promoter of Nipah virus. *Mol. Genet. Genom.* **2020**, *295*, 1501–1516. [CrossRef]

65. Bach, S.; Demper, J.-C.; Biedenkopf, N.; Becker, S.; Hartmann, R.K. RNA secondary structure at the transcription start site influences EBOV transcription initiation and replication in a length- and stability-dependent manner. *RNA Biol.* **2021**, *18*, 523–536. [CrossRef]

66. Luo, M.; Terrell, J.R.; Mcmanus, S.A. Nucleocapsid Structure of Negative Strand RNA Virus. *Viruses* **2020**, *12*, 835. [CrossRef]

67. Wan, W.; Kolesnikova, L.; Clarke, M.; Koehler, A.; Noda, T.; Becker, S.; Briggs, J.A.G. Structure and assembly of the Ebola virus nucleocapsid. *Nature* **2017**, *551*, 394–397. [CrossRef]

68. Green, T.J.; Zhang, X.; Wertz, G.W.; Luo, M. Structure of the vesicular stomatitis virus nucleoprotein-RNA complex. *Science* **2006**, *313*, 357–360. [CrossRef]

69. Albertini, A.A.V.; Wernimont, A.K.; Muziol, T.; Ravelli, R.B.G.; Clapier, C.R.; Schoehn, G.; Weissenhorn, W.; Ruigrok, R.W.H. Crystal structure of the rabies virus nucleoprotein-RNA complex. *Science* **2006**, *313*, 360–363. [CrossRef]

70. Sugita, Y.; Matsunami, H.; Kawaoka, Y.; Noda, T.; Wolf, M. Cryo-EM structure of the Ebola virus nucleoprotein-RNA complex at 3.6 Å resolution. *Nature* **2018**, *563*, 137–140. [CrossRef]

71. Tawar, R.G.; Duquerroy, S.; Vonrhein, C.; Varela, P.F.; Damier-Piolle, L.; Castagné, N.; MacLellan, K.; Bedouelle, H.; Bricogne, G.; Bhella, D.; et al. Crystal structure of a nucleocapsid-like nucleoprotein-RNA complex of respiratory syncytial virus. *Science* **2009**, *326*, 1279–1283. [CrossRef]

72. Tomonaga, K.; Kobayashi, T.; Ikuta, K. Molecular and cellular biology of Borna disease virus infection. *Microbes Infect.* **2002**, *4*, 491–500. [CrossRef]

73. Guo, Y.; Duan, M.; Wang, X.; Gao, J.; Guan, Z.; Zhang, M. Early events in rabies virus infection-Attachment, entry, and intracellular trafficking. *Virus Res.* **2019**, *263*, 217–225. [CrossRef]

74. Fukuhara, H.; Mwaba, M.H.; Maenaka, K. Structural characteristics of measles virus entry. *Curr. Opin. Virol.* **2020**, *41*, 52–58. [CrossRef]

75. Aggarwal, M.; Plemper, R.K. Structural Insight into Paramyxovirus and Pneumovirus Entry Inhibition. *Viruses* **2020**, *12*, 342. [CrossRef] [PubMed]

76. Noton, S.L.; Fearns, R. Initiation and regulation of paramyxovirus transcription and replication. *Virology* **2015**, *479–480*, 545–554. [CrossRef] [PubMed]

77. Biedenkopf, N.; Lier, C.; Becker, S. Dynamic Phosphorylation of VP30 Is Essential for Ebola Virus Life Cycle. *J. Virol.* **2016**, *90*, 4914–4925. [CrossRef] [PubMed]

78. Cao, D.; Gao, Y.; Liang, B. Structural Insights into the Respiratory Syncytial Virus RNA Synthesis Complexes. *Viruses* **2021**, *13*, 834. [CrossRef] [PubMed]

79. Fearns, R.; Plemper, R.K. Polymerases of paramyxoviruses and pneumoviruses. *Virus Res.* **2017**, *234*, 87–102. [CrossRef] [PubMed]

80. Weik, M.; Modrof, J.; Klenk, H.-D.; Becker, S.; Mühlberger, E. Ebola virus VP30-mediated transcription is regulated by RNA secondary structure formation. *J. Virol.* **2002**, *76*, 8532–8539. [CrossRef]

81. Richard, C.-A.; Rincheval, V.; Lassoued, S.; Fix, J.; Cardone, C.; Esneau, C.; Nekhai, S.; Galloux, M.; Rameix-Welti, M.-A.; Sizun, C.; et al. RSV hijacks cellular protein phosphatase 1 to regulate M2-1 phosphorylation and viral transcription. *PLoS Pathog.* **2018**, *14*, e1006920. [CrossRef]

82. Ogino, T.; Green, T.J. RNA Synthesis and Capping by Non-segmented Negative Strand RNA Viral Polymerases: Lessons From a Prototypic Virus. *Front. Microbiol.* **2019**, *10*, 1490. [CrossRef]

83. Ogino, T.; Green, T.J. Transcriptional Control and mRNA Capping by the GDP Polyribonucleotidyltransferase Domain of the Rabies Virus Large Protein. *Viruses* **2019**, *11*, 504. [CrossRef]

84. Whelan, S.P.J.; Barr, J.N.; Wertz, G.W. Transcription and replication of nonsegmented negative-strand RNA viruses. *Curr. Top. Microbiol. Immunol.* **2004**, *283*, 61–119. [CrossRef] [PubMed]

85. Cordey, S.; Roux, L. Transcribing paramyxovirus RNA polymerase engages the template at its 3′ extremity. *J. Gen. Virol.* **2006**, *87*, 665–672. [CrossRef]

86. Mühlberger, E. Filovirus replication and transcription. *Future Virol.* **2007**, *2*, 205–215. [CrossRef] [PubMed]
87. Albariño, C.G.; Wiggleton Guerrero, L.; Chakrabarti, A.K.; Nichol, S.T. Transcriptional analysis of viral mRNAs reveals common transcription patterns in cells infected by five different filoviruses. *PLoS ONE* **2018**, *13*, e0201827. [CrossRef]
88. Shabman, R.S.; Jabado, O.J.; Mire, C.E.; Stockwell, T.B.; Edwards, M.; Mahajan, M.; Geisbert, T.W.; Basler, C.F. Deep sequencing identifies noncanonical editing of Ebola and Marburg virus RNAs in infected cells. *mBio* **2014**, *5*, e02011. [CrossRef]
89. Cowton, V.M.; McGivern, D.R.; Fearns, R. Unravelling the complexities of respiratory syncytial virus RNA synthesis. *J. Gen. Virol.* **2006**, *87*, 1805–1821. [CrossRef]
90. Noda, T.; Hagiwara, K.; Sagara, H.; Kawaoka, Y. Characterization of the Ebola virus nucleoprotein-RNA complex. *J. Gen. Virol.* **2010**, *91*, 1478–1483. [CrossRef] [PubMed]
91. Kolesnikova, L.; Mühlberger, E.; Ryabchikova, E.; Becker, S. Ultrastructural organization of recombinant Marburg virus nucleoprotein: Comparison with Marburg virus inclusions. *J. Virol.* **2000**, *74*, 3899–3904. [CrossRef]
92. Watanabe, S.; Noda, T.; Kawaoka, Y. Functional mapping of the nucleoprotein of Ebola virus. *J. Virol.* **2006**, *80*, 3743–3751. [CrossRef] [PubMed]
93. Liu, B.; Dong, S.; Li, G.; Wang, W.; Liu, X.; Wang, Y.; Yang, C.; Rao, Z.; Guo, Y. Structural Insight into Nucleoprotein Conformation Change Chaperoned by VP35 Peptide in Marburg Virus. *J. Virol.* **2017**, *91*. [CrossRef] [PubMed]
94. Renner, M.; Bertinelli, M.; Leyrat, C.; Paesen, G.C.; de Oliveira, L.F.S.; Huiskonen, J.T.; Grimes, J.M. Nucleocapsid assembly in pneumoviruses is regulated by conformational switching of the N protein. *Elife* **2016**, *5*, e12627. [CrossRef]
95. Qanungo, K.R.; Shaji, D.; Mathur, M.; Banerjee, A.K. Two RNA polymerase complexes from vesicular stomatitis virus-infected cells that carry out transcription and replication of genome RNA. *Proc. Natl. Acad. Sci. USA* **2004**, *101*, 5952–5957. [CrossRef] [PubMed]
96. Gordon, T.B.; Hayward, J.A.; Marsh, G.A.; Baker, M.L.; Tachedjian, G. Host and Viral Proteins Modulating Ebola and Marburg Virus Egress. *Viruses* **2019**, *11*, 25. [CrossRef]
97. El Najjar, F.; Schmitt, A.P.; Dutch, R.E. Paramyxovirus glycoprotein incorporation, assembly and budding: A three way dance for infectious particle production. *Viruses* **2014**, *6*, 3019–3054. [CrossRef] [PubMed]
98. Cox, R.M.; Plemper, R.K. Structure and organization of paramyxovirus particles. *Curr. Opin. Virol.* **2017**, *24*, 105–114. [CrossRef] [PubMed]
99. Ruigrok, R.W.H.; Crépin, T.; Kolakofsky, D. Nucleoproteins and nucleocapsids of negative-strand RNA viruses. *Curr. Opin. Microbiol.* **2011**, *14*, 504–510. [CrossRef]
100. Cifuentes-Muñoz, N.; Branttie, J.; Slaughter, K.B.; Dutch, R.E. Human Metapneumovirus Induces Formation of Inclusion Bodies for Efficient Genome Replication and Transcription. *J. Virol.* **2017**, *91*. [CrossRef] [PubMed]
101. Rincheval, V.; Lelek, M.; Gault, E.; Bouillier, C.; Sitterlin, D.; Blouquit-Laye, S.; Galloux, M.; Zimmer, C.; Eleouet, J.-F.; Rameix-Welti, M.-A. Functional organization of cytoplasmic inclusion bodies in cells infected by respiratory syncytial virus. *Nat. Commun.* **2017**, *8*, 563. [CrossRef]
102. Lahaye, X.; Vidy, A.; Pomier, C.; Obiang, L.; Harper, F.; Gaudin, Y.; Blondel, D. Functional characterization of Negri bodies (NBs) in rabies virus-infected cells: Evidence that NBs are sites of viral transcription and replication. *J. Virol.* **2009**, *83*, 7948–7958. [CrossRef] [PubMed]
103. Heinrich, B.S.; Cureton, D.K.; Rahmeh, A.A.; Whelan, S.P.J. Protein expression redirects vesicular stomatitis virus RNA synthesis to cytoplasmic inclusions. *PLoS Pathog.* **2010**, *6*, e1000958. [CrossRef] [PubMed]
104. Hoenen, T.; Shabman, R.S.; Groseth, A.; Herwig, A.; Weber, M.; Schudt, G.; Dolnik, O.; Basler, C.F.; Becker, S.; Feldmann, H. Inclusion bodies are a site of ebolavirus replication. *J. Virol.* **2012**, *86*, 11779–11788. [CrossRef]
105. Nanbo, A.; Watanabe, S.; Halfmann, P.; Kawaoka, Y. The spatio-temporal distribution dynamics of Ebola virus proteins and RNA in infected cells. *Sci. Rep.* **2013**, *3*, 1206. [CrossRef]
106. Nikolic, J.; Le Bars, R.; Lama, Z.; Scrima, N.; Lagaudrière-Gesbert, C.; Gaudin, Y.; Blondel, D. Negri bodies are viral factories with properties of liquid organelles. *Nat. Commun.* **2017**, *8*, 58. [CrossRef] [PubMed]
107. Heinrich, B.S.; Maliga, Z.; Stein, D.A.; Hyman, A.A.; Whelan, S.P.J. Phase Transitions Drive the Formation of Vesicular Stomatitis Virus Replication Compartments. *mBio* **2018**, *9*. [CrossRef] [PubMed]
108. Galloux, M.; Risso-Ballester, J.; Richard, C.-A.; Fix, J.; Rameix-Welti, M.-A.; Eléouët, J.-F. Minimal Elements Required for the Formation of Respiratory Syncytial Virus Cytoplasmic Inclusion Bodies In Vivo and In Vitro. *mBio* **2020**, *11*. [CrossRef] [PubMed]
109. Zhou, Y.; Su, J.M.; Samuel, C.E.; Ma, D. Measles Virus Forms Inclusion Bodies with Properties of Liquid Organelles. *J. Virol.* **2019**, *93*. [CrossRef]
110. Guseva, S.; Milles, S.; Jensen, M.R.; Salvi, N.; Kleman, J.-P.; Maurin, D.; Ruigrok, R.W.H.; Blackledge, M. Measles virus nucleo- and phosphoproteins form liquid-like phase-separated compartments that promote nucleocapsid assembly. *Sci. Adv.* **2020**, *6*, eaaz7095. [CrossRef]
111. Ringel, M.; Heiner, A.; Behner, L.; Halwe, S.; Sauerhering, L.; Becker, N.; Dietzel, E.; Sawatsky, B.; Kolesnikova, L.; Maisner, A. Nipah virus induces two inclusion body populations: Identification of novel inclusions at the plasma membrane. *PLoS Pathog.* **2019**, *15*, e1007733. [CrossRef] [PubMed]
112. Schudt, G.; Kolesnikova, L.; Dolnik, O.; Sodeik, B.; Becker, S. Live-cell imaging of Marburg virus-infected cells uncovers actin-dependent transport of nucleocapsids over long distances. *Proc. Natl. Acad. Sci. USA* **2013**, *110*, 14402–14407. [CrossRef] [PubMed]

113. Dolnik, O.; Stevermann, L.; Kolesnikova, L.; Becker, S. Marburg virus inclusions: A virus-induced microcompartment and interface to multivesicular bodies and the late endosomal compartment. *Eur. J. Cell Biol.* **2015**, *94*, 323–331. [CrossRef] [PubMed]

114. Chou, Y.; Heaton, N.S.; Gao, Q.; Palese, P.; Singer, R.H.; Singer, R.; Lionnet, T. Colocalization of different influenza viral RNA segments in the cytoplasm before viral budding as shown by single-molecule sensitivity FISH analysis. *PLoS Pathog.* **2013**, *9*, e1003358. [CrossRef]

115. Alenquer, M.; Vale-Costa, S.; Etibor, T.A.; Ferreira, F.; Sousa, A.L.; Amorim, M.J. Influenza A virus ribonucleoproteins form liquid organelles at endoplasmic reticulum exit sites. *Nat. Commun.* **2019**, *10*, 1629. [CrossRef]

116. Nikolic, J.; Lagaudrière-Gesbert, C.; Scrima, N.; Blondel, D.; Gaudin, Y. Structure and Function of Negri Bodies. *Adv. Exp. Med. Biol.* **2019**, *1215*, 111–127. [CrossRef] [PubMed]

117. Pollin, R.; Granzow, H.; Köllner, B.; Conzelmann, K.-K.; Finke, S. Membrane and inclusion body targeting of lyssavirus matrix proteins. *Cell. Microbiol.* **2013**, *15*, 200–212. [CrossRef]

118. Lahaye, X.; Vidy, A.; Fouquet, B.; Blondel, D. Hsp70 protein positively regulates rabies virus infection. *J. Virol.* **2012**, *86*, 4743–4751. [CrossRef]

119. Fouquet, B.; Nikolic, J.; Larrous, F.; Bourhy, H.; Wirblich, C.; Lagaudrière-Gesbert, C.; Blondel, D. Focal adhesion kinase is involved in rabies virus infection through its interaction with viral phosphoprotein P. *J. Virol.* **2015**, *89*, 1640–1651. [CrossRef]

120. Gao, C.; Ma, C.; Wang, H.; Zhong, H.; Zang, J.; Zhong, R.; He, F.; Yang, D. Intrinsic disorder in protein domains contributes to both organism complexity and clade-specific functions. *Sci. Rep.* **2021**, *11*, 2985. [CrossRef]

121. Nikolic, J.; Civas, A.; Lama, Z.; Lagaudrière-Gesbert, C.; Blondel, D. Rabies Virus Infection Induces the Formation of Stress Granules Closely Connected to the Viral Factories. *PLoS Pathog.* **2016**, *12*, e1005942. [CrossRef] [PubMed]

122. Carromeu, C.; Simabuco, F.M.; Tamura, R.E.; Farinha Arcieri, L.E.; Ventura, A.M. Intracellular localization of human respiratory syncytial virus L protein. *Arch. Virol.* **2007**, *152*, 2259–2263. [CrossRef] [PubMed]

123. García-Barreno, B.; Delgado, T.; Melero, J.A. Identification of protein regions involved in the interaction of human respiratory syncytial virus phosphoprotein and nucleoprotein: Significance for nucleocapsid assembly and formation of cytoplasmic inclusions. *J. Virol.* **1996**, *70*, 801–808. [CrossRef] [PubMed]

124. Santangelo, P.J.; Bao, G. Dynamics of filamentous viral RNPs prior to egress. *Nucleic Acids Res.* **2007**, *35*, 3602–3611. [CrossRef]

125. Weber, E.; Humbert, B.; Streckert, H.J.; Werchau, H. Nonstructural protein 2 (NS2) of respiratory syncytial virus (RSV) detected by an antipeptide serum. *Respiration* **1995**, *62*, 27–33. [CrossRef]

126. Ghildyal, R.; Mills, J.; Murray, M.; Vardaxis, N.; Meanger, J. Respiratory syncytial virus matrix protein associates with nucleocapsids in infected cells. *J. Gen. Virol.* **2002**, *83*, 753–757. [CrossRef]

127. Munday, D.C.; Wu, W.; Smith, N.; Fix, J.; Noton, S.L.; Galloux, M.; Touzelet, O.; Armstrong, S.D.; Dawson, J.M.; Aljabr, W.; et al. Interactome analysis of the human respiratory syncytial virus RNA polymerase complex identifies protein chaperones as important cofactors that promote L-protein stability and RNA synthesis. *J. Virol.* **2015**, *89*, 917–930. [CrossRef] [PubMed]

128. Radhakrishnan, A.; Yeo, D.; Brown, G.; Myaing, M.Z.; Iyer, L.R.; Fleck, R.; Tan, B.-H.; Aitken, J.; Sanmun, D.; Tang, K.; et al. Protein analysis of purified respiratory syncytial virus particles reveals an important role for heat shock protein 90 in virus particle assembly. *Mol. Cell. Proteom.* **2010**, *9*, 1829–1848. [CrossRef]

129. Ravi, L.I.; Tan, T.J.; Tan, B.H.; Sugrue, R.J. Virus-induced activation of the rac1 protein at the site of respiratory syncytial virus assembly is a requirement for virus particle assembly on infected cells. *Virology* **2021**, *557*, 86–99. [CrossRef]

130. Jeffree, C.E.; Brown, G.; Aitken, J.; Su-Yin, D.Y.; Tan, B.-H.; Sugrue, R.J. Ultrastructural analysis of the interaction between F-actin and respiratory syncytial virus during virus assembly. *Virology* **2007**, *369*, 309–323. [CrossRef]

131. Kipper, S.; Hamad, S.; Caly, L.; Avrahami, D.; Bacharach, E.; Jans, D.A.; Gerber, D.; Bajorek, M. New host factors important for respiratory syncytial virus (RSV) replication revealed by a novel microfluidics screen for interactors of matrix (M) protein. *Mol. Cell. Proteom.* **2015**, *14*, 532–543. [CrossRef] [PubMed]

132. Lifland, A.W.; Jung, J.; Alonas, E.; Zurla, C.; Crowe, J.E.; Santangelo, P.J. Human respiratory syncytial virus nucleoprotein and inclusion bodies antagonize the innate immune response mediated by MDA5 and MAVS. *J. Virol.* **2012**, *86*, 8245–8258. [CrossRef]

133. Freibaum, B.D.; Messing, J.; Yang, P.; Kim, H.J.; Taylor, J.P. High-fidelity reconstitution of stress granules and nucleoli in mammalian cellular lysate. *J. Cell Biol.* **2021**, *220*. [CrossRef] [PubMed]

134. García, J.; García-Barreno, B.; Vivo, A.; Melero, J.A. Cytoplasmic inclusions of respiratory syncytial virus-infected cells: Formation of inclusion bodies in transfected cells that coexpress the nucleoprotein, the phosphoprotein, and the 22K protein. *Virology* **1993**, *195*, 243–247. [CrossRef]

135. Spehner, D.; Drillien, R.; Howley, P.M. The assembly of the measles virus nucleoprotein into nucleocapsid-like particles is modulated by the phosphoprotein. *Virology* **1997**, *232*, 260–268. [CrossRef] [PubMed]

136. Tawara, J.T.; Goodman, J.R.; Imagawa, D.T.; Adams, J.M. Fine structure of cellular inclusions in experimental measles. *Virology* **1961**, *14*, 410–416. [CrossRef]

137. Pfaller, C.K.; Radeke, M.J.; Cattaneo, R.; Samuel, C.E. Measles virus C protein impairs production of defective copyback double-stranded viral RNA and activation of protein kinase R. *J. Virol.* **2014**, *88*, 456–468. [CrossRef] [PubMed]

138. Ma, D.; George, C.X.; Nomburg, J.L.; Pfaller, C.K.; Cattaneo, R.; Samuel, C.E. Upon Infection, Cellular WD Repeat-Containing Protein 5 (WDR5) Localizes to Cytoplasmic Inclusion Bodies and Enhances Measles Virus Replication. *J. Virol.* **2018**, *92*. [CrossRef]

139. Pfaller, C.K.; Bloyet, L.-M.; Donohue, R.C.; Huff, A.L.; Bartemes, W.P.; Yousaf, I.; Urzua, E.; Clavière, M.; Zachary, M.; de Masson d'Autume, V.; et al. The C Protein Is Recruited to Measles Virus Ribonucleocapsids by the Phosphoprotein. *J. Virol.* **2020**, *94*. [CrossRef]

140. Longhi, S.; Receveur-Bréchot, V.; Karlin, D.; Johansson, K.; Darbon, H.; Bhella, D.; Yeo, R.; Finet, S.; Canard, B. The C-terminal domain of the measles virus nucleoprotein is intrinsically disordered and folds upon binding to the C-terminal moiety of the phosphoprotein. *J. Biol. Chem.* **2003**, *278*, 18638–18648. [CrossRef]

141. Schoehn, G.; Mavrakis, M.; Albertini, A.; Wade, R.; Hoenger, A.; Ruigrok, R.W.H. The 12 A structure of trypsin-treated measles virus N-RNA. *J. Mol. Biol.* **2004**, *339*, 301–312. [CrossRef]

142. Jensen, M.R.; Communie, G.; Ribeiro, E.A.; Martinez, N.; Desfosses, A.; Salmon, L.; Mollica, L.; Gabel, F.; Jamin, M.; Longhi, S.; et al. Intrinsic disorder in measles virus nucleocapsids. *Proc. Natl. Acad. Sci. USA* **2011**, *108*, 9839–9844. [CrossRef] [PubMed]

143. Desfosses, A.; Goret, G.; Farias Estrozi, L.; Ruigrok, R.W.H.; Gutsche, I. Nucleoprotein-RNA orientation in the measles virus nucleocapsid by three-dimensional electron microscopy. *J. Virol.* **2011**, *85*, 1391–1395. [CrossRef] [PubMed]

144. Gely, S.; Lowry, D.F.; Bernard, C.; Jensen, M.R.; Blackledge, M.; Costanzo, S.; Bourhis, J.-M.; Darbon, H.; Daughdrill, G.; Longhi, S. Solution structure of the C-terminal X domain of the measles virus phosphoprotein and interaction with the intrinsically disordered C-terminal domain of the nucleoprotein. *J. Mol. Recognit.* **2010**, *23*, 435–447. [CrossRef] [PubMed]

145. Milles, S.; Jensen, M.R.; Lazert, C.; Guseva, S.; Ivashchenko, S.; Communie, G.; Maurin, D.; Gerlier, D.; Ruigrok, R.W.H.; Blackledge, M. An ultraweak interaction in the intrinsically disordered replication machinery is essential for measles virus function. *Sci. Adv.* **2018**, *4*, eaat7778. [CrossRef]

146. Abdella, R.; Aggarwal, M.; Okura, T.; Lamb, R.A.; He, Y. Structure of a paramyxovirus polymerase complex reveals a unique methyltransferase-CTD conformation. *Proc. Natl. Acad. Sci. USA* **2020**, *117*, 4931–4941. [CrossRef] [PubMed]

147. Guryanov, S.G.; Liljeroos, L.; Kasaragod, P.; Kajander, T.; Butcher, S.J. Crystal Structure of the Measles Virus Nucleoprotein Core in Complex with an N-Terminal Region of Phosphoprotein. *J. Virol.* **2015**, *90*, 2849–2857. [CrossRef]

148. Du Pont, V.; Jiang, Y.; Plemper, R.K. Bipartite interface of the measles virus phosphoprotein X domain with the large polymerase protein regulates viral polymerase dynamics. *PLoS Pathog.* **2019**, *15*, e1007995. [CrossRef]

149. Longhi, S.; Bloyet, L.-M.; Gianni, S.; Gerlier, D. How order and disorder within paramyxoviral nucleoproteins and phosphoproteins orchestrate the molecular interplay of transcription and replication. *Cell. Mol. Life Sci.* **2017**, *74*, 3091–3118. [CrossRef]

150. Brunel, J.; Chopy, D.; Dosnon, M.; Bloyet, L.-M.; Devaux, P.; Urzua, E.; Cattaneo, R.; Longhi, S.; Gerlier, D. Sequence of events in measles virus replication: Role of phosphoprotein-nucleocapsid interactions. *J. Virol.* **2014**, *88*, 10851–10863. [CrossRef]

151. Bloyet, L.-M.; Brunel, J.; Dosnon, M.; Hamon, V.; Erales, J.; Gruet, A.; Lazert, C.; Bignon, C.; Roche, P.; Longhi, S.; et al. Modulation of Re-initiation of Measles Virus Transcription at Intergenic Regions by PXD to NTAIL Binding Strength. *PLoS Pathog.* **2016**, *12*, e1006058. [CrossRef]

152. Katoh, H.; Kubota, T.; Kita, S.; Nakatsu, Y.; Aoki, N.; Mori, Y.; Maenaka, K.; Takeda, M.; Kidokoro, M. Heat shock protein 70 regulates degradation of the mumps virus phosphoprotein via the ubiquitin-proteasome pathway. *J. Virol.* **2015**, *89*, 3188–3199. [CrossRef] [PubMed]

153. Zhang, S.; Jiang, Y.; Cheng, Q.; Zhong, Y.; Qin, Y.; Chen, M. Inclusion Body Fusion of Human Parainfluenza Virus Type 3 Regulated by Acetylated α-Tubulin Enhances Viral Replication. *J. Virol.* **2017**, *91*. [CrossRef]

154. Li, Y.; Zhang, C.; Lu, N.; Deng, X.; Zang, G.; Zhang, S.; Tang, H.; Zhang, G. Involvement of Actin-Regulating Factor Cofilin in the Inclusion Body Formation and RNA Synthesis of Human Parainfluenza Virus Type 3 via Interaction With the Nucleoprotein. *Front. Microbiol.* **2019**, *10*, 95. [CrossRef]

155. Carlos, T.S.; Young, D.F.; Schneider, M.; Simas, J.P.; Randall, R.E. Parainfluenza virus 5 genomes are located in viral cytoplasmic bodies whilst the virus dismantles the interferon-induced antiviral state of cells. *J. Gen. Virol.* **2009**, *90*, 2147–2156. [CrossRef]

156. Goldsmith, C.S.; Whistler, T.; Rollin, P.E.; Ksiazek, T.G.; Rota, P.A.; Bellini, W.J.; Daszak, P.; Wong, K.; Shieh, W.-J.; Zaki, S.R. Elucidation of Nipah virus morphogenesis and replication using ultrastructural and molecular approaches. *Virus Res.* **2003**, *92*, 89–98. [CrossRef]

157. Becker, S.; Rinne, C.; Hofsäss, U.; Klenk, H.D.; Mühlberger, E. Interactions of Marburg virus nucleocapsid proteins. *Virology* **1998**, *249*, 406–417. [CrossRef]

158. Bamberg, S.; Kolesnikova, L.; Möller, P.; Klenk, H.-D.; Becker, S. VP24 of Marburg virus influences formation of infectious particles. *J. Virol.* **2005**, *79*, 13421–13433. [CrossRef] [PubMed]

159. Groseth, A.; Charton, J.E.; Sauerborn, M.; Feldmann, F.; Jones, S.M.; Hoenen, T.; Feldmann, H. The Ebola virus ribonucleoprotein complex: A novel VP30-L interaction identified. *Virus Res.* **2009**, *140*, 8–14. [CrossRef] [PubMed]

160. Shu, T.; Gan, T.; Bai, P.; Wang, X.; Qian, Q.; Zhou, H.; Cheng, Q.; Qiu, Y.; Yin, L.; Zhong, J.; et al. Ebola virus VP35 has novel NTPase and helicase-like activities. *Nucleic Acids Res.* **2019**, *47*, 5837–5851. [CrossRef]

161. Woolsey, C.; Menicucci, A.R.; Cross, R.W.; Luthra, P.; Agans, K.N.; Borisevich, V.; Geisbert, J.B.; Mire, C.E.; Fenton, K.A.; Jankeel, A.; et al. A VP35 Mutant Ebola Virus Lacks Virulence but Can Elicit Protective Immunity to Wild-Type Virus Challenge. *Cell Rep.* **2019**, *28*, 3032–3046.e6. [CrossRef]

162. Huang, Y.; Xu, L.; Sun, Y.; Nabel, G.J. The Assembly of Ebola Virus Nucleocapsid Requires Virion-Associated Proteins 35 and 24 and Posttranslational Modification of Nucleoprotein. *Mol. Cell* **2002**, *10*, 307–316. [CrossRef]

163. Mateo, M.; Carbonnelle, C.; Martinez, M.J.; Reynard, O.; Page, A.; Volchkova, V.A.; Volchkov, V.E. Knockdown of Ebola virus VP24 impairs viral nucleocapsid assembly and prevents virus replication. *J. Infect. Dis.* **2011**, *204* (Suppl. 3), S892–S896. [CrossRef]

164. Zhang, A.P.P.; Bornholdt, Z.A.; Liu, T.; Abelson, D.M.; Lee, D.E.; Li, S.; Woods, V.L.; Saphire, E.O. The ebola virus interferon antagonist VP24 directly binds STAT1 and has a novel, pyramidal fold. *PLoS Pathog.* **2012**, *8*, e1002550. [CrossRef] [PubMed]

165. Watt, A.; Moukambi, F.; Banadyga, L.; Groseth, A.; Callison, J.; Herwig, A.; Ebihara, H.; Feldmann, H.; Hoenen, T. A novel life cycle modeling system for Ebola virus shows a genome length-dependent role of VP24 in virus infectivity. *J. Virol.* **2014**, *88*, 10511–10524. [CrossRef]

166. Miyake, T.; Farley, C.M.; Neubauer, B.E.; Beddow, T.P.; Hoenen, T.; Engel, D.A. Ebola Virus Inclusion Body Formation and RNA Synthesis Are Controlled by a Novel Domain of Nucleoprotein Interacting with VP35. *J. Virol.* **2020**, *94*. [CrossRef]

167. Modrof, J.; Möritz, C.; Kolesnikova, L.; Konakova, T.; Hartlieb, B.; Randolf, A.; Mühlberger, E.; Becker, S. Phosphorylation of Marburg virus VP30 at serines 40 and 42 is critical for its interaction with NP inclusions. *Virology* **2001**, *287*, 171–182. [CrossRef] [PubMed]

168. Martinez, M.J.; Volchkova, V.A.; Raoul, H.; Alazard-Dany, N.; Reynard, O.; Volchkov, V.E. Role of VP30 phosphorylation in the Ebola virus replication cycle. *J. Infect. Dis.* **2011**, *204* (Suppl. 3), S934–S940. [CrossRef]

169. Lier, C.; Becker, S.; Biedenkopf, N. Dynamic phosphorylation of Ebola virus VP30 in NP-induced inclusion bodies. *Virology* **2017**, *512*, 39–47. [CrossRef] [PubMed]

170. Kruse, T.; Biedenkopf, N.; Hertz, E.P.T.; Dietzel, E.; Stalmann, G.; López-Méndez, B.; Davey, N.E.; Nilsson, J.; Becker, S. The Ebola Virus Nucleoprotein Recruits the Host PP2A-B56 Phosphatase to Activate Transcriptional Support Activity of VP30. *Mol. Cell* **2018**, *69*, 136–145.e6. [CrossRef] [PubMed]

171. Noda, T.; Ebihara, H.; Muramoto, Y.; Fujii, K.; Takada, A.; Sagara, H.; Kim, J.H.; Kida, H.; Feldmann, H.; Kawaoka, Y. Assembly and budding of Ebolavirus. *PLoS Pathog.* **2006**, *2*, e99. [CrossRef]

172. Dolnik, O.; Kolesnikova, L.; Welsch, S.; Strecker, T.; Schudt, G.; Becker, S. Interaction with Tsg101 is necessary for the efficient transport and release of nucleocapsids in marburg virus-infected cells. *PLoS Pathog.* **2014**, *10*, e1004463. [CrossRef]

173. Watanabe, S.; Noda, T.; Halfmann, P.; Jasenosky, L.; Kawaoka, Y. Ebola virus (EBOV) VP24 inhibits transcription and replication of the EBOV genome. *J. Infect. Dis.* **2007**, *196* (Suppl. 2), S284–S290. [CrossRef]

174. Noda, T.; Kolesnikova, L.; Becker, S.; Kawaoka, Y. The importance of the NP: VP35 ratio in Ebola virus nucleocapsid formation. *J. Infect. Dis.* **2011**, *204* (Suppl. 3), S878–S883. [CrossRef]

175. Noda, T.; Halfmann, P.; Sagara, H.; Kawaoka, Y. Regions in Ebola virus VP24 that are important for nucleocapsid formation. *J. Infect. Dis.* **2007**, *196* (Suppl. 2), S247–S250. [CrossRef]

176. Banadyga, L.; Hoenen, T.; Ambroggio, X.; Dunham, E.; Groseth, A.; Ebihara, H. Ebola virus VP24 interacts with NP to facilitate nucleocapsid assembly and genome packaging. *Sci. Rep.* **2017**, *7*, 7698. [CrossRef]

177. Takamatsu, Y.; Kolesnikova, L.; Schauflinger, M.; Noda, T.; Becker, S. The Integrity of the YxxL Motif of Ebola Virus VP24 Is Important for the Transport of Nucleocapsid-Like Structures and for the Regulation of Viral RNA Synthesis. *J. Virol.* **2020**, *94*. [CrossRef] [PubMed]

178. Schudt, G.; Dolnik, O.; Kolesnikova, L.; Biedenkopf, N.; Herwig, A.; Becker, S. Transport of Ebolavirus Nucleocapsids Is Dependent on Actin Polymerization: Live-Cell Imaging Analysis of Ebolavirus-Infected Cells. *J. Infect. Dis.* **2015**, *212* (Suppl. 2), S160–S166. [CrossRef]

179. Su, J.M.; Wilson, M.Z.; Samuel, C.E.; Ma, D. Formation and Function of Liquid-Like Viral Factories in Negative-Sense Single-Stranded RNA Virus Infections. *Viruses* **2021**, *13*, 126. [CrossRef] [PubMed]

180. Nevers, Q.; Albertini, A.A.; Lagaudrière-Gesbert, C.; Gaudin, Y. Negri bodies and other virus membrane-less replication compartments. *Biochim. Biophys. Acta Mol. Cell Res.* **2020**, *1867*, 118831. [CrossRef] [PubMed]

181. Geisbert, T.W.; Jahrling, P.B. Differentiation of filoviruses by electron microscopy. *Virus Research* **1995**, *39*, 129–150. [CrossRef]

182. Kolesnikova, L.; Bugany, H.; Klenk, H.-D.; Becker, S. VP40, the matrix protein of Marburg virus, is associated with membranes of the late endosomal compartment. *J. Virol.* **2002**, *76*, 1825–1838. [CrossRef]

183. Dolnik, O.; Volchkova, V.A.; Escudero-Perez, B.; Lawrence, P.; Klenk, H.-D.; Volchkov, V.E. Shedding of Ebola Virus Surface Glycoprotein Is a Mechanism of Self-regulation of Cellular Cytotoxicity and Has a Direct Effect on Virus Infectivity. *J. Infect. Dis.* **2015**, *212* (Suppl. 2), S322–S328. [CrossRef]

184. Wendt, L.; Brandt, J.; Bodmer, B.S.; Reiche, S.; Schmidt, M.L.; Traeger, S.; Hoenen, T. The Ebola Virus Nucleoprotein Recruits the Nuclear RNA Export Factor NXF1 into Inclusion Bodies to Facilitate Viral Protein Expression. *Cells* **2020**, *9*, 187. [CrossRef] [PubMed]

185. Brandt, J.; Wendt, L.; Bodmer, B.S.; Mettenleiter, T.C.; Hoenen, T. The Cellular Protein CAD is Recruited into Ebola Virus Inclusion Bodies by the Nucleoprotein NP to Facilitate Genome Replication and Transcription. *Cells* **2020**, *9*, 1126. [CrossRef] [PubMed]

186. Nelson, E.V.; Schmidt, K.M.; Deflubé, L.R.; Doğanay, S.; Banadyga, L.; Olejnik, J.; Hume, A.J.; Ryabchikova, E.; Ebihara, H.; Kedersha, N.; et al. Ebola Virus Does Not Induce Stress Granule Formation during Infection and Sequesters Stress Granule Proteins within Viral Inclusions. *J. Virol.* **2016**, *90*, 7268–7284. [CrossRef] [PubMed]

187. Le Sage, V.; Cinti, A.; McCarthy, S.; Amorim, R.; Rao, S.; Daino, G.L.; Tramontano, E.; Branch, D.R.; Mouland, A.J. Ebola virus VP35 blocks stress granule assembly. *Virology* **2017**, *502*, 73–83. [CrossRef] [PubMed]

188. Lee, J.E.; Cathey, P.I.; Wu, H.; Parker, R.; Voeltz, G.K. Endoplasmic reticulum contact sites regulate the dynamics of membraneless organelles. *Science* **2020**, *367*. [CrossRef] [PubMed]

189. Patel, A.; Malinovska, L.; Saha, S.; Wang, J.; Alberti, S.; Krishnan, Y.; Hyman, A.A. ATP as a biological hydrotrope. *Science* **2017**, *356*, 753–756. [CrossRef]

190. Zhao, Y.G.; Zhang, H. Phase Separation in Membrane Biology: The Interplay between Membrane-Bound Organelles and Membraneless Condensates. *Dev. Cell* **2020**, *55*, 30–44. [CrossRef] [PubMed]

191. Noda, T.; Murakami, S.; Nakatsu, S.; Imai, H.; Muramoto, Y.; Shindo, K.; Sagara, H.; Kawaoka, Y. Importance of the 1+7 configuration of ribonucleoprotein complexes for influenza A virus genome packaging. *Nat. Commun.* **2018**, *9*, 54. [CrossRef]

192. Muramoto, Y.; Takada, A.; Fujii, K.; Noda, T.; Iwatsuki-Horimoto, K.; Watanabe, S.; Horimoto, T.; Kida, H.; Kawaoka, Y. Hierarchy among viral RNA (vRNA) segments in their role in vRNA incorporation into influenza A virions. *J. Virol.* **2006**, *80*, 2318–2325. [CrossRef]

193. Fournier, E.; Moules, V.; Essere, B.; Paillart, J.-C.; Sirbat, J.-D.; Cavalier, A.; Rolland, J.-P.; Thomas, D.; Lina, B.; Isel, C.; et al. Interaction network linking the human H3N2 influenza A virus genomic RNA segments. *Vaccine* **2012**, *30*, 7359–7367. [CrossRef]

194. Fournier, E.; Moules, V.; Essere, B.; Paillart, J.-C.; Sirbat, J.-D.; Isel, C.; Cavalier, A.; Rolland, J.-P.; Thomas, D.; Lina, B.; et al. A supramolecular assembly formed by influenza A virus genomic RNA segments. *Nucleic Acids Res.* **2012**, *40*, 2197–2209. [CrossRef]

195. Lakdawala, S.S.; Wu, Y.; Wawrzusin, P.; Kabat, J.; Broadbent, A.J.; Lamirande, E.W.; Fodor, E.; Altan-Bonnet, N.; Shroff, H.; Subbarao, K. Influenza a virus assembly intermediates fuse in the cytoplasm. *PLoS Pathog.* **2014**, *10*, e1003971. [CrossRef]

196. De Castro Martin, I.F.; Fournier, G.; Sachse, M.; Pizarro-Cerda, J.; Risco, C.; Naffakh, N. Influenza virus genome reaches the plasma membrane via a modified endoplasmic reticulum and Rab11-dependent vesicles. *Nat. Commun.* **2017**, *8*, 1396. [CrossRef]

197. Wu, X.; Qi, X.; Liang, M.; Li, C.; Cardona, C.J.; Li, D.; Xing, Z. Roles of viroplasm-like structures formed by nonstructural protein NSs in infection with severe fever with thrombocytopenia syndrome virus. *FASEB J.* **2014**, *28*, 2504–2516. [CrossRef] [PubMed]

198. Fontana, J.; López-Montero, N.; Elliott, R.M.; Fernández, J.J.; Risco, C. The unique architecture of Bunyamwera virus factories around the Golgi complex. *Cell. Microbiol.* **2008**, *10*, 2012–2028. [CrossRef] [PubMed]

199. Baird, N.L.; York, J.; Nunberg, J.H. Arenavirus infection induces discrete cytosolic structures for RNA replication. *J. Virol.* **2012**, *86*, 11301–11310. [CrossRef]

200. Bartok, E.; Hartmann, G. Immune Sensing Mechanisms that Discriminate Self from Altered Self and Foreign Nucleic Acids. *Immunity* **2020**, *53*, 54–77. [CrossRef] [PubMed]

201. Chatterjee, S.; Basler, C.F.; Amarasinghe, G.K.; Leung, D.W. Molecular Mechanisms of Innate Immune Inhibition by Non-Segmented Negative-Sense RNA Viruses. *J. Mol. Biol.* **2016**, *428*, 3467–3482. [CrossRef] [PubMed]

202. Gerlier, D.; Lyles, D.S. Interplay between innate immunity and negative-strand RNA viruses: Towards a rational model. *Microbiol. Mol. Biol. Rev.* **2011**, *75*, 468–490. [CrossRef]

203. Chelbi-Alix, M.K.; Vidy, A.; El Bougrini, J.; Blondel, D. Rabies viral mechanisms to escape the IFN system: The viral protein P interferes with IRF-3, Stat1, and PML nuclear bodies. *J. Interferon Cytokine Res.* **2006**, *26*, 271–280. [CrossRef] [PubMed]

204. Brzózka, K.; Finke, S.; Conzelmann, K.-K. Identification of the rabies virus alpha/beta interferon antagonist: Phosphoprotein P interferes with phosphorylation of interferon regulatory factor 3. *J. Virol.* **2005**, *79*, 7673–7681. [CrossRef] [PubMed]

205. Basler, C.F.; Mikulasova, A.; Martinez-Sobrido, L.; Paragas, J.; Mühlberger, E.; Bray, M.; Klenk, H.-D.; Palese, P.; García-Sastre, A. The Ebola virus VP35 protein inhibits activation of interferon regulatory factor 3. *J. Virol.* **2003**, *77*, 7945–7956. [CrossRef] [PubMed]

206. Basler, C.F.; Amarasinghe, G.K. Evasion of interferon responses by Ebola and Marburg viruses. *J. Interferon Cytokine Res.* **2009**, *29*, 511–520. [CrossRef]

207. Prins, K.C.; Cárdenas, W.B.; Basler, C.F. Ebola virus protein VP35 impairs the function of interferon regulatory factor-activating kinases IKKepsilon and TBK-1. *J. Virol.* **2009**, *83*, 3069–3077. [CrossRef]

208. Hartman, A.L.; Bird, B.H.; Towner, J.S.; Antoniadou, Z.-A.; Zaki, S.R.; Nichol, S.T. Inhibition of IRF-3 activation by VP35 is critical for the high level of virulence of ebola virus. *J. Virol.* **2008**, *82*, 2699–2704. [CrossRef]

209. Cárdenas, W.B.; Loo, Y.-M.; Gale, M.; Hartman, A.L.; Kimberlin, C.R.; Martínez-Sobrido, L.; Saphire, E.O.; Basler, C.F. Ebola virus VP35 protein binds double-stranded RNA and inhibits alpha/beta interferon production induced by RIG-I signaling. *J. Virol.* **2006**, *80*, 5168–5178. [CrossRef]

210. Schümann, M.; Gantke, T.; Mühlberger, E. Ebola virus VP35 antagonizes PKR activity through its C-terminal interferon inhibitory domain. *J. Virol.* **2009**, *83*, 8993–8997. [CrossRef] [PubMed]

211. Wu, X.; Qi, X.; Qu, B.; Zhang, Z.; Liang, M.; Li, C.; Cardona, C.J.; Li, D.; Xing, Z. Evasion of antiviral immunity through sequestering of TBK1/IKKε/IRF3 into viral inclusion bodies. *J. Virol.* **2014**, *88*, 3067–3076. [CrossRef] [PubMed]

212. Jobe, F.; Simpson, J.; Hawes, P.; Guzman, E.; Bailey, D. Respiratory Syncytial Virus Sequesters NF-κB Subunit p65 to Cytoplasmic Inclusion Bodies To Inhibit Innate Immune Signaling. *J. Virol.* **2020**, *94*. [CrossRef]

213. Fricke, J.; Koo, L.Y.; Brown, C.R.; Collins, P.L. p38 and OGT sequestration into viral inclusion bodies in cells infected with human respiratory syncytial virus suppresses MK2 activities and stress granule assembly. *J. Virol.* **2013**, *87*, 1333–1347. [CrossRef]

214. Onomoto, K.; Jogi, M.; Yoo, J.-S.; Narita, R.; Morimoto, S.; Takemura, A.; Sambhara, S.; Kawaguchi, A.; Osari, S.; Nagata, K.; et al. Critical role of an antiviral stress granule containing RIG-I and PKR in viral detection and innate immunity. *PLoS ONE* **2012**, *7*, e43031. [CrossRef]

215. Dinh, P.X.; Beura, L.K.; Das, P.B.; Panda, D.; Das, A.; Pattnaik, A.K. Induction of stress granule-like structures in vesicular stomatitis virus-infected cells. *J. Virol.* **2013**, *87*, 372–383. [CrossRef] [PubMed]

216. Hong, Y.; Bai, M.; Qi, X.; Li, C.; Liang, M.; Li, D.; Cardona, C.J.; Xing, Z. Suppression of the IFN-α and -β Induction through Sequestering IRF7 into Viral Inclusion Bodies by Nonstructural Protein NSs in Severe Fever with Thrombocytopenia Syndrome Bunyavirus Infection. *J. Immunol.* **2019**, *202*, 841–856. [CrossRef] [PubMed]

217. Kitagawa, Y.; Sakai, M.; Shimojima, M.; Saijo, M.; Itoh, M.; Gotoh, B. Nonstructural protein of severe fever with thrombocytopenia syndrome phlebovirus targets STAT2 and not STAT1 to inhibit type I interferon-stimulated JAK-STAT signaling. *Microbes Infect.* **2018**, *20*, 360–368. [CrossRef]
218. Ning, Y.-J.; Feng, K.; Min, Y.-Q.; Cao, W.-C.; Wang, M.; Deng, F.; Hu, Z.; Wang, H. Disruption of type I interferon signaling by the nonstructural protein of severe fever with thrombocytopenia syndrome virus via the hijacking of STAT2 and STAT1 into inclusion bodies. *J. Virol.* **2015**, *89*, 4227–4236. [CrossRef] [PubMed]
219. Santiago, F.W.; Covaleda, L.M.; Sanchez-Aparicio, M.T.; Silvas, J.A.; Diaz-Vizarreta, A.C.; Patel, J.R.; Popov, V.; Yu, X.; García-Sastre, A.; Aguilar, P.V. Hijacking of RIG-I signaling proteins into virus-induced cytoplasmic structures correlates with the inhibition of type I interferon responses. *J. Virol.* **2014**, *88*, 4572–4585. [CrossRef]

 *cells*

Review

Post-Transcriptional Regulation of Viral RNA through Epitranscriptional Modification

David G. Courtney 

Wellcome-Wolfson Institute for Experimental Medicine, Queen's University Belfast, Belfast BT9 7BL, UK; david.courtney@qub.ac.uk; Tel.: +44-2890972713

Abstract: The field of mRNA modifications has been steadily growing in recent years as technologies have improved and the importance of these residues became clear. However, a subfield has also arisen, specifically focused on how these modifications affect viral RNA, with the possibility that viruses can also be used as a model to best determine the role that these modifications play on cellular mRNAs. First, virologists focused on the most abundant internal mRNA modification, m^6A, mapping this modification and elucidating its effects on the RNA of a wide range of RNA and DNA viruses. Next, less common RNA modifications including m^5C, Nm and ac^4C were investigated and also found to be present on viral RNA. It now appears that viral RNA is littered with a multitude of RNA modifications. In biological systems that are under constant evolutionary pressure to out compete both the host as well as newly arising viral mutants, it poses an interesting question about what evolutionary benefit these modifications provide as it seems evident, at least to this author, that these modifications have been selected for. In this review, I discuss how RNA modifications are identified on viral RNA and the roles that have now been uncovered for these modifications in regard to viral replication. Finally, I propose some interesting avenues of research that may shed further light on the exact role that these modifications play in viral replication.

Keywords: virus; RNA; modification; epitranscriptomic; N6-methyladenosine; mapping; 5-methylcytosine; pseudouridine; HIV-1

Citation: Courtney, D.G. Post-Transcriptional Regulation of Viral RNA through Epitranscriptional Modification. *Cells* **2021**, *10*, 1129. https://doi.org/10.3390/cells10051129

Academic Editors: Thomas Hoenen and Allison Groseth

Received: 13 April 2021
Accepted: 5 May 2021
Published: 7 May 2021

1. Introduction

The post-transcriptional regulation of mRNA function by the covalent modification of individual nucleotides, referred to as epitranscriptomic gene regulation, has attracted increasing interest in recent years. Through the development of better mapping techniques to identify the sites of RNA modification, the ability to quantify these modified residues through mass spectrometry, and the identification of modification writer and reader proteins (Table 1), researchers have been better able to understand the role that these modifications play across the cellular landscape.

RNA modifications include the addition of generally small biochemical groups to adenosine, cytosine, uracil and guanosine, with a methyl group being the most common addition. The most common mRNA modification, m^6A, comprises an additional methyl group at the N6 position of adenosine and constitutes ~0.4% of all adenosine residues on human cellular mRNA [1]. The second most common modification appears to be a modified uracil residue called pseudouridine, at ~0.3% of all uracil residues on human mRNAs. Recent data have demonstrated that 5-methylcytosine (m^5C) at ~0.05% is also quite prevalent on cellular mRNAs [1]. 2′O-methylated base modifications (N_m) are also widespread, where the methyl group is added to the ribose base as opposed to the nucleoside as is the case for the previous three modifications. 2′O-methylation occurs on all four nucleosides on cellular mRNAs. Common 5′cap modification 7-methylguanosine (m^7G) has also been recently reported to be found internally on cellular mRNAs, though not a great deal is known about the abundance at this early stage [2]. The dimethyl modification N6, N6-dimethyladenosine ($N^{6,6}A$), and N1-methyladenosine (m^1A) have also been proposed to be present on cellular mRNAs, though if they are indeed present, they are at extremely low levels [3]. Finally, N4-acetylcytidine (ac^4C) rounds out the list of common mRNA modifications and this arises from the addition of an acetyl group to cytosine. Interestingly, this is the first and, so far, only acetylation event reported on eukaryotic mRNAs [4].

Table 1. A summary of writer and reader proteins for common mRNA modifications and the proposed roles for these modifications on human cellular mRNAs.

Modification	Writers	Readers	*Roles on mRNAs*
N6-methyladenosine (m^6A)	METTL3 METTL4 METTL16	YTHDF1-3 YTHDC1-2	Splicing Stability Translation Localisation
5-methylcytidine (m^5C)	NSUN2 DNMT2	YBX1	Splicing Translation
2'O-methylated nucleosides (Nm)	FTSJ3	Unknown	Structure Stability
N^4-acetylcytidine (ac^4C)	NAT10	Unknown	Stability Translation
Pseudouridine (ψ)	PUS7 TRUB1 DKC1	Unknown	Codon misreading
7-methylguanosine (m^7G)	METTL1	Unknown	Translation
N^1-methyladenosine (m^1A)	TRMT6/61A	YTHDF1-3	Translation
1-methylguanosine (m^1G)	TRMT10A/B	Unknown	Unknown
N6,N6-dimethyladenosine ($m^{6,6}A$)	Unknown	Unknown	Unknown

The abundance of these modifications on viral RNA is less well understood. However, in recent years, there have been a growing number of publications investigating the presence of RNA modifications, mostly m^6A, on RNA from a variety of RNA and DNA viruses, as detailed in Table 2. It appears that m^6A modifications are highly prevalent on viral RNA, seemingly present on RNA from every virus that has been investigated. Many of the studies compiled in Table 2 identify the presence of RNA modifications by mapping their locations. Three publications, however, performed ultra high-performance liquid chromatography and tandem mass spectrometry (UPLC–MS/MS) to identify pools of RNA modifications in highly purified RNA isolated from HIV-1/MLV virions [5,6], or RNA isolated from positive-strand RNA virus-infected cells [7]. All these studies listed in Table 2 provide us with a clear picture that RNA modifications litter the viral RNA of infected cells.

The presence of these modifications has been found to increase the replication of HIV-1, SV40, HBV and IAV [8–13]. However, the exact nature of how RNA modifications confer this advantage remains a mystery. Viral genomes are constantly under an intense selective pressure. Therefore, RNA modifications that confer a selective advantage to viral replication will be selected for and theoretically a virus will quickly evolve to acquire a level of modification that maximizes this advantage. I and others have recently proven this to be the case in two retroviruses, HIV-1 and MLV, where a number of RNA modifications were found on viral RNAs at greater levels than have been observed on cellular mRNAs [5,6,9,10,14]. This conclusion is further supported by a previous study that found an enrichment of a range of RNA modifications on RNA viruses including Zika, Dengue and hepatitis C viruses [4]. The modifications m^6A, m^5C, 2'O-methylation and ac^4C have now been mapped on HIV-1 RNAs [6,9–11,14,15], with m^6A having additionally been mapped on RNA from IAV, MLV, SV40 and a number of flaviviruses [5,6,12,13,16,17].

While the function of these modifications on mRNA is slowly being elucidated, with roles in splicing [18–20], translation [21–27], trafficking [28] and stability [11,29–31] all being proposed, whether they have any unique and distinct roles on viral RNA is still as yet unclear. This review will attempt to summarise the current understanding in the field surrounding RNA modifications and their roles in viral replication. I will review the current methods in the field for identifying sites of modification on viral RNA, the stages of the replication cycle seemingly most susceptible to epitranscriptomic mediated alterations, and finally future possible areas of research that could proceed to answer important questions surrounding viral epitranscriptomics.

Table 2. A summary of viruses that have previously been reported to carry some of these nine common RNA modifications on virally encoded mRNAs.

Virus	Genome	m^6A	m^5C	ψ	m^1A	ac^4C	Nm	m^7G	m^1G	$m^{6,6}A$
Adenovirus serotype 5	DNA	[18]								
Dengue virus	RNA	[7,16]	[7]	[7]	[7]	[7]	[7]	[7]	[7]	[7]
Enterovirus 71	RNA	[32]								
Epstein–Barr virus	DNA	[33]								
Hepatitis B virus	DNA	[8,34]								
Hepatitis C virus	RNA	[7,16,34]	[7]	[7]	[7]	[7]	[7]	[7]	[7]	[7]
HIV-1	RNA	[6,7,9,10,14,35]	[6,7,36]	[7]	[6,7]	[7,11]	[6,7,15]	[6,7]	[6,7]	[6,7]
Human metapneumovirus	RNA	[37]								
Influenza A virus	RNA	[12,38,39]								
Kaposi's sarcoma-associated herpesvirus	DNA	[40,41]								
Measles virus	RNA	[42]								
Murine leukaemia virus	RNA	[5]	[5,43]		[5]	[5]	[5]	[5]	[5]	[5]
Poliovirus	RNA	[7]	[7]	[7]	[7]	[7]	[7]	[7]	[7]	[7]
Respiratory syncytial virus	RNA	[44]								
SARS-CoV-2	RNA	[45]	[46]							
Sendai virus	RNA	[42]								
Simian virus 40	DNA	[13]								
Vesicular stomatitis virus	RNA	[42]								
West Nile virus	RNA	[16]								
Yellow fever virus	RNA	[16]								
Zika virus	RNA	[7,16,17]	[7]	[7]	[7]	[7]	[7]	[7]	[7]	[7]

2. Viral Modification Mapping

2.1. Antibody Mapping

With the commercial availability of modified nucleoside-specific antibodies, novel mapping techniques started to be developed [47]. This was initially focused on m^6A identification, before methods were adapted to identify m^5C, m^1A and ac^4C [3,23,48]. In short, these methods generally involve extraction of RNA from the target cells, poly(A) purification of mRNA when it is the target of interest, fragmentation of the RNA, capture of RNA fragments containing a given modification by the modification-specific antibody, capture of the antibody on beads and then isolation of the captured RNAs followed by deep sequencing (Figure 1).

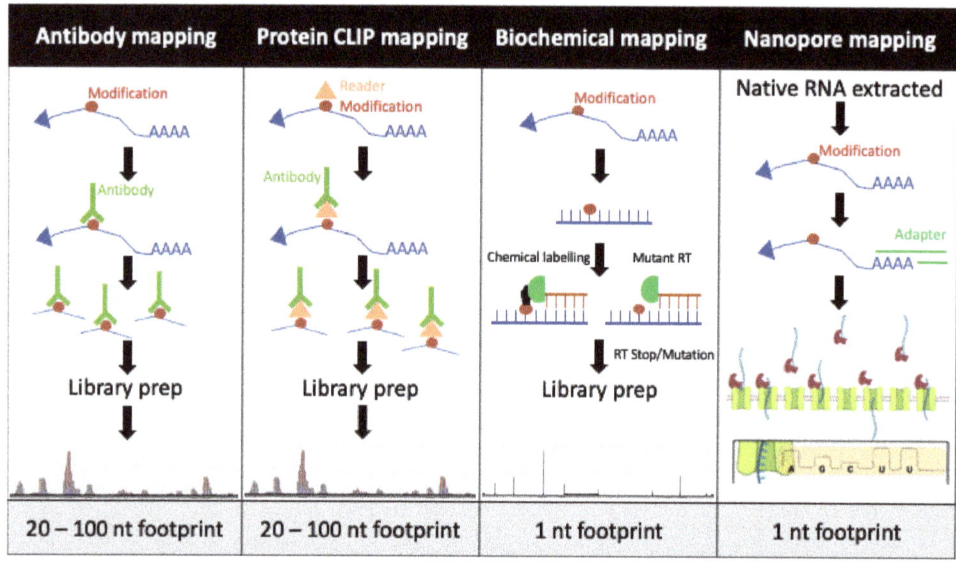

Figure 1. Schematic of the four main methods of mapping RNA modifications. Antibody mapping and protein clip mapping are straightforward techniques involving capture of modified RNA fragments by antibodies before elution and next-generation sequencing, which yields footprints of 20–100 nt. Biochemical mapping generally involves either chemical labelling of a modified residue to block reverse transcription, or a mutant reverse transcriptase that spontaneously stops upon encountering a modified residue. Again, these products undergo next-generation sequencing, but the resultant footprint of these methods is 1 nt. Finally, Nanopore mapping uses a new technique of nucleotide detection by calculating electrical current as the RNA passes through a pore. Each nucleotide alters the electrical current differently, with minor fluctuations also detectable when modified nucleotides are present. This method also results in a 1 nt footprint and is capable of sequencing native RNA.

These antibody-based methods, or slightly altered versions of them, have since been used to great effect to map a large number of modifications on RNAs from a whole host of viruses. HIV-1 has been the most extensively studied by this method, with antibody mapping having been used for m^6A, m^5C and ac^4C modification identification [6,9–11,14]. Mapping of modifications on MLV, another retrovirus, has been performed for both m^6A and m^5C using these methods [5], while influenza A virus (IAV) [12], SV40 [13], Zika virus [17], hepatitis B virus [8] and hepatitis C virus [16] m^6A modifications have also been mapped in this way.

However, antibody-based mapping of modifications is inherently noisy, with input or IgG controls being required for a number of these methods to remove background signal (Table 3). In addition, this form of mapping results in large footprints of around 20–100 nucleotides, making it practically impossible to determine the exact modified residue on viral or cellular RNA (Table 3). Although antibody-based mapping is relatively quick to perform, unlike some of the other methods, and can at least train a researcher's eye to a region of interest, it should now be complemented with additional mapping methods to validate any proposed sites of modification.

Table 3. A summary of the advantages and disadvantages of the forms of mapping techniques described in this review.

Mapping Method	Advantages	Disadvantages
Antibody mapping	1. Fast, straightforward technique 2. Can be used to map modifications on lowly expressed RNA	1. Large footprint of ~20–100 nucleotides 2. Can generate mapping artifacts
Protein CLIP mapping	1. Quite straightforward 2. Modification specific	1. Large footprint of ~20–100 nucleotides 2. Must know the writer or reader protein of interest prior to mapping 3. Can generate mapping artifacts
Biochemical mapping	1. Single-nucleotide resolution 2. Can be used to quantify modification occupancy at specific residues	1. Can require very large read-depth 2. May not pick up lowly expressed RNAs 3. lower, more technically difficult technique 4. Can generate mapping artifacts
Nanopore mapping	1. Can map modifications on native RNA	1. Difficult to differentiate between modifications at present 2. Can generate mapping artifacts

2.2. Protein CLIP Mapping

The use of crosslinked immunoprecipitation can be a useful method for RNA modification mapping if the researcher is aware of modification-specific writer or reader proteins. For example, the YTH domain-containing family of protein including YTHDF1, YTHDF2, YTHDF3, YTHDC1 and YTHDC2 (Table 1) are known to be m^6A-specific RNA-binding proteins, known as m^6A 'readers' first described by Dominissini et al. [47]. Performing CLIP of these specific proteins can help identify RNA footprints containing an m^6A residue, to a similar resolution to some antibody-based approaches such as PA-m^6A-seq (Figure 1) [49]. This CLIP-based approach has been used to good effect for mapping m^6A modifications on viral RNAs including those of HIV-1 [10,14], IAV [12] and SV40 [13].

In addition to CLIP-seq that uses modification-specific reader proteins, for m^5C, a highly novel CLIP-based method using the NSUN family of writer proteins has also been described and used for mapping m^5C on viral RNAs. The m^5C RNA modification is mediated by the seven members of the NSUN family of methyltransferases, NSUN1 through NSUN7, in addition to the DNA methyltransferase homolog DNMT2 (Table 1) [50,51]. NSUN protein mediated methylation of cytosine uses two highly conserved cysteine residues. One cysteine residue (C321 in NSUN2) forms a transient covalent bond to the pyrimidine base, while the second conserved cysteine residue (C271 in NSUN2) is essential for release of the RNA [52]. Hussain et al. 2013 [53] very cleverly exploited this phenomenon to generate a spontaneously crosslinking NSUN2 mutant (C271A) and then proceeded to overexpress this protein to map NSUN2 targeted cytosine residues by immunoprecipitation and deep sequencing without any need for an actual crosslinking step in the procedure. This also avoids any off-target crosslinking issues as only RNA bound by the NSUN protein, in this case NSUN2, will be covalently bound and appear in the deep sequencing analysis. Colleagues and I followed up on this previous study and exploited the same phenomenon with NSUN proteins to identify NSUN2 as the primary m^5C writer for HIV-1 RNA and to map sites of m^5C modification on HIV-1 [3]. This may prove to be a powerful method in the future for both writer identification and m^5C site validation on both viral and cellular mRNAs.

2.3. Biochemical Mapping

Biochemical mapping methods are at present the best option for mapping RNA modifications at single-nucleotide resolution (Figure 1). Although these methods have generally only been utilised for mapping modifications on cellular RNA transcripts, they could theoretically be exploited to map modifications on viral RNAs. This would in turn allow for the quantification of the level of modification occupancy at each residue (Table 3). This is a key attribute of biochemical methods that the above mapping techniques fail to deliver. Below I will describe one commonly utilised biochemical technique for different modifications of increasing interest in the area of viral epitranscriptomics.

One method, termed miCLIP, has been described for the identification of m^6A sites at single-nucleotide resolution on cellular RNAs [54]. This technique uses an m^6A-specific antibody UV crosslinked to m^6A-containing RNA, similar to those described above. This results in the introduction of a single polymorphism by reverse transcriptase, which can be detected and quantified by deep sequencing. This method could easily be translated to the study of viral epitranscriptomics.

RNA bisulfite sequencing is another well-used biochemical method, which has been used to map m^5C modifications to single-nucleotide resolution on cellular RNAs. By this method, RNA is denatured and incubated at a high temperature with sodium bisulfite to chemically deaminate all unmethylated cytosine residues to uracil [55]. This is due to the low reactivity of m^5C with HSO_3. Cytosine residues that are 'protected' from deamination can then be detected by standard sequencing techniques. However, one drawback of this technique is that protection from deamination can be due to the presence of not only m^5C residues but further oxidised forms of cytidine including 5-hydroxymethylcytidine (hm^5C) and 5-formycytidine (f^5C). Unfortunately, the efficiency of bisulfite conversion is affected by RNA secondary structure. This may be a problem for the use of this technique to map modifications on viral RNAs, which are notoriously rich in secondary structures and may produce too many artifacts [56]. This will have to be tested experimental before one can know for certain.

The ψ-seq technique has been well described for the mapping of ψ residues on eukaryotic cellular RNA [57,58]. This protocol uses N-cyclohexyl-N'-(2-morpholinoethyl) carbodiimide metho-p-toluenesulfonate (CMC) to selectively modify ψ residues. This large CMC modification on each ψ results in a total block to reverse transcription and these prematurely stopped cDNA fragments can be deep sequenced and identified bioinformatically (Figure 1). An identical non-CMC control sample is processed in parallel to determine background levels of premature stopping. This technique again provides the researcher with single-nucleotide resolution mapping of ψ residues; however, this method is still to be tested for mapping ψ modifications on viral RNAs.

Two methods to accurately map the location of Nm base modifications are RiboMethSeq [59] and Nm-seq [60]. RiboMethSeq is a straightforward method of using alkaline fragmentation on an RNA pool, where Nm residues are generally resistant to fragmentation. In short, this fragmented pool is then ligated to adapters and processed for Illumina sequencing following standard protocols. If the sequencing is performed to a great enough depth, underrepresented sites of fragmentation can be identified bioinformatically and it can be surmised that they have arisen due to the presence of Nm residues [59]. By this method, single-nucleotide resolution can be achieved as well as modification occupancy frequency (Table 3). Nm-seq is a more time-consuming approach, though requiring much less read depth. Nm-seq relies on performing multiple oxidation–elimination–dephosphorylation cycles, where, every cycle, an unmodified nucleotide is eliminated from an RNA string unless it is Nm modified and thus protected from elimination. This approach is then coupled with Illumina sequencing where an adapter is ligated to the immediate 3′ residue of the RNA string, which has been enriched for Nm modified nucleotides [60]. In this way Nm residues can be identified by deep sequencing followed by bioinformatic detection of overrepresented residues at the 3′ end of reads, indicating protection from oxidation and thus likely the presence of an Nm residue. At present, only RiboMethSeq has been shown to be effective in detecting Nm residues in viruses, with this approach having been used to great effect on HIV-1, as will be described later [15]. However, I do not foresee any issues with exploiting either sequencing method for the detection of Nm modifications on viral RNAs in future studies.

In addition to those described above, a number of additional biochemical methods have recently been published for mapping modifications including ac4C [23,61,62], m^{7}G [2] and m^{1}A [63], all of which it is possible to imagine can be translated to the study of viral epitranscriptomics.

2.4. Nanopore Mapping

The advent of direct RNA sequencing through Nanopore technology is a particularly exciting advance in the field of epitranscriptomics, and particularly in regard to viral RNA as one can now sequence native viral RNA harvested from isolated cellular compartments as well as purified virions (Figure 1; Table 3). This allows researchers the opportunity to determine whether the modification landscape of viral RNA is consistent throughout the cell, or whether these sites of modification are dynamic. For instance, are modified residues modified throughout the entire viral replication cycle, such as with an influenza vRNA transcribed in the nucleus, trafficked through the cytoplasm, packaged at the cellular membrane, encapsulated in a virion, and upon infection trafficked again through the cytoplasm to the nucleus? Or perhaps only a subpopulation of viral RNA is modified, and this aids in distinguishing viral RNA to be trafficked and packaged versus translated, as could be the case for some positive-strand RNA viruses.

In fact, nanopore-based direct RNA sequencing was recently used for modification identification early in the COVID-19 pandemic by Kim et al. 2020, where the authors identified potentially at least 41 sites of modification on SARS-CoV-2 viral RNAs [46]. One particularly interesting observation in this study regarding viral transcripts was the discovery that modified viral RNA had shorter poly(A) tails than their unmodified counterparts, with the authors proceeding to speculate that the presence of internal modifications could affect viral RNA stability, but further work will be required to elucidate such a proposed mechanism [46].

3. Viral RNA Trafficking

The correct trafficking of viral RNA to sites of replication, translation and packaging is critical to the successful completion of the viral replication cycle. RNA modifications have been identified in a number of cases as having an important role in this process of RNA trafficking. Gokhale et al. [16] investigated the role of m^{6}A modification on hepatitis C virus (HCV), while also mapping sites of modification on a range of other flaviviruses. These authors found that m^{6}A modification had a direct effect on viral RNA retention in virus replication factories, effectively slowing down the infection and potentially leading to a prolonged chronic infection as is characteristic of HCV infection in the liver. However, when m^{6}A was depleted, these viral transcripts are more readily bound by viral Core protein and are successfully trafficked to sites of virion packaging within the cell [16]. Two further studies investigated similar dynamics of viral RNA trafficking, but with retroviruses, and found that, in this instance, RNA modifications contributed a positive effect to viral RNA trafficking. Lichinchi et al. [9] explored the effect of m^{6}A on HIV-1 and found that the presence of two m^{6}A sites in the Rev-response element (RRE) increased the affinity of Rev for the RRE that in turn increased the nuclear export of RRE containing HIV-1 RNA. While Eckwahl et al. [43] focused on m^{5}C modification of MLV RNA, which the authors found, through an association with ALYREF, also increased nuclear export of viral RNA and thus increased viral replication. The differing roles for RNA modifications in the trafficking of HCV and retrovirus RNA further adds a fascinating layer to the complexity of modification mediated post-transcriptional regulation of viral RNA.

4. Degradation of Viral RNA

RNA modifications, most notably m^{6}A, have been shown to dysregulate the stability of cellular mRNAs generally through interaction with the YTH domain-containing family of proteins [30]. This has been shown to be the case with HBV and KSHV RNA, where both sets of authors demonstrated through global depletion of m^{6}A by methyltransferase knockdown or simply depletion of YTHDF proteins by siRNAs, that m^{6}A contributed to the destabilisation of viral RNA mostly likely through interactions with YTHDF proteins [8,41]. However, in the case of HIV-1 and IAV, this m^{6}A induced RNA destabilization does not appear to be the case. The presence of m^{6}A residues on the RNA of HIV-1 [10] and IAV [12] have been shown to increase RNA stability and in both these studies authors suspect this is due to an interaction with YTHDF proteins, primarily YTHDF2. This stabilization was most apparent in YTHDF2 tethering experiments in a mammalian expression system by Kennedy et al. where YTHDF1, 2 and 3 proteins where tethered to a luciferase reporter mRNA by MS2 hairpins in the 3'UTR [10]. This interaction was found to increase the luciferase activity by approximately 3–4 fold for each YTHDF protein.

Aside from m^{6}A, a recent study by Tsai et al. investigating the role of ac^{4}C modifications in HIV-1 RNA found that these modifications to increase the stability of HIV-1 RNA [11]. This finding is supported by previous work surrounding ac^{4}C that found a similar phenotype on cellular mRNAs [23]. The authors demonstrate this finding through acetyltransferase knockout and mutagenesis of modified sites on viral RNA. At present, no RNA-binding proteins specific to ac^{4}C are known so the authors were unable to speculate as to whether this increase in RNA stability is due to RNA structural changes or RNA-protein interactions.

5. Splicing of Viral RNA

RNA modifications have been implicated in the alternation of splicing events for both HIV-1 and adenovirus RNA [6,18]. Regarding HIV-1, colleagues and I reported that m^{5}C is generally present at specific locations across the HIV-1 mRNA genome [6]. However, when m^{5}C modification was perturbed due to writer knockout or mutagenesis to prevent modification, alternative splicing at one specific site, namely the D1/A2 splice junction, was altered. Interestingly, this reduction in splice acceptor usage was found for both early (~1.8 kb) and late (~4 kb) HIV-1 classes of transcripts. For adenoviral RNA, Price et al. investigated the role of m^{6}A modifications on viral splicing [18]. These authors found that depletion of m^{6}A modifications globally, by siRNA mediated knockdown of METTL3 expression, significantly reduced the expression of specifically late adenoviral transcripts. They went on to determine that this phenotype was caused by a reduction in splicing efficiency. These studies imply that the presence of RNA modifications, which are already known to affect splicing on cellular mRNA [64], are being utilised to also alter viral RNA splicing patterns.

6. Immune Evasion by Viral RNA

This idea of RNA modifications preventing innate immune sensors from recognising foreign RNAs is not a new concept [65,66]. Karikó et al., and more recently Durbin et al., published studies detailing mechanisms by which innate immune sensors may be blocked from recognising foreign RNA if nucleosides within the RNA are modified. Karikó et al. focused on Toll-like receptors (TLRs), while Durbin et al. investigated the immuno-activating conformational change of RIG-I. Both studies found that the presence of m^{6}A or ψ diminished the innate immune signalling by TLRs and RIG-I, respectively. However, it should be noted that both these studies used RNAs with high levels of modified nucleosides much greater than would be physiologically relevant for viral RNAs.

However, in what feels like a seamless follow on to this thought-provoking work, one exciting study into the role of RNA modifications in the viral replication cycle was recently published by Ringeard et al. [15] looking into 2'-O-methyl modifications on HIV-1 RNA. This research provided clear evidence that 2'-O-methylation of HIV-1 gRNA by the methyltransferase FTSJ3 prevents recognition of gRNA by the innate immune sensor MDA5 [15]. Through preventing the addition of 2'-O-methyl marks to HIV-1 gRNA by siRNA mediated FTSJ3 knockdown, the authors show that incoming gRNA induces IFN-α and IFN-β expression. Since this work was published four further studies exploring innate immune sensing of viral RNA were published by Lu et al. [37], Chen et al. [35], Kim et al. [34] and

Lu et al. [42] demonstrating, in a similar manner to Ringeard et al., that HMPV, HIV-1, HBV/HCV, VSV, MeV and SeV RNAs are also modified to avoid detection by the host cell. In each of these cases, the viral RNA of each virus is m^6A modified and this is found to prevent recognition by RIG-I, in validation of the phenomenon described previously by Durbin et al. These studies perform experiments to reduce the modification of virally encoded adenosines using methods such as mutation of the viral genome or treatment with 3-deazaadenosine (DAA), an inhibitor of S-Adenosylhomocysteine (SAH) hydrolase and find that this in turn increases the cellular type 1 interferon response to infection.

7. Future Avenues of Research

As more research is performed and published surrounding viral epitranscriptomics there are several ways in which scientists in the field can improve on our current knowledge base. First, we need to expand our interests beyond m^6A and into other common modifications already found on cellular mRNAs. Lead candidates would include m^5C, ψ and Nm residues, of which very little is currently known about in regard to viral RNA [5,6,15,36]. When investigating how prevalent these modifications are and where they are on viral RNA, we should seriously consider applying multiple mapping methods to increase our confidence in every site of modification being identified. Antibody-based mapping is being used readily as it is a fast and relatively straightforward approach, but it is also error prone due to off target binding leading to faulty mapping data. As stated above, antibody-based mapping also can only provide a footprint of where a modification may be located of approximately 20–100 nucleotides. However, if coupled with biochemical methods, such as those described above, we can map modifications on viral RNA to single base resolution and at the same time be inherently more confident that these are indeed sites of modification. With this data in hand, we can more reliably design hypo-modified viruses through silent mutagenesis to better grasp the phenotypes that arise due to the presence of individual modifications on a viral RNA.

Another methodology for the study of viral epitranscriptomics that may be better utilised in the future is the generation of viral stocks where the genomic RNA is entirely unmodified. This would help answer important questions about the role of modifications during the initial stages of infection, prior to transcription. This concept has rarely been utilised so far [15], but this author believes it could be an extremely useful technique for all RNA viruses, especially as the role of RNA modifications in immune evasion is better investigated.

As the research progresses, we also need to do better at identifying the exact writer protein for the viral RNA being investigated (Table 1). Regarding m^6A, the writer complex for viral RNA is almost always METTL3/METTL14/WTAP, but for the likes of m^5C, it is not as straightforward as the writer may be one out of a family of NSUN proteins or even DNMT2. Through the identification of the correct writer protein, the sites of modification can be better validated, by knocking out or knocking down expression of the writer and then remapping the modification on viral RNA. If the correct writer is no longer present, it would be expected that modification would be ablated. The identification of the primary writer protein will also allow researchers to better investigate the relationship between these proteins and the viral polymerases. A number of RNA modifications have been found to be added co-transcriptionally to cellular mRNAs. Therefore, it is a valid hypothesis that the polymerases of viruses that exploit RNA modifications would associate with writer machinery. This is an area of study that is significantly lacking but may reveal surprising discoveries in the coming years that could impact on our knowledge of both viral and cellular RNA modification utilisation. If these potential interactions are proven to be essential to viral replication, this would also be an interesting avenue of research into the development of antivirals.

The field of viral epitranscriptomics is just at the cusp of a research explosion over the next few years as the line that separates molecular virologists and RNA biologists begins to blur. Exciting questions, as discussed above, will be answered in the coming years and we will hopefully establish the key roles for individual RNA modifications in the replication cycles of a wide variety of viruses. Technological advances will only aid in our understanding of these modifications, with enhanced mapping techniques and better quantification of virus modified sites hopefully not far off.

Funding: D.G.C. is supported by a European Research Council Starting Grant (PTFLU—949506).

Institutional Review Board Statement: Not applicable.

Informed Consent Statement: Not applicable.

Data Availability Statement: Data sharing not applicable.

Conflicts of Interest: The authors declare no conflict of interest.

References

1. Li, X.; Xiong, X.; Yi, C. Epitranscriptome Sequencing Technologies: Decoding RNA Modifications. *Nat. Methods* **2016**, *14*, 23–31. [CrossRef] [PubMed]
2. Zhang, L.-S.; Liu, C.; Ma, H.; Dai, Q.; Sun, H.-L.; Luo, G.; Zhang, Z.; Zhang, L.; Hu, L.; Dong, X.; et al. Transcriptome-Wide Mapping of Internal N7-Methylguanosine Methylome in Mammalian MRNA. *Mol. Cell* **2019**, 1–13. [CrossRef]
3. Dominissini, D.; Eyal, E.; Hershkovitz, V.; Salmon-Divon, M.; Clark, W.C.; Dai, Q.; Ben-Haim, M.S.; Solomon, O.; Amariglio, N.; Rechavi, G.; et al. The Dynamic N1-Methyladenosine Methylome in Eukaryotic Messenger RNA. *Nature* **2016**, *530*, 441–446. [CrossRef]
4. Suzuki, T.; Ito, S.; Horikawa, S.; Suzuki, T.; Kawauchi, H.; Tanaka, Y.; Suzuki, T. Human NAT10 Is an ATP-Dependent Rna Acetyltransferase Responsible for N4-Acetylcytidine Formation in 18 S Ribosomal RNA (RRNA). *J. Biol. Chem.* **2014**, *289*, 35724–35730. [CrossRef]
5. Courtney, D.G.; Chalem, A.; Bogerd, H.P.; Law, B.A.; Kennedy, E.M.; Holley, C.L.; Cullen, B.R. Extensive Epitranscriptomic Methylation of A and C Residues on Murine Leukemia Virus Transcripts Enhances Viral Gene Expression. *mBio* **2019**, *10*, e01209-19. [CrossRef]
6. Courtney, D.G.; Tsai, K.; Bogerd, H.P.; Kennedy, E.M.; Law, B.A.; Emery, A.; Swanstrom, R.; Holley, C.L.; Cullen, B.R. Epitranscriptomic Addition of M5C to HIV-1 Transcripts Regulates Viral Gene Expression. *Cell Host Microbe* **2019**, *26*, 217–227. [CrossRef]
7. McIntyre, W.; Netzband, R.; Bonenfant, G.; Biegel, J.M.; Miller, C.; Fuchs, G.; Henderson, E.; Arra, M.; Canki, M.; Fabris, D.; et al. Positive-Sense RNA Viruses Reveal the Complexity and Dynamics of the Cellular and Viral Epitranscriptomes during Infection. *Nucleic Acids Res.* **2018**, *46*, 5776–5791. [CrossRef]
8. Imam, H.; Khan, M.; Gokhale, N.S.; McIntyre, A.B.R.; Kim, G.-W.; Jang, J.Y.; Kim, S.-J.; Mason, C.E.; Horner, S.M.; Siddiqui, A. N6-Methyladenosine Modification of Hepatitis B Virus RNA Differentially Regulates the Viral Life Cycle. *Proc. Natl. Acad. Sci. USA* **2018**, *115*, 201808319. [CrossRef] [PubMed]
9. Lichinchi, G.; Gao, S.; Saletore, Y.; Gonzalez, G.M.; Bansal, V.; Wang, Y.; Mason, C.E.; Rana, T.M. Dynamics of the Human and Viral M6A RNA Methylomes during HIV-1 Infection of T Cells. *Nat. Microbiol.* **2016**, *1*, 16011. [CrossRef] [PubMed]
10. Kennedy, E.M.; Bogerd, H.P.; Kornepati, A.V.R.; Kang, D.; Ghoshal, D.; Marshall, J.B.; Poling, B.C.; Tsai, K.; Gokhale, N.S.; Horner, S.M.; et al. Posttranscriptional M6A Editing of HIV-1 MRNAs Enhances Viral Gene Expression. *Cell Host Microbe* **2016**, *19*, 675–685. [CrossRef]
11. Tsai, K.; Jaguva Vasudevan, A.A.; Martinez Campos, C.; Emery, A.; Swanstrom, R.; Cullen, B.R. Acetylation of Cytidine Residues Boosts HIV-1 Gene Expression by Increasing Viral RNA Stability. *Cell Host Microbe* **2020**. [CrossRef] [PubMed]
12. Courtney, D.G.; Kennedy, E.M.; Dumm, R.E.; Bogerd, H.P.; Tsai, K.; Heaton, N.S.; Cullen, B.R. Epitranscriptomic Enhancement of Influenza A Virus Gene Expression and Replication. *Cell Host Microbe* **2017**, *22*, 377–386. [CrossRef]
13. Tsai, K.; Courtney, D.G.; Cullen, B.R. Addition of M6A to SV40 Late MRNAs Enhances Viral Structural Gene Expression and Replication. *PLoS Pathog.* **2018**, *14*, e1006919. [CrossRef] [PubMed]
14. Tirumuru, N.; Simen Zhao, B.; Lu, W.; Lu, Z.; He, C.; Wu, L.; Zhao, B.S.; Lu, W.; Lu, Z.; He, C.; et al. N6-Methyladenosine of HIV-1 RNA Regulates Viral Infection and HIV-1 Gag Protein Expression. *eLife* **2016**, *5*, e15528. [CrossRef]
15. Ringeard, M.; Marchand, V.; Decroly, E.; Motorin, Y.; Bennasser, Y. FTSJ3 Is an RNA 2′-O-Methyltransferase Recruited by HIV to Avoid Innate Immune Sensing. *Nature* **2019**, *565*, 500–504. [CrossRef] [PubMed]
16. Gokhale, N.S.; McIntyre, A.B.R.; McFadden, M.J.; Roder, A.E.; Kennedy, E.M.; Gandara, J.A.; Hopcraft, S.E.; Quicke, K.M.; Vazquez, C.; Willer, J.; et al. N6-Methyladenosine in Flaviviridae Viral RNA Genomes Regulates Infection. *Cell Host Microbe* **2016**, *20*, 654–665. [CrossRef]
17. Lichinchi, G.; Zhao, B.S.; Wu, Y.; Lu, Z.; Qin, Y.; He, C.; Rana, T.M. Dynamics of Human and Viral RNA Methylation during Zika Virus Infection. *Cell Host Microbe* **2016**, *20*, 666–673. [CrossRef] [PubMed]
18. Price, A.M.; Hayer, K.E.; McIntyre, A.B.R.; Gokhale, N.S.; Abebe, J.S.; della Fera, A.N.; Mason, C.E.; Horner, S.M.; Wilson, A.C.; Depledge, D.P.; et al. Direct RNA Sequencing Reveals M6A Modifications on Adenovirus RNA Are Necessary for Efficient Splicing. *Nat. Commun.* **2020**, *11*, 1–17. [CrossRef]

19. Roundtree, I.A.; Luo, G.Z.; Zhang, Z.; Wang, X.; Zhou, T.; Cui, Y.; Sha, J.; Huang, X.; Guerrero, L.; Xie, P.; et al. YTHDC1 Mediates Nuclear Export of N6-Methyladenosine Methylated MRNAs. *eLife* **2017**. [CrossRef] [PubMed]

20. Wojtas, M.N.; Pandey, R.R.; Mendel, M.; Homolka, D.; Sachidanandam, R.; Pillai, R.S. Regulation of M6A Transcripts by the 3′→5′ RNA Helicase YTHDC2 Is Essential for a Successful Meiotic Program in the Mammalian Germline. *Mol. Cell* **2017**. [CrossRef]

21. Amort, T.; Rieder, D.; Wille, A.; Khokhlova-Cubberley, D.; Riml, C.; Trixl, L.; Jia, X.Y.; Micura, R.; Lusser, A. Distinct 5-Methylcytosine Profiles in Poly(A) RNA from Mouse Embryonic Stem Cells and Brain. *Genome Biol.* **2017**. [CrossRef] [PubMed]

22. Choe, J.; Lin, S.; Zhang, W.; Liu, Q.; Wang, L.; Ramirez-Moya, J.; Du, P.; Kim, W.; Tang, S.; Sliz, P.; et al. MRNA Circularization by METTL3–EIF3h Enhances Translation and Promotes Oncogenesis. *Nature* **2018**, *561*, 556–560. [CrossRef] [PubMed]

23. Arango, D.; Sturgill, D.; Alhusaini, N.; Dillman, A.A.; Sweet, T.J.; Hanson, G.; Hosogane, M.; Sinclair, W.R.; Nanan, K.K.; Mandler, M.D.; et al. Acetylation of Cytidine in MRNA Promotes Translation Efficiency. *Cell* **2018**, 1–15. [CrossRef]

24. Coots, R.A.; Liu, X.M.; Mao, Y.; Dong, L.; Zhou, J.; Wan, J.; Zhang, X.; Qian, S.B. M6A Facilitates EIF4F-Independent MRNA Translation. *Mol. Cell* **2017**. [CrossRef]

25. Meyer, K.D.; Patil, D.P.; Zhou, J.; Zinoviev, A.; Skabkin, M.A.; Elemento, O.; Pestova, T.V.; Qian, S.B.; Jaffrey, S.R. 5′ UTR M6A Promotes Cap-Independent Translation. *Cell* **2015**, *163*, 999–1010. [CrossRef]

26. Wang, X.; Zhao, B.S.; Roundtree, I.A.; Lu, Z.; Han, D.; Ma, H.; Weng, X.; Chen, K.; Shi, H.; He, C. N6-Methyladenosine Modulates Messenger RNA Translation Efficiency. *Cell* **2015**. [CrossRef]

27. Lin, S.; Choe, J.; Du, P.; Triboulet, R.; Gregory, R.I. The M6A Methyltransferase METTL3 Promotes Translation in Human Cancer Cells. *Mol. Cell* **2016**, *62*, 335–345. [CrossRef]

28. Yang, X.; Yang, Y.; Sun, B.F.; Chen, Y.S.; Xu, J.W.; Lai, W.Y.; Li, A.; Wang, X.; Bhattarai, D.P.; Xiao, W.; et al. 5-Methylcytosine Promotes MRNA Export-NSUN2 as the Methyltransferase and ALYREF as an m 5 C Reader. *Cell Res.* **2017**, *27*, 606–625. [CrossRef]

29. Schwartz, S.; Mumbach, M.R.; Jovanovic, M.; Wang, T.; Maciag, K.; Bushkin, G.G.; Mertins, P.; Ter-Ovanesyan, D.; Habib, N.; Cacchiarelli, D.; et al. Perturbation of M6A Writers Reveals Two Distinct Classes of MRNA Methylation at Internal and 5′ Sites. *Cell Rep.* **2014**. [CrossRef]

30. Wang, X.; Lu, Z.; Gomez, A.; Hon, G.C.; Yue, Y.; Han, D.; Fu, Y.; Parisien, M.; Dai, Q.; Jia, G.; et al. N 6-Methyladenosine-Dependent Regulation of Messenger RNA Stability. *Nature* **2014**. [CrossRef] [PubMed]

31. Woo, H.-H.; Chambers, S.K. Human ALKBH3-Induced M1A Demethylation Increases the CSF-1 MRNA Stability in Breast and Ovarian Cancer Cells. *Biochim. Et Biophys. Acta (Bba) - Gene Regul. Mech.* **2018**, *1862*, 35–46. [CrossRef]

32. Yao, M.; Dong, Y.; Wang, Y.; Liu, H.; Ma, H.; Zhang, H.; Zhang, L.; Cheng, L.; Lv, X.; Xu, Z.; et al. N6-Methyladenosine Modifications Enhance Enterovirus 71 ORF Translation through METTL3 Cytoplasmic Distribution. *Biochem. Biophys. Res. Commun.* **2020**, *527*, 297–304. [CrossRef]

33. Zheng, X.; Wang, J.; Zhang, X.; Fu, Y.; Peng, Q.; Lu, J.; Wei, L.; Li, Z.; Liu, C.; Wu, Y.; et al. RNA M6 A Methylation Regulates Virus-Host Interaction and EBNA2 Expression during Epstein-Barr Virus Infection. *Immun. Inflamm. Dis.* **2021**. [CrossRef]

34. Kim, G.W.; Imam, H.; Khan, M.; Siddiqui, A. N6-Methyladenosine Modification of Hepatitis B and C Viral RNAs Attenuates Host Innate Immunity via RIG-I Signaling. *J. Biol. Chem.* **2020**, *295*, 13123–13133. [CrossRef]

35. Chen, S.; Kumar, S.; Espada, C.E.; Tirumuru, N.; Cahill, M.P.; Hu, L.; He, C.; Wu, L. N6-Methyladenosine Modification of HIV-1 RNA Suppresses Type-I Interferon Induction in Differentiated Monocytic Cells and Primary Macrophages. *PLoS Pathog.* **2021**, *17*, e1009421. [CrossRef] [PubMed]

36. Kong, W.; Biswas, A.; Zhou, D.; Fiches, G.; Fujinaga, K.; Santoso, N.; Zhu, J. Nucleolar Protein NOP2/NSUN1 Suppresses HIV-1 Transcription and Promotes Viral Latency by Competing with Tat for TAR Binding and Methylation. *PLoS Pathog.* **2020**, *16*, e1008430. [CrossRef] [PubMed]

37. Lu, M.; Zhang, Z.; Xue, M.; Zhao, B.S.; Harder, O.; Li, A.; Liang, X.; Gao, T.Z.; Xu, Y.; Zhou, J.; et al. N6-Methyladenosine Modification Enables Viral RNA to Escape Recognition by RNA Sensor RIG-I. *Nat. Microbiol.* **2020**, 1–15. [CrossRef] [PubMed]

38. Narayan, P.; Ayers, D.F.; Rottman, F.M.; Maroney, P.A.; Nilsen, T.W. Unequal Distribution of N6-Methyladenosine in Influenza Virus MRNAs. *Mol. Cell. Biol.* **1987**, *7*, 1572–1575. [CrossRef]

39. Krug, R.M.; Morgan, M.A.; Shatkin, A.J. Influenza Viral MRNA Contains Internal N6-Methyladenosine and 5′-Terminal 7-Methylguanosine in Cap Structures. *J. Virol.* **1976**, *20*, 45–53. [CrossRef]

40. Hesser, C.R.; Karijolich, J.; Dominissini, D.; He, C.; Glaunsinger, B.A. N6-Methyladenosine Modification and the YTHDF2 Reader Protein Play Cell Type Specific Roles in Lytic Viral Gene Expression during Kaposi's Sarcoma-Associated Herpesvirus Infection. *PLoS Pathog.* **2018**. [CrossRef]

41. Tan, B.; Liu, H.; Zhang, S.; da Silva, S.R.; Zhang, L.; Meng, J.; Cui, X.; Yuan, H.; Sorel, O.; Zhang, S.W.; et al. Viral and Cellular N6-Methyladenosine and N6,2′-O-Dimethyladenosine Epitranscriptomes in the KSHV Life Cycle. *Nat. Microbiol.* **2017**, *3*, 108–120. [CrossRef] [PubMed]

42. Lu, M.; Xue, M.; Wang, H.-T.; Kairis, E.L.; Ahmad, S.; Wei, J.; Zhang, Z.; Liu, Q.; Zhang, Y.; Gao, Y.; et al. Nonsegmented Negative-Sense RNA Viruses Utilize N6-Methyladenosine (m 6 A) as a Common Strategy To Evade Host Innate Immunity. *J. Virol.* **2021**, *95*. [CrossRef]

43. Eckwahl, M.; Xu, R.; Michalkiewicz, J.; Zhang, W.; Patel, P.; Cai, Z.; Pan, T. 5-Methylcytosine RNA Modifications Promote Retrovirus Replication in an ALYREF Reader Protein-Dependent Manner. *J. Virol.* **2020**. [CrossRef]

44. Xue, M.; Zhao, B.S.; Zhang, Z.; Lu, M.; Harder, O.; Chen, P.; Lu, Z.; Li, A.; Ma, Y.; Xu, Y.; et al. Viral N 6-Methyladenosine Upregulates Replication and Pathogenesis of Human Respiratory Syncytial Virus. *Nat. Commun.* **2019**, *10*, 1–18. [CrossRef] [PubMed]

45. Liu, J.; Xu, Y.-P.; Li, K.; Ye, Q.; Zhou, H.-Y.; Sun, H.; Li, X.; Yu, L.; Deng, Y.-Q.; Li, R.-T.; et al. The M6A Methylome of SARS-CoV-2 in Host Cells. *Cell Res.* **2021**. [CrossRef]

46. Kim, D.; Lee, J.Y.; Yang, J.S.; Kim, J.W.; Kim, V.N.; Chang, H. The Architecture of SARS-CoV-2 Transcriptome. *Cell* **2020**, *181*, 914–921.e10. [CrossRef] [PubMed]

47. Dominissini, D.; Moshitch-Moshkovitz, S.; Schwartz, S.; Salmon-Divon, M.; Ungar, L.; Osenberg, S.; Cesarkas, K.; Jacob-Hirsch, J.; Amariglio, N.; Kupiec, M.; et al. Topology of the Human and Mouse M6A RNA Methylomes Revealed by M6A-Seq. *Nature* **2012**, *485*, 201–206. [CrossRef]

48. Edelheit, S.; Schwartz, S.; Mumbach, M.R.; Wurtzel, O.; Sorek, R. Transcriptome-Wide Mapping of 5-Methylcytidine RNA Modifications in Bacteria, Archaea, and Yeast Reveals M5C within Archaeal MRNAs. *PLoS Genet.* **2013**. [CrossRef]

49. Chen, K.; Lu, Z.; Wang, X.; Fu, Y.; Luo, G.Z.; Liu, N.; Han, D.; Dominissini, D.; Dai, Q.; Pan, T.; et al. High-Resolution N6-Methyladenosine (M6A) Map Using Photo-Crosslinking-Assisted M6A Sequencing. *Angew. Chem. Int. Ed.* **2015**, *54*, 1587–1590. [CrossRef] [PubMed]

50. Reid, R.; Greene, P.J.; Santi, D.V. Exposition of a Family of RNA M5C Methyltransferases from Searching Genomic and Proteomic Sequences. *Nucleic Acids Res.* **1999**, *27*, 3138–3145. [CrossRef] [PubMed]

51. Bohnsack, K.E.; Höbartner, C.; Bohnsack, M.T. Eukaryotic 5-Methylcytosine (M 5 C) RNA Methyltransferases: Mechanisms, Cellular Functions, and Links to Disease. *Genes* **2019**, *10*, 102. [CrossRef]

52. King, M.Y.; Redman, K.L. RNA Methyltransferases Utilize Two Cysteine Residues in the Formation of 5-Methylcytosine. *Biochemistry* **2002**, *41*, 11218–11225. [CrossRef]

53. Hussain, S.; Sajini, A.A.; Blanco, S.; Dietmann, S.; Lombard, P.; Sugimoto, Y.; Paramor, M.; Gleeson, J.G.; Odom, D.T.; Ule, J.; et al. NSun2-Mediated Cytosine-5 Methylation of Vault Noncoding RNA Determines Its Processing into Regulatory Small RNAs. *Cell Rep.* **2013**, *4*, 255–261. [CrossRef]

54. Linder, B.; Grozhik, A.V.; Olarerin-George, A.O.; Meydan, C.; Mason, C.E.; Jaffrey, S.R. Single-Nucleotide-Resolution Mapping of M6A and M6Am throughout the Transcriptome. *Nat. Methods* **2015**, *12*, 767–772. [CrossRef] [PubMed]

55. Schaefer, M.; Pollex, T.; Hanna, K.; Lyko, F. RNA Cytosine Methylation Analysis by Bisulfite Sequencing. *Nucleic Acids Res.* **2009**, *37*. [CrossRef] [PubMed]

56. Huang, T.; Chen, W.; Liu, J.; Gu, N.; Zhang, R. Genome-Wide Identification of MRNA 5-Methylcytosine in Mammals. *Nat. Struct. Mol. Biol.* **2019**, *26*. [CrossRef] [PubMed]

57. Carlile, T.M.; Rojas-Duran, M.F.; Zinshteyn, B.; Shin, H.; Bartoli, K.M.; Gilbert, W.V. Pseudouridine Profiling Reveals Regulated MRNA Pseudouridylation in Yeast and Human Cells. *Nature* **2014**, *515*, 143–146. [CrossRef] [PubMed]

58. Schwartz, S.; Bernstein, D.A.; Mumbach, M.R.; Jovanovic, M.; Herbst, R.H.; León-Ricardo, B.X.; Engreitz, J.M.; Guttman, M.; Satija, R.; Lander, E.S.; et al. Transcriptome-Wide Mapping Reveals Widespread Dynamic-Regulated Pseudouridylation of NcRNA and MRNA. *Cell* **2014**, *159*, 148–162. [CrossRef]

59. Marchand, V.; Blanloeil-Oillo, F.; Helm, M.; Motorin, Y. Illumina-Based RiboMethSeq Approach for Mapping of 2′-O-Me Residues in RNA. *Nucleic Acids Res.* **2016**, *44*, e135. [CrossRef]

60. Hsu, P.J.; Fei, Q.; Dai, Q.; Shi, H.; Dominissini, D.; Ma, L.; He, C. Single Base Resolution Mapping of 2′-O-Methylation Sites in Human MRNA and in 3′ Terminal Ends of Small RNAs. *Methods* **2018**. [CrossRef]

61. Thalalla Gamage, S.; Sas-Chen, A.; Schwartz, S.; Meier, J.L. Quantitative Nucleotide Resolution Profiling of RNA Cytidine Acetylation by Ac4C-Seq. *Nat. Protoc.* **2021**, *16*, 2286–2307. [CrossRef]

62. Sas-Chen, A.; Thomas, J.M.; Matzov, D.; Taoka, M.; Nance, K.D.; Nir, R.; Bryson, K.M.; Shachar, R.; Liman, G.L.S.; Burkhart, B.W.; et al. Dynamic RNA Acetylation Revealed by Quantitative Cross-Evolutionary Mapping. *Nature* **2020**, *583*, 638–643. [CrossRef] [PubMed]

63. Zhou, H.; Rauch, S.; Dai, Q.; Cui, X.; Zhang, Z.; Nachtergaele, S.; Sepich, C.; He, C.; Dickinson, B.C. Evolution of a Reverse Transcriptase to Map N 1-Methyladenosine in Human Messenger RNA. *Nat. Methods* **2019**, *16*, 1281–1288. [CrossRef]

64. Shi, H.; Wei, J.; He, C. Where, When, and How: Context-Dependent Functions of RNA Methylation Writers, Readers, and Erasers. *Mol. Cell* **2019**, *74*, 640–650. [CrossRef] [PubMed]

65. Durbin, A.F.; Wang, C.; Marcotrigiano, J.; Gehrke, L. RNAs Containing Modified Nucleotides Fail to Trigger RIG-I Conformational Changes for Innate Immune Signaling. *mBio* **2016**, *7*, 1–11. [CrossRef] [PubMed]

66. Karikó, K.; Buckstein, M.; Ni, H.; Weissman, D. Suppression of RNA Recognition by Toll-like Receptors: The Impact of Nucleoside Modification and the Evolutionary Origin of RNA. *Immunity* **2005**, *23*, 165–175. [CrossRef] [PubMed]

Article

Interferon-Induced HERC5 Inhibits Ebola Virus Particle Production and Is Antagonized by Ebola Glycoprotein

Ermela Paparisto [1], Nina R. Hunt [1], Daniel S. Labach [1], Macon D. Coleman [1], Eric J. Di Gravio [1], Mackenzie J. Dodge [1], Nicole J. Friesen [1], Marceline Côté [2], Andreas Müller [3], Thomas Hoenen [3] and Stephen D. Barr [1,*]

[1] Department of Microbiology and Immunology, Schulich School of Medicine and Dentistry, Western University, Dental Sciences Building Room 3007, London, ON N6A 5C1, Canada; epaparis@uwo.ca (E.P.); nhunt8@uwo.ca (N.R.H.); dlabach@uwo.ca (D.S.L.); mcolem5@uwo.ca (M.D.C.); edigravi@uwo.ca (E.J.D.G.); mdodge@uwo.ca (M.J.D.); nfriese5@uwo.ca (N.J.F.)

[2] Department of Biochemistry, Microbiology, and Immunology, Ottawa Institute of Systems Biology, University of Ottawa, Roger-Guindon Hall Room 4214, Ottawa, ON K1H 8M5 , Canada; Marceline.Cote@uottawa.ca

[3] Friedrich-Loeffler-Institut, Institute of Molecular Virology and Cell Biology, Südufer 10, 17493 Greifswald—Insel Riems, Germany; andreas.mueller@seracell.de (A.M.); Thomas.Hoenen@fli.de (T.H.)

* Correspondence: stephen.barr@uwo.ca

Abstract: Survival following Ebola virus (EBOV) infection correlates with the ability to mount an early and robust interferon (IFN) response. The host IFN-induced proteins that contribute to controlling EBOV replication are not fully known. Among the top genes with the strongest early increases in expression after infection in vivo is IFN-induced HERC5. Using a transcription- and replication-competent VLP system, we showed that HERC5 inhibits EBOV virus-like particle (VLP) replication by depleting EBOV mRNAs. The HERC5 RCC1-like domain was necessary and sufficient for this inhibition and did not require zinc finger antiviral protein (ZAP). Moreover, we showed that EBOV (Zaire) glycoprotein (GP) but not Marburg virus GP antagonized HERC5 early during infection. Our data identify a novel 'protagonist–antagonistic' relationship between HERC5 and GP in the early stages of EBOV infection that could be exploited for the development of novel antiviral therapeutics.

Keywords: Ebola virus; Marburg virus; HERC5; antiviral; interferon

Citation: Paparisto, E.; Hunt, N.R.; Labach, D.S.; Coleman, M.D.; Di Gravio, E.J.; Dodge, M.J.; Friesen, N.J.; Côté, M.; Müller, A.; Hoenen, T.; et al. Interferon-Induced HERC5 Inhibits Ebola Virus Particle Production and Is Antagonized by Ebola Glycoprotein. *Cells* **2021**, *10*, 2399. https://doi.org/10.3390/cells10092399

Academic Editors: Reinhild Prange and Alexander E. Kalyuzhny

Received: 29 April 2021
Accepted: 31 August 2021
Published: 13 September 2021

Publisher's Note: MDPI stays neutral with regard to jurisdictional claims in published maps and institutional affiliations.

1. Introduction

Ebola virus (EBOV) is a member of the *Filoviridae* family of single-stranded negative-sense RNA viruses with a filamentous morphology. EBOV infection results in severe hemorrhagic fever and can lead to death 6-16 days after the onset of symptoms in up to 90% of cases, making EBOV one of the most virulent pathogens to infect humans [1]. Studies involving primate models, and human studies carried out during the 2013–2016 outbreak, showed that EBOV exposure results in an early and robust immune response, largely characterized by the up-regulation of IFN-stimulated genes [2–12]. A contributing factor to the pathophysiology of EBOV infection is the ability of the virus to evade the host IFN response [7,13–16]. Using in vitro models of infection, it was shown that EBOV is able to evade the innate immune response through various IFN antagonisms, notably involving VP24 and VP35 proteins [17–19]. The key mediators of this early cellular IFN response to EBOV and how EBOV withstands this early response are not fully characterized.

Restriction factors are key intrinsic mediators of the early IFN response and potently inhibit different steps in the life cycle of evolutionarily diverse viruses in the absence of viral antagonists [20]. Bone marrow stromal cell antigen 2 (BST-2)/tetherin is one such factor that potently inhibits the release of EBOV from cells by tethering virions to the surface of cells [21,22]. This inhibition is counteracted by EBOV GP [23–26]. IFN-inducible trans-membrane proteins 1–3 (IFITM1–3) comprise another family of factors that restrict the cellular entry of EBOV, although an EBOV antagonist to these proteins has yet to be identified [27,28]. HECT and RCC1-like containing domain 5 (HERC5) are some of the genes with the strongest early increases in expression in multiple tissues after EBOV infection [3,5,6,29]. HERC5 is an evolutionarily ancient restriction factor that inhibits the replication of diverse viruses [30–36]. By virtue of its C-terminal HECT domain, HERC5 is the main cellular E3 ligase for conjugating ISG15 to substrates and localizes to polyribosomes to modify newly translated viral proteins, thereby disrupting key aspects of viral particle production [31,35,37,38]. E3 ligase-independent antiviral activity has also been demonstrated towards HIV-1, where it inhibits the nuclear export of incompletely-spliced viral RNAs by a mechanism requiring its N-terminal RCC1-like domain (RLD) [30].

Here, we examined the antiviral activity of HERC5 towards EBOV VLP production and replication. We identified a novel E3 ligase-independent mechanism by which HERC5 inhibits viral particle production involving the depletion of EBOV mRNAs. In addition, we demonstrated that EBOV GP antagonizes HERC5 activity and rescues EBOV VLP production and replication.

2. Materials and Methods

2.1. Cell Lines

293T and HeLa cells were obtained from American Type Culture Collection. 293T ZC3HAV1 (ZAP) knockout cells were obtained from Dr. Takaoka (Hokkaido University, Japan) via Dr. Li (University of California, Los Angeles, CA, USA) and Dr. MacDonald (The Rockefeller University, New York, NY, USA). Cells were maintained in standard growth medium (Dulbecco's Modified Eagle's Medium (DMEM)), supplemented with 10% heat-inactivated Fetal Bovine Serum (FBS), 100 U/mL Penicillin and 100 µg/mL Streptomycin) at 37 °C with 5% CO_2.

2.2. Plasmids, Transfections, Antibodies and Quantitative Western Blotting

Expression plasmids carrying FLAG-tagged HERC5, HERC5-ΔRLD, HERC5-ΔHECT and HERC5-C994A, and HERC4 have been described previously [36]. The plasmid carrying FLAG-tagged RLD only (pFLAG-RLDonly) was generated by standard restriction enzyme cloning of the HERC5 RLD (containing a 3′ stop codon) into p3xFLAG-CMV-10 (Sigma). The promoterless empty vector plasmid pGL3, pEGFP-C1 (pEGFP) and pZAP (short isoform) were obtained from Promega, Clontech and Dharmacon, respectively. pLKO.1/scrambled shRNA and pLKO.1/HERC5 shRNA were previously described [30,31]. VP40 and GP were cloned into p3xFLAG-CMV-10 (Sigma) to generate pFLAG-VP40, pFLAG GP and pEGFP-C1 (containing a CMV promoter) (Clontech) to generate pVP40-EGFP using standard restriction enzyme cloning. EBOV expression plasmids: pCAGGS plasmids (containing a CMV enhancer, chicken beta-actin promoter and beta-actin intron sequence) carrying only EBOV (Zaire) VP40, VP30, VP35, L, NP, or GP were obtained from Dr. Kawaoka (University of Wisconsin) [39]. Plasmids for the trVLP assay were provided by Dr. Hoenen (Friedrich-Loeffler-Institut, Germany): Plasmids carrying NP, VP35, VP30, L, Tim-1, T7 and the tetracistronic minigenomes (p4cisvRNA-hrLuc, p4cis-vRNA-EGFP) have been previously described [40,41]. All EBOV gene sequences in the minigenomes and plasmids carrying NP, VP35, VP30, and L originated from the Zaire EBOV isolate *H. sapiens-tc/COD/1976/Yambuku-Mayinga*. The EBOV GP and MARV GP expression plasmids were kind gifts of Dr. Cunningham (Brigham and Women's Hospital) [42,43]. Transfections were performed using Lipofectamine 2000 (Invitrogen) per manufacturer's instructions unless otherwise stated. Co-transfections of HERC5 plasmids with pVP40 were performed at a ratio of 10:1, respectively, unless otherwise noted. VP40 VLPs were purified from cell supernatants by centrifugation over a 20% sucrose cushion at $21,000\times g$ for 2 h. Cell lysates and VP40 VLP pellets were subjected to quantitative Western blot analyses using LI-COR, as previously described [30]. Densitometric analysis was performed using ImageJ 1.53e 64-bit version software. Antibodies: Anti-FLAG was purchased from Sigma, anti-ZAP from AbCam (Cat. #ab154680), anti-VP40 from GeneTex (Cat. #GTX134034), anti-MARV GP from Alpha Diagnostic International (Cat. #MVGP12-A), anti-EBOV GP from Bio-Techne (Cat. #MAB9016), anti-β-actin from Rockland, anti-EGFP from Clontech and anti-GAPDH (clone 6C5) from EMD/Millipore.

2.3. Confocal Immunofluorescence Microscopy

HeLa cells were cultured in 12-well plates on 18 mm coverslips and co-transfected with either pFLAG-HERC5 and pVP40-EGFP (10:1 ratio) or pGL3 and pVP40-EGFP (10:1 ratio). Twenty-four hours after transfection, the coverslips containing the cells were washed twice with PF buffer ($1\times$ PBS + 1% FBS), fixed for 10 min in $1\times$ PBS containing 4% formaldehyde and 2% sucrose, permeabilized in $1\times$ PBS containing 0.1% Triton X 100 (Sigma) and then washed twice more with PF buffer. Coverslips were incubated with primary antibody rabbit anti-FLAG (1:500 dilution) for 1 h, washed $3\times$ with PF buffer and incubated with either secondary antibody anti-rabbit 594 (1:1000) for 1 h. Coverslips were washed $3\times$, incubated in Hoechst 33342 (1:10,000 dilution) (Life Technologies) for 5 min and washed $6\times$ with PF buffer. Coverslips were then mounted on glass slides with 10 μL Vectashield mounting media (Vector Laboratories Inc., Burlingame, CA, USA) and sealed with nail polish. Confocal micrographs were obtained using a Leica TCS SP8 (Leica Microsystems) microscope, and Leica Application Software X was used for image acquisition.

2.4. Transmission Electron Microscopy

Cells were co-transfected with empty vector or pFLAG-HERC5 and pVP40-EGFP at a 10:1 ratio. After 48 h, cells were resuspended in media, fixed in 2.5% glutaraldehyde in 0.1 M sodium cacodylate (pH 7.4) for 2 h, and washed $3\times$ in 0.1 M sodium cacodylate. Cells were pelleted and fixed with 2% osmium tetroxide in sodium cacodylate. After ~1 h in the dark, cells were washed $3\times$ in ddH$_2$O. Water was discarded, and samples were left at 4 °C overnight. Samples were dehydrated by adding 1 mL 20% acetone in ddH$_2$O, mixed and incubated for 10 min at room temperature. Cells were pelleted, acetone removed, and the procedure was repeated with 50%, 70%, 90%, 100%, 100% and 100% acetone. Cells were embedded in resin by adding 1 mL of a 2:1 mix of acetone:resin (Epon) and incubated for ~4 h at room temperature in a rotating tube shaker. Cells were pelleted, acetone:resin mix was discarded and repeated with a 1:1 mix overnight, 1:2 mix overnight, and finally, resin only overnight. Samples were cut in 70 nm slices using a Sorval Ultracut ultramicrotome and placed onto 400 mesh nickel grids (Embra). Grids were placed on drops of 2% uranyl acetate in ddH$_2$O to stain for 20 min in the dark and washed 5–6$\times$ in ddH$_2$O for 1 min. Samples were then stained in drops of Sato's lead citrate (5 mM calcined lead citrate, 11 mM lead nitrate, 11 mM lead acetate, 95 mM sodium citrate) for 1 min and washed using ddH$_2$O. Samples were imaged using a Phillips CM10 Transmission Electron Microscope. The AMT Advantage digital imaging system was used for image acquisition.

2.5. Quantitative PCR

The total RNA was extracted using the PureLink RNA mini kit (Ambion, Life Technologies). Using the M-MLV reverse transcriptase and Oligo(dT) primers (Eurofins), 500 ng of RNA was reverse transcribed to cDNA. Prior to qPCR, cDNA samples were diluted 1:5 with water. Each PCR reaction consisted of 10 μL of SYBR Green Master Mix, 1.6 μL of gene-specific primers (0.8 μL of 10 μM forward primer and 0.8 μL of 10 μM reverse primer), 4 μL of diluted cDNA, and water to a total volume of 20 μL. Quantification of endogenous mRNA was run on the QuantStudio5 qPCR machine (Applied Biosystems) under the following cycling conditions: 2 min at 95 °C and 40 cycles of 5 sec at 95 °C, 10 s at 60 °C, and 20 s at 72 °C. The QuantStudio Design and Analysis Desktop Software (version 1.4) was used to determine the C$_T$ for each PCR reaction. Primer pairs were as follows: HERC5- (fwd: 5′ ATG AGC TAA GAC CCT GTT TGG 3′; rev: 5′ CCC AAA TCA GAA ACA TAG GCA AG 3′); ZAP- (fwd: 5′ CGCTTAATGGTAGCTGCAGC 3′; rev: 5′ CCTACAGAACA-GAGGTGGATTCC 3′); GAPDH- (fwd: 5′ CAT GTT CGT CAT GGG TGT GAA CCA 3′; rev: 5′ AGT GAT GGC ATG GAC TGT GGT CAT 3′); EGFP- (fwd: 5′ GACAAC-

CACTACCTGAGCAC 3′; rev: 5′ CAGGACCATGTGATCGCG3′); EBOV VP40- (fwd: 5′GCTTCCTCTAGGTGTCGCTG3′ ; rev 5′GGTTGCCTTGCCGAAATGG3′); EBOV GP- (fwd: 5′GTGAATGGGCTGAAAACTGC3′ ; rev 5′CCGTTCCTGATACTTTGTGC3′); EBOV VP30- (fwd: 5′CCAGACAGCATTCAAGGG3′; rev 5′GCTGGAGGAACTGTTAATGG3′); EBOV VP35- (fwd: 5′CGACTCAAAACGACAGAATGC3′ ; rev 5′GGTTTGGCTTCGTTTGT TGC3′); EBOV NP- (fwd: 5′GCCAACTTATCATACAGGCC3′ ; rev 5′CCAAATACTTGACT GCGCC3′); EBOV L- (fwd: 5′CCTAGTCACTAGAGCTTGCG3′ ; rev 5′GGCTCAACAGGA CAGAATCC3′). To ensure no carry-over of DNA into each total purified RNA sample, 100 ng of RNA was used directly as a template without reverse transcription for qPCR using the primer sets described above.

2.6. trVLP Assay

Expression plasmids carrying tim-1, T7, NP, VP35, VP30, L, and the tetracistronic minigenome (p4cis-vRNA-hrLuc) carrying luciferase, VP40, GP and VP24 have been previously described [40,44]. trVLP assays were performed as previously described, with the following changes [40,41]. Passage zero (p0) cells were seeded in 12-well plates and transfected at 50% confluency using Transit LT-1 (Mirus Bio LLC, Madison, WI, USA) with expression plasmids carrying T7-polymerase (125 ng; all amounts per well), the viral proteins NP (62.5 ng), VP35 (62.5 ng), VP30 (37.5 ng), L (500 ng), a tetracistronic minigenome (125 ng), and Firefly luciferase (100 ng) following the manufacturer's instructions. Twenty-four hours prior to infection of p1, p2, p3 and p4 cells, target cells were pre-transfected with expression plasmids carrying NP (62.5 ng), VP35 (62.5 ng), VP30 (37.5 ng), L (500 ng), Tim-1 (125 ng) and either HERC5 (125 ng) or empty vector (125 ng).

2.7. Cell Viability Assay

293T cells were co-transfected with pFLAG-VP40, GFP-VP40 or GFP alone, as well as increasing concentrations of pFLAG-HERC5 or empty vector control plasmid. Forty-eight hours post-transfection Cell Counting Kit-8 (CCK-8) (GLPBIO) was used to measure cell viability as per the manufacturer's instructions.

2.8. Statistical Analyses

GraphPad Prism v9 was used for all statistical analyses stated in the text. *p* values and statistical tests used are stated in the text where appropriate. *p* values less than 0.05 were deemed significant. Quantification of immunogold labelling for statistical analysis was performed as described [31].

3. Results

3.1. HERC5 Inhibits EBOV trVLP Replication

Previous studies have identified HERC5 as a potent inhibitor of diverse viruses [30–36]. To determine if HERC5 restricts EBOV particle production and replication, we used an EBOV (Zaire) transcription- and replication-competent VLP (trVLP) system. This system utilizes a tetracistronic minigenome ('4cis') carrying a *luciferase* reporter gene together with *VP40*, *VP24*, and *GP* (Figure 1A) [40,45]. The advantage of this system over conventional VLP assays is that the viral proteins VP40, GP and VP24 are encoded by the minigenome and expressed from the EBOV promoter in a more natively regulated fashion [40]. The co-expression of this minigenome with NP, VP35, VP30, and L drive genome replication and transcription, synthesis of the minigenome-encoded proteins, and formation of infectious trVLPs. These trVLPs incorporate minigenomes and are capable of undergoing multiple rounds of replication and infection in target cells that express NP, VP35, VP30, L and Tim-1 (Figure 1B). The replication of these trVLPs was quantified over multiple passages (every three days) by measuring the luciferase reporter activity within cells. As a negative control, the plasmid carrying the Ebola L gene was omitted from the transfections, which abrogated the trVLP formation. Compared to the control cells transfected with an empty vector plasmid, cells expressing HERC5 exhibited a significant reduction in trVLP replication over four passages (Figure 1C). The reduction in luciferase reporter activity also correlated with a reduction in GP and VP40 mRNA levels (Figure 1D).

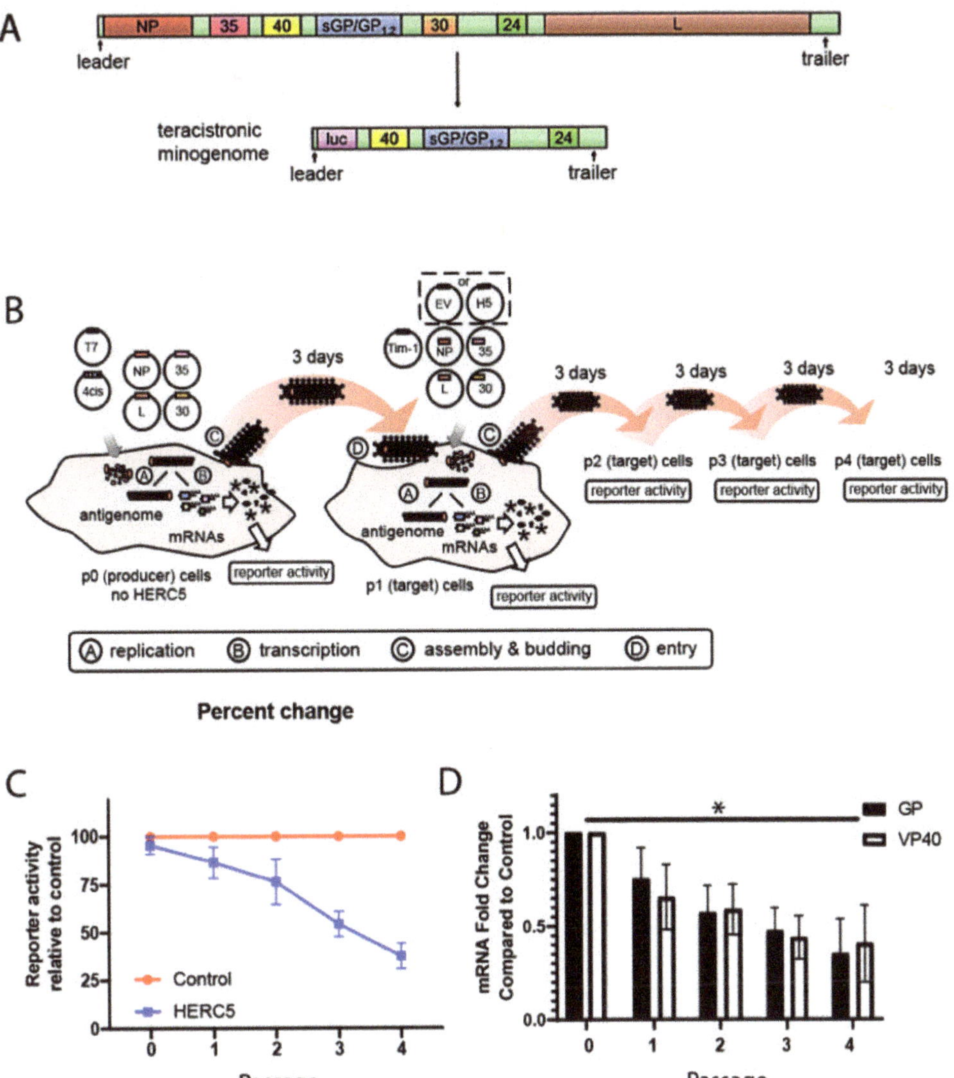

Figure 1. HERC5 inhibits EBOV trVLP replication. (**A**) Schematic depicting EBOV full-length genome and the derived tetracistronic minigenome. (**B**) The trVLP propagation assay. A tetracistronic EBOV minigenome (4cis) is expressed in cells together with the viral ribonucleoprotein complex (RNP) proteins (NP, VP35, VP30 and L). After the initial transcription by a co-expressed T7 polymerase, the minigenome is replicated and transcribed by the RNP proteins. Expression of VP40, GP and VP24 from the minigenome leads to the formation of infectious trVLPs containing minigenomes, which can infect target cells. Multiple infectious cycles can be modeled in cells expressing NP, VP35, L, VP30 and Tim-1 without the need for additional transfections of plasmids carrying VP40, GP and VP24. The figure was adapted from (Watt et al., 2014), copyright © American Society for Microbiology, J. Virol. 88, 2014, 10,511–10,524, doi:10.1128/JVI.01272-14. (**C**) Quantification of trVLP

propagation in the presence and absence of HERC5. The trVLP propagation assay was performed using tetracistronic minigenomes carrying a luciferase reporter, EBOV VP40, VP24 and EBOV GP over four passages (spanning 12 days). All EBOV minigenomes and plasmids carrying the EBOV proteins are based on EBOV *H. sapiens*-tc/COD/1976/Yambuku-Mayinga. Luciferase reporter activity relative to the control (trVLPs propagated in the absence of HERC5) is shown. The data shown represent the average (+/− S.E.M.) of four independent experiments. Linear regression analysis, F = 39.14. DFn = 1, DFd = 36; $p < 0.0001$. (**D**) The mRNA of GP and VP40 was measured using qRT-PCR at each passage. The data shown represent the average (+/− S.E.M.) of the four independent experiments represented in part C. * $p < 0.05$; One-way ANOVA with Dunnet's multiple comparisons test compared to the control.

3.2. HERC5 Inhibits EBOV VP40 Particle Production

Previous studies showed that HERC5 interferes with the function of key viral structural proteins [30,31,35,36]. The EBOV structural protein VP40 is necessary and sufficient for the assembly and budding of virus particles. When expressed in the absence of any other viral protein, VP40 can form VLPs that bud and are released from cells similar to wild-type EBOV [46–48]. To determine if HERC5 targets VP40, we co-transfected 293T cells with a plasmid carrying VP40 and increasing concentrations of plasmids carrying either empty vector control or FLAG-tagged HERC5. VP40 protein levels within cells and in VLPs were measured using quantitative Western blotting. HERC5 transfection did not alter cell viability (Figure S1A). As shown in Figure 2A and Figure S1B,C, HERC5 inhibited the production of VP40 VLPs in a dose-dependent manner when VP40 is tagged with either GFP or with FLAG but had no effect on intracellular GFP levels. As a control, transfection with *HERC4*, a closely related member of the small *HERC* family, did not significantly alter cell viability, VP40 or GFP levels (Figure S1D–F). In contrast, when *HERC5* mRNA levels were reduced using RNA interference, an increase in intracellular VP40 protein levels and an increase in the production of VP40 VLPs were observed compared to the control cells (Figure 2B,C).

We also assessed the impact of HERC5 expression on VLPs using confocal microscopy and transmission electron microscopy (TEM). As expected, cells expressing VP40 with enhanced green fluorescent protein fused at its amino-terminus (VP40-EGFP) exhibited punctate fluorescence at the cell surface (Figure 2D). In contrast, cells co-expressing VP40-EGFP and HERC5 exhibited substantially less punctate fluorescence at the cell surface compared to the control cells. The presence of VP40 protein at the cell surface was also confirmed using TEM and immunogold TEM (Figure 2E,F). In cells expressing VP40-EGFP alone, an accumulation of immunogold particles was observed in budding structures at the cell surface, which was significantly different from a random distribution (Tables S1 and S2). Cells expressing HERC5 exhibited markedly fewer VP40-EGFP-containing structures at the cell surface compared to the control cells. In addition, cells expressing HERC5 exhibited on average eight-fold fewer immunogold particles per cell compared to the control cells (Figure 2G). Notably, the few VP40-EGFP-containing structures that were observed in cells expressing HERC5 were located predominantly in a region under the plasma membrane.

We then asked whether the reduced VP40 protein levels correlated with reduced intracellular VP40 mRNA levels. The quantitative PCR showed that 293T cells co-expressing HERC5 and FLAG-tagged VP40 exhibited reduced intracellular levels of VP40 mRNA (nine-fold) compared to the control cells not expressing HERC5 (Figure 2H). Similar results were obtained when HERC5 was co-expressed with a VP40-EGFP fusion protein (Figure S2). As a control, HERC5 expression had no significant effect on EGFP mRNA levels when EGFP was expressed alone (Figure 2H and Figure S2). To determine if the effect of HERC5 is specific for VP40 mRNA, we assessed the impact of HERC5 expression on the level of other EBOV mRNAs. Cells co-expressing HERC5 and either VP30, VP35, L or NP exhibited a two- to five-fold reduction in mRNA levels compared to the control cells (Figure 2H). Together, these data show that HERC5 inhibits EBOV VP40 particle production by a mechanism involving the depletion of EBOV mRNAs.

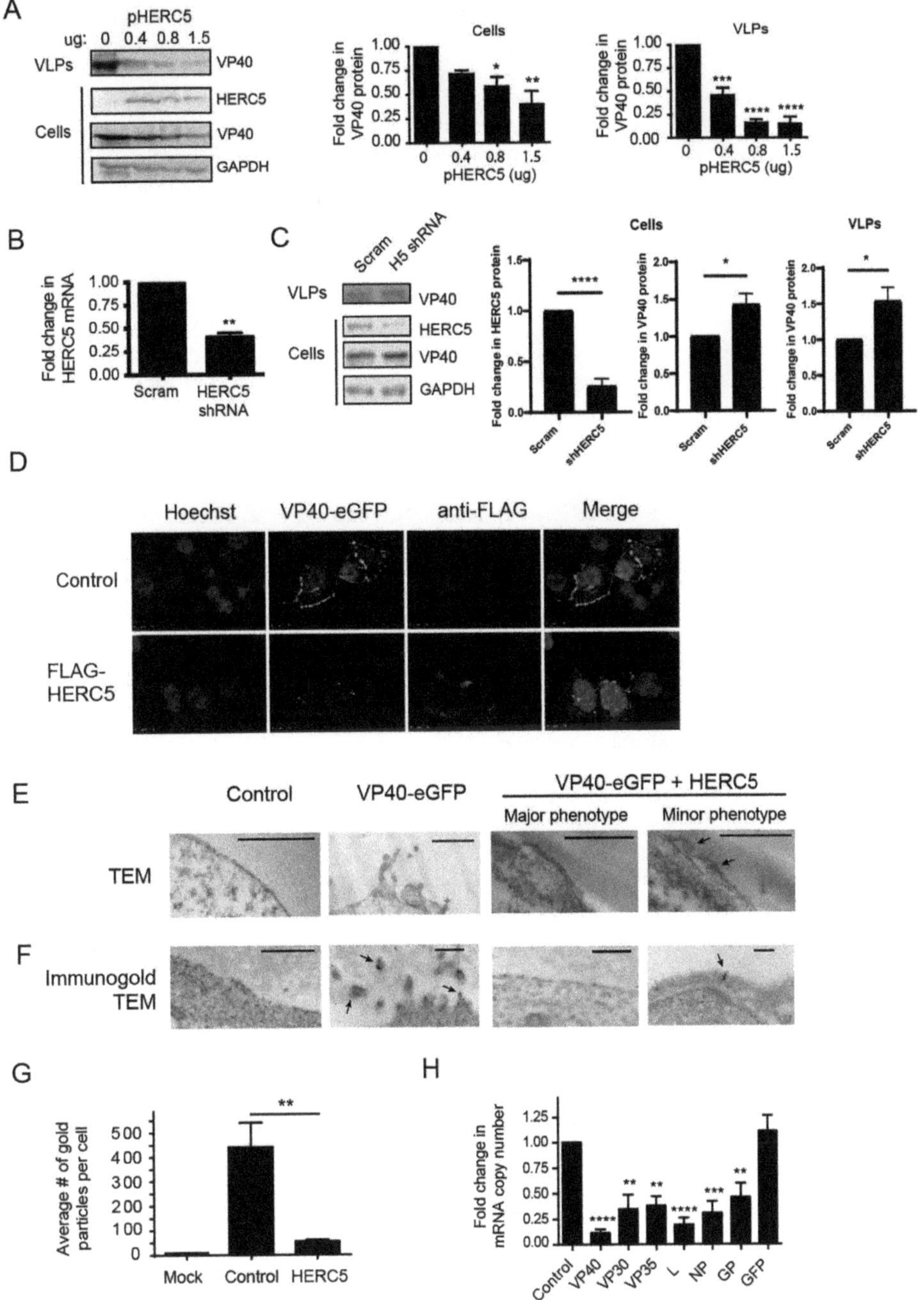

Figure 2. HERC5 inhibits EBOV VP40 particle production. (**A**) 293T cells were co-transfected with plasmids carrying FLAG-tagged VP40 (pFLAG-VP40) and increasing concentrations of FLAG-tagged HERC5 (pFLAG-HERC5). Empty vector

plasmid was transfected in the condition with no HERC5 and used to ensure equal amounts of DNA were transfected in each condition. Forty-eight hours post-transfection, purified VLPs released into the cell supernatant and intracellular protein were subjected to quantitative Western blot analysis using anti-FLAG, anti-VP40 and anti-GAPDH. The average densitometric quantification of VP40 protein bands is shown to the right after normalization to GAPDH levels ($+/-$ S.E.M.). A representative Western blot of four independent experiments is shown. (**B**) 293T cells were co-transfected with pFLAG-VP40 and either scrambled short-hairpin RNA (shRNA) (scram) or HERC5$_{shRNA}$ (shHERC5). Forty-eight hours after transfection, intracellular levels of HERC5 mRNA were quantified via qPCR. Data shown is the average ($+/-$ S.E.M.) of three independent experiments. (**C**) 293T cells were transfected with either scrambled short-hairpin RNA (shRNA) (scram) or HERC5$_{shRNA}$ (shHERC5) for 24 h and then with pFLAG-HERC5 and pFLAG-VP40 for forty-eight hours. Purified VLPs released into the cell supernatant and intracellular protein were subjected to quantitative Western blot analysis using anti-FLAG and anti-GAPDH. The average densitometric quantification of VP40 protein bands is shown to the right after normalization to GAPDH levels ($+/-$ S.E.M.). A representative Western blot of four independent experiments is shown. (**D**) HeLa cells were co-transfected with pVP40-EGFP and either empty vector (control) or pFLAG-HERC5 and visualized using confocal microscopy 48 h post-transfection. (**E**) 293T cells were "mock" transfected (control), transfected with empty vector and pVP40-EGFP, or transfected with pFLAG-HERC5 and pVP40-EGFP and analyzed via transmission electron microscopy (TEM) after 48 h. Virus particles beneath the plasma membrane are indicated with arrows. (**F**) Representative immunogold TEM images of 293T cells transfected as in (**E**) and labelled with 5 ($+/-$ 2) nm anti-GFP immunogold particles. Immunogold-labelled VLPs are indicated with arrows. Scale bars = 500 nm. (**G**) The number of gold particles per positive cell was counted and presented as the average number of particles per cell ($+/-$ S.E.M). (**H**) 293T cells were co-transfected with plasmids carrying FLAG-HERC5 (or empty vector) and either EBOV VP40, VP30, VP35, L, NP, GP or GFP at a ratio of 10:1 (HERC5: EBOV plasmids). Forty-eight hours post-transfection viral mRNA was measured using qPCR after normalization to GAPDH mRNA levels. Data shown are representative of three independent experiments ($+/-$ S.E.M.). **** $p < 0.0001$, *** $p < 0.001$, ** $p < 0.01$, * $p < 0.05$; One-way ANOVA with Dunnet's multiple comparisons test compared to the control (**A**,**G**); Student's paired t-test (**B**,**C**,**H**).

3.3. HERC5 RLD Is Necessary and Sufficient for Inhibition of VP40 Particle Production

To determine if the RLD or HECT domains of HERC5 are required for inhibition, we tested the ability of several HERC5 mutants to inhibit VP40 particle production. 293T cells were co-transfected with plasmids carrying VP40 and either empty vector (control), wild type HERC5 or HERC5 mutants lacking the RCC1-like domain (HERC5-ΔRLD), spacer region (HERC5-Δspacer) or HECT domain (HERC5-ΔHECT). We also tested the HERC5 RLD alone (HERC5-RLDonly) or HERC5 containing a cysteine to an alanine point mutation of residue 994 (HERC5-C994A), which specifically inactivates its E3 ligase activity (Figure 3A). Each of the FLAG-tagged mutant proteins was expressed at similar levels in 293T cells (Figure 3B).

As shown in Figure 3C, cells expressing wild type HERC5, HERC5-ΔHECT or HERC5-C994A reduced VP40 protein levels, which also correlated with reduced VP40 VLP production. In contrast, cells expressing HERC5-ΔRLD, and to a lesser extent HERC5-Δspacer, exhibited a diminished capacity to reduce VP40 protein levels and VP40 VLP production. Notably, expression of the HERC5 RLD alone (HERC5-RLDonly) reduced VP40 protein levels and VP40 VLP production similar to wild-type HERC5 (Figure 3D). We also examined the ability of the different HERC5 mutants to reduce VP40 mRNA levels. All HERC5 mutants except for HERC5-ΔRLD significantly reduced VP40 mRNA levels (Figure 3E). Taken together, these data show that the HERC5 RLD is necessary and sufficient to reduce VP40 mRNA levels and VP40 particle production.

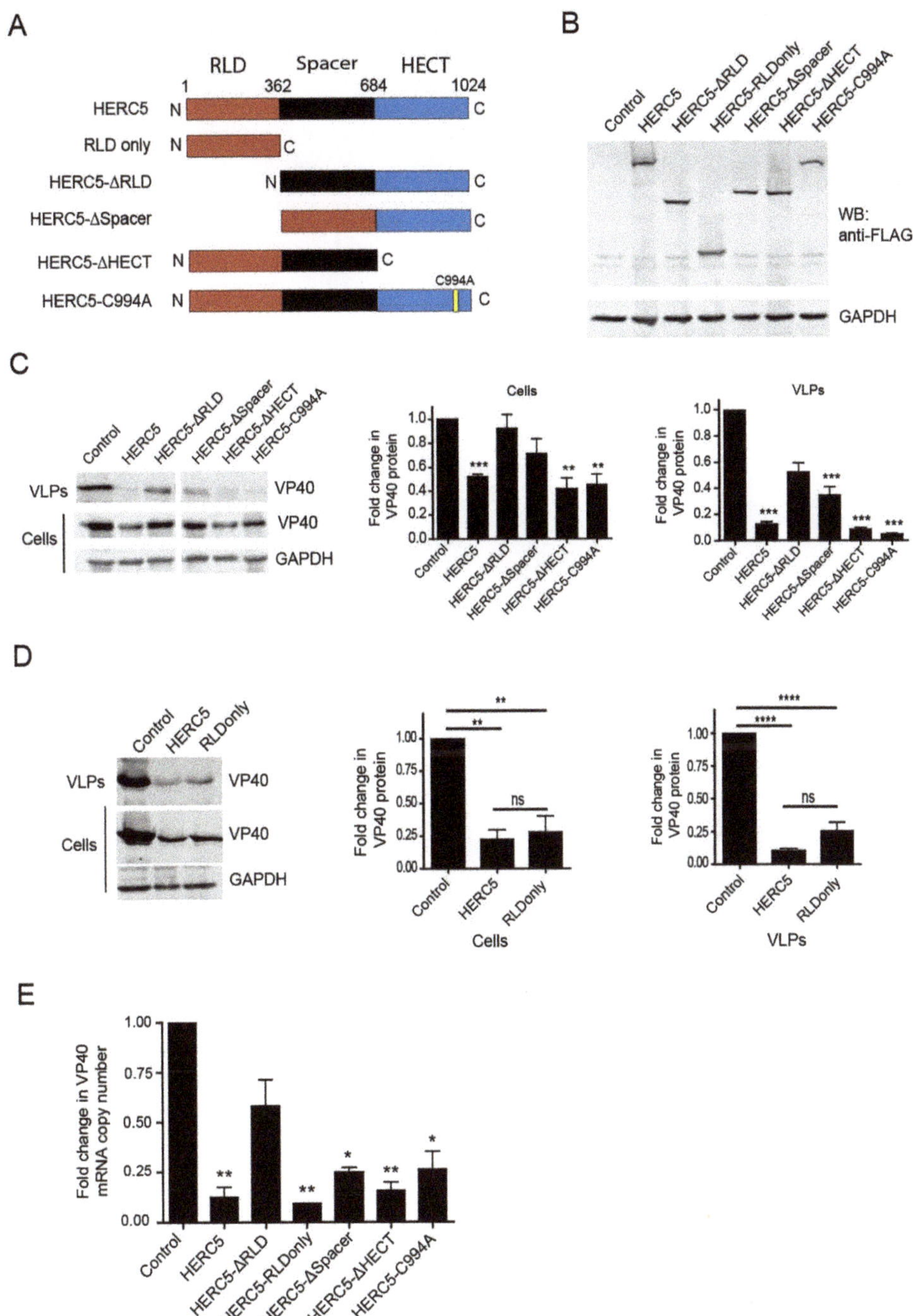

Figure 3. The RLD is necessary and sufficient for HERC5-mediated restriction. (**A**) Schematic of the different HERC5 mutant constructs. (**B**) Representative Western blot showing consistent expression of wild-type HERC5 and mutant forms of HERC5.

293T cells were transfected with either empty vector or plasmids carrying FLAG-tagged HERC5, HERC5-ΔRLD, HERC5-RLDonly, HERC5-ΔSpacer, HERC5-ΔHECT or HERC5-C994A. Forty-eight hours after transfection, cell lysate was subjected to Western blot analysis using anti-FLAG and anti-GAPDH. (**C**) 293T cells were co-transfected with plasmids carrying FLAG-tagged VP40 and either empty vector, wild-type HERC5 or one of the HERC5 mutants listed in (**A**). Forty-eight hours post-transfection, purified VLPs released into the supernatant and intracellular protein were examined by Western blotting using anti-FLAG and anti-GAPDH. VP40 protein levels were quantified densitometrically after normalization to GAPDH levels (graphs on the right). (**D**) 293T cells were co-transfected with plasmids carrying VP40-EGFP and either empty vector, HERC5 or HERC5-RLDonly. Cell lysates and VLPs were analyzed via Western blotting using anti-GFP and anti-GAPDH. VP40-EGFP protein levels were quantified densitometrically (graphs on the right). (**E**) 293T cells were co-transfected with plasmids carrying FLAG-tagged VP40 and either empty vector, HERC5, HERC5-ΔRLD, HERC5-RLDonly, HERC5-ΔSpacer, HERC5-ΔHECT or HERC5-C994A. Forty-eight hours post-transfection, mRNA was isolated and used to measure intracellular VP40 mRNA levels using qPCR. All data shown are representative of three independent experiments (+/− S.E.M.). **** $p < 0.0001$, *** $p < 0.001$, ** $p < 0.01$, * $p \leq 0.05$, ns (not significant) $p > 0.05$; One-way ANOVA with Dunnet's multiple comparisons test compared to the control.

3.4. HERC5 Depletes VP40 mRNA Independently of ZAP

ZAP (also called Zinc finger CCCH-type, antiviral 1, ZC3HAV1, and Poly (ADP-ribose) polymerase 13, PARP13) is an antiviral protein that causes significant loss of viral mRNAs from evolutionarily diverse RNA viruses, including *Filoviridae, Retroviridae, Togaviridae* and *Hepadnaviridae* [49–55]. We, therefore, asked if ZAP was required for HERC5-mediated depletion of EBOV mRNA. We co-expressed VP40 and HERC5 in 293T cells that were knocked out for all ZAP isoforms and measured VP40 mRNA and protein levels using qPCR and Western blotting [56,57]. Cells expressing HERC5 in the absence of ZAP significantly reduced VP40 mRNA levels (Figure 4A). Exogenous expression of ZAP (short isoform) in the ZAP knockout cells reduced VP40 mRNA levels as previously shown [52,56]. Co-expression of HERC5 and ZAP together resulted in an enhanced loss of VP40 mRNA (Figure 4A). In support of this observation, cells expressing HERC5 in the absence of ZAP significantly reduced intracellular VP40 protein and VP40 VLPs the cell supernatant (Figure 4B,C). Together, these data show that ZAP is not required for HERC5-mediated reduction of VP40 mRNA.

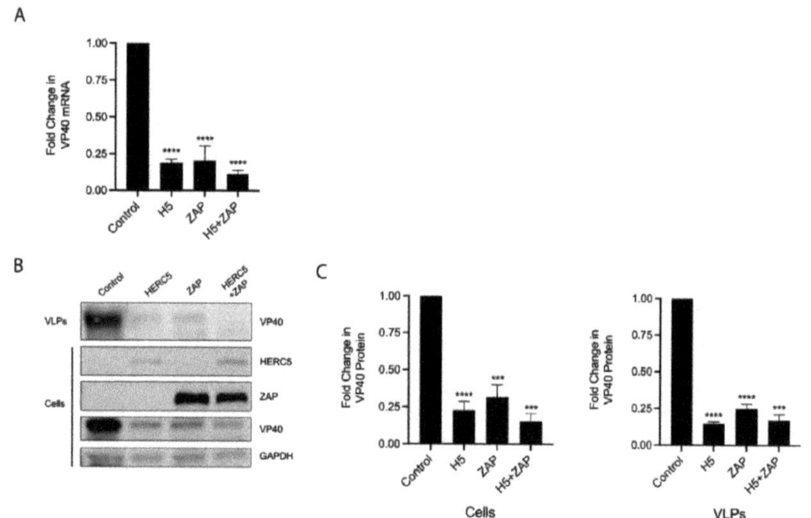

Figure 4. HERC5 restricts VP40 independently of ZAP. 293T ZAP knockout cells were co-transfected with plasmids carrying FLAG-tagged VP40 and either empty vector control, HERC5, ZAP (short isoform),

or HERC5 and ZAP (short isoform). Twenty-four hours post-transfection, cell lysates and VLP-containing supernatants were harvested. (**A**) Intracellular VP40 mRNA levels were measured using qPCR and normalized to GAPDH. The data shown are representative of four independent experiments. (**B**) Purified VLPs released into the cell supernatant and intracellular proteins were subjected to Western blot analysis using anti-FLAG and anti-GAPDH. Representative Western blot of three independent experiments is shown. (**C**) The average densitometric quantification of VP40 protein bands from B is shown after normalization to GAPDH levels. Results are presented as mean (± SEM) fold changes in VP40 protein or mRNA. **** $p < 0.001$, *** $p < 0.001$, One-way ANOVA with Tukey's multiple comparisons test.

3.5. EBOV GP and L Proteins Antagonize HERC5

Despite an early and robust IFN-signaling response to EBOV infection, EBOV proteins ultimately suppress this response leading to pathogenesis [2–12]. Given the potent antiviral activity of HERC5 towards EBOV mRNAs, we asked if any of the EBOV proteins could antagonize this activity. VP40 mRNA levels in cells co-expressing HERC5 and various EBOV proteins were measured by qPCR. As shown in Figure 5A, VP40 mRNA levels were rescued in cells co-expressing GP or L protein, but not VP30, VP35, NP or the non-EBOV protein vesicular stomatitis virus-G (VSV-G) protein. Western blot analysis of cell lysates correlated with the qPCR data where only L and GP proteins rescued intracellular VP40 protein levels (Figure 5B). Western blot analysis of VP40 VLPs in the supernatant revealed that GP but not L protein rescued VLP production, indicating that only GP was able to fully rescue VLP production.

To determine if the ability of EBOV GP to antagonize HERC5 is specific to the *Ebolavirus* genus, we tested the ability of Marburg virus (MARV) GP, which belongs to the *Marburgvirus* genus, to antagonize HERC5. In contrast with EBOV GP, co-expression of MARV GP failed to rescue VP40 VLP production (Figure 5C). Together these data show that EBOV GP antagonizes HERC5 activity and that this antagonism does not appear to be conserved between filovirus genera.

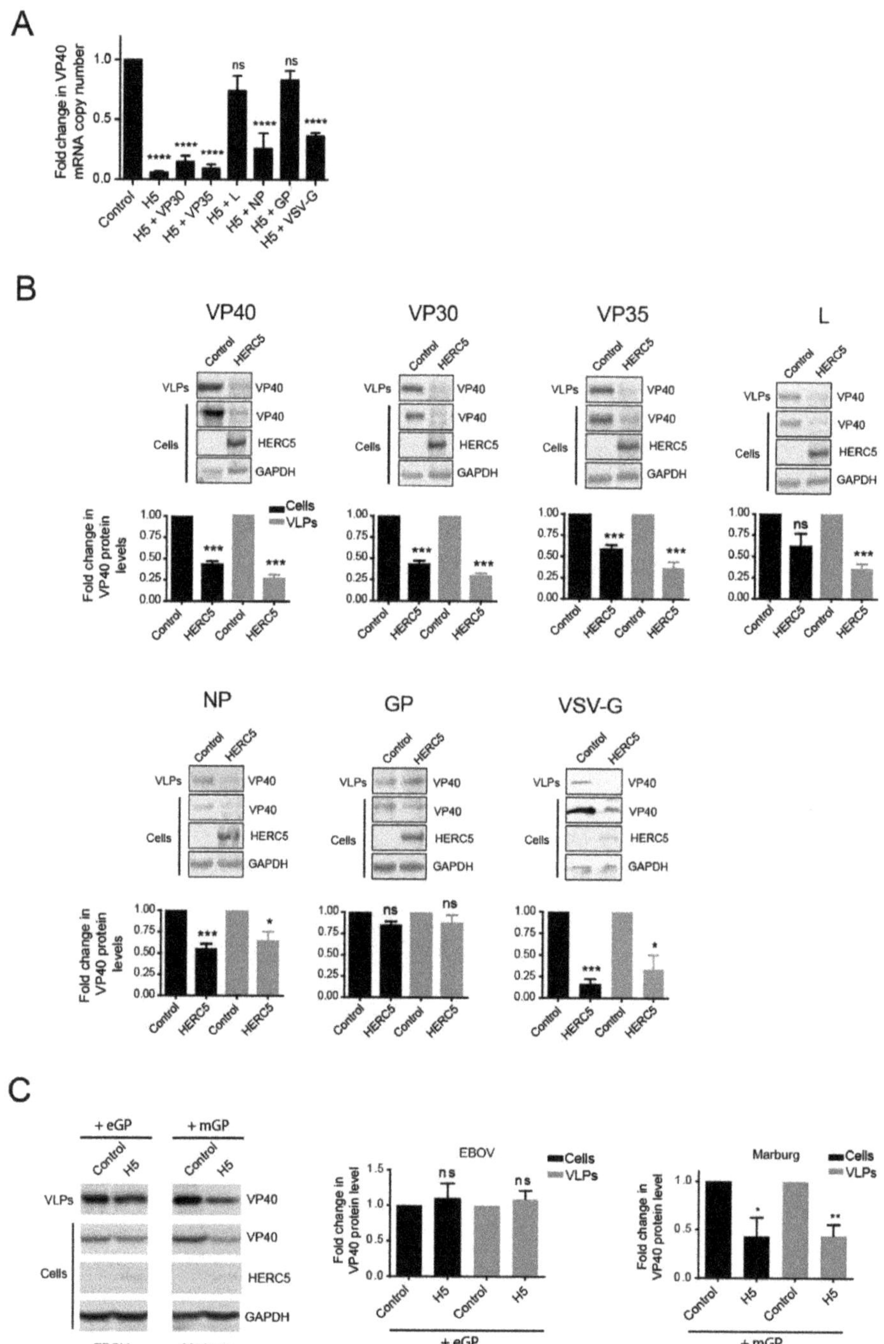

Figure 5. EBOV GP and L antagonize HERC5. 293T cells were co-transfected with plasmids carrying FLAG-tagged VP40 and either empty vector or HERC5 and one plasmid carrying either EBOV VP30, VP35, L NP, GP or VSV-G. Forty-eight hours

post-transfection, VP40 mRNA was measured using qPCR (**A**) and VP40 protein levels in cell lysates and VLPs released into supernatant were analyzed by quantitative Western blotting and quantified densitometrically after normalization to GAPDH levels (**B**). (**C**) 293T cells were co-transfected with plasmids carrying FLAG-tagged VP40 and either empty vector or HERC5, and one of EBOV GP (eGP) or MARV GP (mGP). Forty-eight hours post-transfection, VP40 protein levels in cell lysates and VLPs released into the supernatant were analyzed via Western blotting using anti-FLAG and anti-GAPDH. The data shown represent the average (+/− S.E.M.) of three independent experiments. * $p < 0.05$, ** $p < 0.01$, *** $p < 0.001$, **** $p < 0.0001$, ns (not significant) $p > 0.05$; One-way ANOVA with Dunnett's multiple comparisons test compared to the control (**A**); Student's paired t-test (**B**,**C**).

3.6. EBOV and MARV GP Differentially Antagonize HERC5 Inhibition of EBOV trVLP Replication

We utilized the EBOV trVLP system described in Figure 1 to determine if genus-specific GP (EBOV or MARV) could antagonize the ability of HERC5 to inhibit trVLP replication. To test the effect of different GPs on trVLP replication, two different sets of trVLP particles were generated at P0. One set contained EBOV GP (trVLP$_{\text{EBOV GP}}$) and was generated as described in Figure 1A. The second set was generated in an identical way except that the EBOV *GP* gene in the '4cis' plasmid minigenome was substituted with the MARV *GP* gene (trVLP$_{\text{MARV GP}}$). This allowed us to test the impact of different GPs in the VLPs while maintaining the same background of EBOV proteins. As a negative control, the plasmid carrying the Ebola *L* gene was omitted from the transfections, which abrogates trVLP formation. Compared to the control cells not expressing HERC5, cells expressing HERC5 exhibited significantly reduced levels of trVLP$_{\text{MARV GP}}$ and trVLP$_{\text{EBOV GP}}$ replication over four passages (spanning 12 days) (Figure 6). Notably, HERC5 inhibited trVLP$_{\text{MARV GP}}$ replication significantly more than trVLP$_{\text{EBOV GP}}$ replication over two passages ($p < 0.01$, Two-way ANOVA). Together, these data show that EBOV GP and MARV GP differentially antagonize HERC5 inhibition of EBOV trVLP replication.

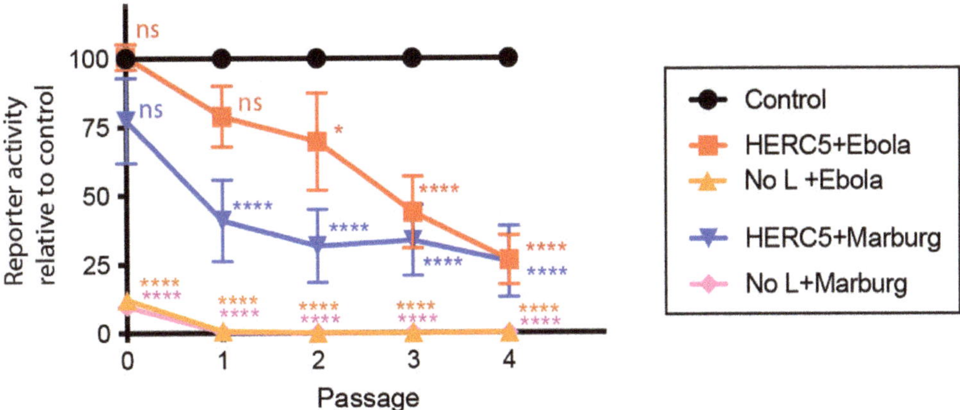

Figure 6. EBOV GP and MARV GP differentially antagonize HERC5. Quantification of trVLP propagation in the presence and absence of HERC5. The trVLP propagation assay was performed using tetracistronic minigenomes carrying a luciferase reporter, EBOV VP40, VP24 and either EBOV GP or MARV GP over four passages (spanning 12 days). All EBOV minigenomes and plasmids carrying the EBOV proteins are based on EBOV *H. sapiens*-tc/COD/1976/Yambuku-Mayinga. As a negative control ('No L'), the plasmid carrying the Ebola *L* gene was omitted from the transfections. Luciferase reporter activity relative to the control (trVLPs propagated in the absence of HERC5) is shown. The data shown represent the average (+/− S.E.M.) of at least six independent experiments. **** $p < 0.0001$, * $p \leq 0.05$, ns (not significant) $p > 0.05$; Two-way ANOVA with Dunnett's multiple comparisons test compared to the no HERC5 control.

4. Discussion

Hundreds of IFN-induced proteins are part of the early and robust immune response to EBOV infection in primates [2–12]. Characterization of the key effector proteins of this defense and how EBOV overcomes them will provide a better understanding of the virus–host interactions that occur early in infection. HERC5 is one of the most up-regulated antiviral proteins in the early response to EBOV infection in vivo; however, its role in EBOV replication was previously unknown [3,5,6,29].

In this study, we showed that HERC5 inhibits EBOV VLP replication via a novel E3 ligase-independent mechanism. This mechanism involves the depletion of viral mRNAs and requires the RLD domain of HERC5. We previously showed that HERC5 inhibits the nuclear export of HIV-1 RNA genomes by a different E3 ligase-independent mechanism, one that also requires the RLD domain of HERC5 [30]. These E3 ligase-independent antiviral activities, together with its well-documented E3 ligase-dependent antiviral activities [58], identifies HERC5 as a multifunctional antiviral protein. It is perhaps not surprising that HERC5 has evolved multiple mechanisms of restriction of viruses. The ancestral *HERC* gene is believed to have arisen from a gene fusion event between an *RCC1*-like gene and a *HECT* gene [59,60]. This fusion event gave rise to a family of small HERC proteins containing an amino-terminal RLD and a carboxyl-terminal HECT domain that is highly conserved among vertebrates spanning >595 million years of evolution [36,59,60]. Moreover, HERC5 has been evolving under strong positive selection, which is characteristic of many host restriction factors involved in an evolutionary struggle with viruses [30,60,61]. The ability of HERC5 to inhibit viruses via both E3 ligase-dependent and -independent mechanisms would confer a strong evolutionary advantage to its host, making it more difficult for viruses to evolve countermeasures to HERC5.

Like HERC5, ZAP is present in evolutionarily diverse vertebrates and has evolved under strong positive selection [30,36,62]. ZAP targets diverse viruses such as HIV-1, MoLV and XMRV (*Retroviridae*), Ebola and Marburg viruses (*Filoviridae*), alphavirus, Sindbis, Semliki Forest and Ross River viruses (*Togaviridae*), hepatitis B virus (*Hepadnaviridae*) and double-stranded DNA murine gamma herpesvirus (*Herpesviridae*) [49–55,63]. ZAP is known to inhibit a wide range of antiviral activities, including recruiting the exosome complex to target viral RNAs for degradation [49,51,53,55,57,64–68]. ZAP also exhibits virus specificity since it has no antiviral effect on vesicular stomatitis, poliovirus, yellow fever and herpes simplex I viruses [49]. We showed here that HERC5 depletes EBOV mRNAs in a ZAP-independent manner. Our finding that the HERC5 RLD is necessary and sufficient for EBOV mRNA depletion further supports an E3 ligase-independent mechanism of restriction. It was previously shown that the RLD is required for the association of HERC5 with polyribosomes [35]. It is possible that HERC5 exploits this interaction to recruit other RNA degradation machinery to EBOV mRNAs.

Although we showed that the RLD alone was necessary and sufficient to inhibit particle production, HERC5 lacking the RLD failed to completely inhibit VP40 VLP particle production. Since the RLD is important but not essential for its E3 ligase activity, it is possible that the E3 ligase activity of HERC5 also confers some antiviral activity towards VLP production via ISGylation of viral and/or host proteins involved in particle production [31,35,38]. It was previously shown that over-expression of ISG15 alone inhibited budding of EBOV VP40 VLPs by disrupting Nedd4 function and subsequent ubiquitination of VP40, which is necessary for viral egress [69]. It is unknown whether HERC5 was involved in this activity since it was not investigated. Although our data show that the predominant mechanism by which HERC5 inhibits EBOV VLP production involves the depletion of EBOV mRNAs, visual inspection of cells co-expressing EBOV VP40 and HERC5 by TEM and confocal microscopy revealed an accumulation of the VP40 protein at the localized regions in the plasma membrane in some cells, consistent with the idea of a second mechanism of inhibition acting later in particle production. HERC5-induced trapping of virus particles at the plasma membrane has also been observed with HIV-1 [31]. However, it is also possible that these accumulations represent particles in the process of budding that have escaped HERC5 restriction. HERC5 reduced intracellular mRNA levels of viral protein expressed both from a plasmid system (Figure 2H) and of viral mRNA expressed from a tetracistronic minigenome. It is unknown how HERC5 can target viral RNAs but not non-viral RNA such as GFP. Perhaps virus-specific RNA sequences recruit HERC5 and/or RNA depletion machinery similar to how ZAP selectively recognizes high CpG-containing viral RNA. Further studies are needed to decipher this novel antiviral function of HERC5.

Animal model studies have suggested that the Type I IFN response plays an important role in restricting EBOV replication and that the ability of EBOV to overcome this response may be a requirement for lethal infection [70,71]. Although EBOV VP24 and VP35 can act broadly to dampen the IFN response, several IFN-induced antiviral proteins, including HERC5, are also highly upregulated early in response to other stimuli associated with infection, such as pro-inflammatory cytokines [72–74]. As such, it is likely that EBOV evolved additional antagonists of such antiviral proteins. Indeed, EBOV GP can directly antagonize the restriction factor BST-2/tetherin without altering BST-2/tetherin expression levels or cellular localization [24,75–82]. As shown herein, EBOV GP also antagonizes HERC5 without altering HERC5 expression levels. Although controversial, GP sequence diversity has been shown to affect EBOV transmission and virulence, as demonstrated in the 2013-2016 EBOV epidemic [83,84]. We showed here that variations in GP sequence, such as those found between different filovirus genera (e.g., EBOV and MARV), also influence the potency of antagonism of HERC5 during the early stages of EBOV trVLP replication. It is unclear how GP, which is predominantly localized to the plasma membrane, can rescue EBOV mRNA levels. GP expression is known to alter the expression and trafficking of select cellular proteins; therefore, it is possible that proteins involved in viral RNA stability are affected by GP expression [85–87]. Important next steps will be to characterize the mechanism of GP antagonism and to test the importance of this HERC5-GP axis early in infection using animal models.

It is interesting that EBOV L protein was also able to rescue HERC5-induced VP40 mRNA depletion but unable to antagonize the release of VP40 VLPs into the cell supernatant. The mechanism underlying this antagonism is not fully understood; however, it was previously shown that L protein antagonizes ZAP [52]. It is possible that L protein also specifically antagonizes HERC5-induced depletion of mRNAs. However, we speculate that the E3 ligase activity of HERC5 remains functional, leading to the ISGylation of viral and/or host proteins and subsequent arrest of later steps in viral particle production.

In conclusion, we showed that HERC5 inhibits EBOV virus particle production by a mechanism involving the depletion of EBOV mRNAs. Our data also identifies a novel 'protagonist–antagonistic' relationship between HERC5 and GP early in EBOV infection. With the ability to inhibit HERC5 and other restriction factors, GP is an attractive target for the development of small molecule compounds that interfere with this antagonism.

Supplementary Materials: The following are available online at https://www.mdpi.com/article/10.3390/cells10092399/s1, Figure S1: HERC5 depletes GFP- and FLAG-tagged VP40 mRNA but not GFP mRNA, Figure S2: HERC5 depletes GFP- and FLAG-tagged VP40 mRNA but not GFP mRNA, Table S1: Quantification of 5nm gold particle-labeled anti-GFP in cells expressing empty vector and VP40-EGFP, Table S2: Quantification of 5nm gold particle-labeled anti-GFP in cells expressing HERC5 and VP40-EGFP.

Author Contributions: Conceptualization, S.D.B., E.P. and N.R.H.; methodology, E.P., N.R.H., D.S.L., M.D.C., E.J.D.G., M.J.D. and N.J.F.; formal analysis, E.P., N.R.H., D.S.L., M.D.C., E.J.D.G., M.J.D., N.J.F. and S.D.B.; investigation, E.P., N.R.H., D.S.L., M.D.C., E.J.D.G., M.J.D. and N.J.F.; resources, E.P., N.R.H., D.S.L., M.D.C., E.J.D.G., M.J.D., M.C., A.M., T.H. and S.D.B.; writing—original draft preparation, S.D.B., E.P. and N.R.H.; writing—review and editing, S.D.B., E.P., N.R.H., D.S.L., M.D.C., M.C. and T.H.; supervision, S.D.B.; project administration, S.D.B.; funding acquisition, S.D.B. All authors have read and agreed to the published version of the manuscript.

Funding: This research was funded by the Canadian Institutes of Health Research (CIHR) to S.D.B., grant numbers HBF134179 and HBF137693.

Institutional Review Board Statement: Not applicable.

Informed Consent Statement: Not applicable.

Data Availability Statement: Not applicable.

Acknowledgments: We thank Melody Li (University of California, Los Angeles, CA, USA), Margaret MacDonald (The Rockefeller University, New York, NY, USA) and Akinori Takaoka (Hokkaido University, Hokkaido, Japan) for kindly providing the ZAP knockout cell line. We thank Tirthankar Ray for their help in constructing the VP40 plasmid. We thank Karen Nygard and Richard Gardiner for their help with electron microscopy at the Western University Biotron Facility, and Claudia Seah and Flavio Beraldo at the London Regional Microscopy Facility, Western University) for their help with confocal microscopy.

Conflicts of Interest: The authors declare no conflict of interest.

References

1. Feldmann, H.; Geisbert, T.W. Ebola haemorrhagic fever. *Lancet Lond.* **2011**, *377*, 849–862. [CrossRef]
2. Rubins, K.H.; Hensley, L.; Wahl-Jensen, V.; DiCaprio, K.M.D.; Young, H.; Reed, D.S.; Jahrling, P.B.; O Brown, P.; Relman, D.; Geisbert, T.W. The temporal program of peripheral blood gene expression in the response of nonhuman primates to Ebola hemorrhagic fever. *Genome Biol.* **2007**, *8*, R174. [CrossRef]
3. Caballero, I.S.; Honko, A.N.; Gire, S.K.; Winnicki, S.M.; Melé, M.; Gerhardinger, C.; Lin, A.E.; Rinn, J.L.; Sabeti, P.C.; Hensley, L.E.; et al. In vivo Ebola virus infection leads to a strong innate response in circulating immune cells. *BMC Genom.* **2016**, *17*, 707. [CrossRef] [PubMed]
4. Kash, J.C.; Walters, K.-A.; Kindrachuk, J.; Baxter, D.; Scherler, K.; Janosko, K.B.; Adams, R.D.; Herbert, A.S.; James, R.M.; Stonier, S.W.; et al. Longitudinal peripheral blood transcriptional analysis of a patient with severe Ebola virus disease. *Sci. Transl. Med.* **2017**, *9*, eaai9321. [CrossRef] [PubMed]
5. Speranza, E.; Bixler, S.L.; Altamura, L.A.; Arnold, C.E.; Pratt, W.D.; Taylor-Howell, C.; Burrows, C.; Aguilar, W.; Rossi, F.; Shamblin, J.D.; et al. A conserved transcriptional response to intranasal Ebola virus exposure in nonhuman primates prior to onset of fever. *Sci. Transl. Med.* **2018**, *10*, eaaq1016. [CrossRef]
6. Versteeg, K.; Menicucci, A.R.; Woolsey, C.; Mire, C.E.; Geisbert, J.B.; Cross, R.W.; Agans, K.N.; Jeske, D.; Messaoudi, I.; Geisbert, T.W. Infection with the Makona variant results in a delayed and distinct host immune response compared to previous Ebola virus variants. *Sci. Rep.* **2017**, *7*, 9730. [CrossRef]
7. Cilloniz, C.; Ebihara, H.; Ni, C.; Neumann, G.; Korth, M.J.; Kelly, S.M.; Kawaoka, Y.; Feldmann, H.; Katze, M.G. Functional genomics reveals the induction of inflammatory response and metalloproteinase gene expression during lethal ebola virus infection. *J. Virol.* **2011**, *85*, 9060–9068. [CrossRef]
8. Speranza, E.; Altamura, L.A.; Kulcsar, K.; Bixler, S.L.; Rossi, C.A.; Schoepp, R.J.; Nagle, E.; Aguilar, W.; Douglas, C.E.; Delp, K.L.; et al. Comparison of transcriptomic platforms for analysis of whole blood from ebola-infected cynomolgus macaques. *Sci. Rep.* **2017**, *7*, 14756. [CrossRef]
9. Garamszegi, S.; Yen, J.Y.; Honko, A.N.; Geisbert, J.B.; Rubins, K.H.; Geisbert, T.W.; Xia, Y.; Hensley, L.E.; Connor, J.H. Transcriptional correlates of disease outcome in anticoagulant-treated non-human primates infected with ebolavirus. *PLoS Negl. Trop. Dis.* **2014**, *8*, e3061. [CrossRef]
10. Liu, X.; Speranza, E.; Muñoz-Fontela, C.; Haldenby, S.; Rickett, N.Y.; Garcia-Dorival, I.; Fang, Y.; Hall, Y.; Zekeng, E.-G.; Lüdtke, A.; et al. Transcriptomic signatures differentiate survival from fatal outcomes in humans infected with Ebola virus. *Genome Biol.* **2017**, *18*, 4. [CrossRef]
11. Eisfeld, A.J.; Halfmann, P.J.; Wendler, J.P.; Kyle, J.E.; Burnum-Johnson, K.E.; Peralta, Z.; Maemura, T.; Walters, K.B.; Watanabe, T.; Fukuyama, S.; et al. Multi-platform 'omics analysis of human Ebola virus disease pathogenesis. *Cell Host Microbe* **2017**, *22*, 817–829. [CrossRef] [PubMed]
12. Yen, J.Y.; Garamszegi, S.; Geisbert, J.B.; Rubins, K.H.; Geisbert, T.W.; Honko, A.; Xia, Y.; Connor, J.H.; Hensley, L.E. Therapeutics of Ebola hemorrhagic fever: Whole-genome transcriptional analysis of successful disease mitigation. *J. Infect. Dis.* **2011**, *204*, S1043–S1052. [CrossRef]
13. Kash, J.C.; Mühlberger, E.; Carter, V.; Grosch, M.; Perwitasari, O.; Proll, S.C.; Thomas, M.J.; Weber, F.; Klenk, H.-D.; Katze, M.G. Global suppression of the host antiviral response by Ebola- and Marburgviruses: Increased antagonism of the type I interferon response is associated with enhanced virulence. *J. Virol.* **2006**, *80*, 3009–3020. [CrossRef] [PubMed]
14. Hartman, A.L.; Ling, L.; Nichol, S.T.; Hibberd, M.L. Whole-genome expression profiling reveals that inhibition of host innate immune response pathways by Ebola virus can be reversed by a single amino acid change in the VP35 protein. *J. Virol.* **2008**, *82*, 5348–5358. [CrossRef]

15. Hartman, A.L.; Bird, B.H.; Towner, J.S.; Antoniadou, Z.-A.; Zaki, S.R.; Nichol, S.T. Inhibition of IRF-3 activation by VP35 is critical for the high level of virulence of ebola virus. *J. Virol.* **2008**, *82*, 2699–2704. [CrossRef] [PubMed]

16. Prins, K.C.; Delpeut, S.; Leung, D.W.; Reynard, O.; Volchkova, V.A.; Reid, S.P.; Ramanan, P.; Cárdenas, W.B.; Amarasinghe, G.K.; Volchkov, V.E.; et al. Mutations abrogating VP35 interaction with double-stranded RNA render Ebola virus avirulent in guinea pigs. *J. Virol.* **2010**, *84*, 3004–3015. [CrossRef] [PubMed]

17. Basler, C.F.; Amarasinghe, G.K. Evasion of interferon responses by Ebola and Marburg viruses. *J. Interf. Cytokine Res.* **2009**, *29*, 511–520. [CrossRef]

18. Messaoudi, I.; Amarasinghe, G.K.; Basler, C.F. Filovirus pathogenesis and immune evasion: Insights from Ebola virus and Marburg virus. *Nat. Rev. Microbiol.* **2015**, *13*, 663–676. [CrossRef]

19. Luthra, P.; Ramanan, P.; Mire, C.E.; Weisend, C.; Tsuda, Y.; Yen, B.; Liu, G.; Leung, D.W.; Geisbert, T.W.; Ebihara, H.; et al. Mutual antagonism between the Ebola Virus VP35 protein and the RIG-I activator PACT determines infection outcome. *Cell Host Microbe* **2013**, *14*, 74–84. [CrossRef]

20. Bieniasz, P.D. Intrinsic immunity: A front-line defense against viral attack. *Nat. Immunol.* **2004**, *5*, 1109–1115. [CrossRef]

21. Neil, S.J.; Zang, T.; Bieniasz, P.D. Tetherin inhibits retrovirus release and is antagonized by HIV-1 Vpu. *Nature* **2008**, *451*, 425–430. [CrossRef]

22. Van Damme, N.; Goff, D.; Katsura, C.; Jorgenson, R.L.; Mitchell, R.; Johnson, M.C.; Stephens, E.B.; Guatelli, J. The interferon-induced protein BST-2 restricts HIV-1 release and is downregulated from the cell surface by the viral Vpu protein. *Cell Host Microbe* **2008**, *3*, 245–252. [CrossRef]

23. Jouvenet, N.; Neil, S.J.D.; Zhadina, M.; Zang, T.; Kratovac, Z.; Lee, Y.; McNatt, M.; Hatziioannou, T.; Bieniasz, P.D. Broad-spectrum inhibition of retroviral and filoviral particle release by tetherin. *J. Virol.* **2009**, *83*, 1837–1844. [CrossRef] [PubMed]

24. Kaletsky, R.L.; Francica, J.R.; Agrawal-Gamse, C.; Bates, P. Tetherin-mediated restriction of filovirus budding is antagonized by the Ebola glycoprotein. *Proc. Natl. Acad. Sci. USA* **2009**, *106*, 2886–2891. [CrossRef] [PubMed]

25. Sakuma, T.; Noda, T.; Urata, S.; Kawaoka, Y.; Yasuda, J. Inhibition of Lassa and Marburg virus production by tetherin. *J. Virol.* **2009**, *83*, 2382–2385. [CrossRef] [PubMed]

26. Radoshitzky, S.R.; Dong, L.; Chi, X.; Clester, J.C.; Retterer, C.; Spurgers, K.; Kuhn, J.H.; Sandwick, S.; Ruthel, G.; Kota, K.; et al. Infectious Lassa virus, but not filoviruses, is restricted by BST-2/tetherin. *J. Virol.* **2010**, *84*, 10569–10580. [CrossRef] [PubMed]

27. Huang, I.-C.; Bailey, C.C.; Weyer, J.L.; Radoshitzky, S.R.; Becker, M.M.; Chiang, J.J.; Brass, A.L.; Ahmed, A.A.; Chi, X.; Dong, L.; et al. Distinct Patterns of IFITM-Mediated Restriction of Filoviruses, SARS Coronavirus, and Influenza A Virus. *PLoS Pathog.* **2011**, *7*, e1001258. [CrossRef]

28. Wrensch, F.; Karsten, C.B.; Gnirß, K.; Hoffmann, M.; Lu, K.; Takada, A.; Winkler, M.; Simmons, G.; Pöhlmann, S. Interferon-Induced transmembrane protein–mediated inhibition of host cell entry of Ebolaviruses. *J. Infect. Dis.* **2015**, *212*, S210–S218. [CrossRef] [PubMed]

29. Menicucci, A.R.; Versteeg, K.; Woolsey, C.; Mire, C.E.; Geisbert, J.B.; Cross, R.W.; Agans, K.N.; Jankeel, A.; Geisbert, T.W.; Messaoudi, I. Transcriptome analysis of circulating immune cell subsets highlight the role of Monocytes in zaire Ebola virus Makona pathogenesis. *Front. Immunol.* **2017**, *8*, 1372. [CrossRef]

30. Woods, M.W.; Tong, J.G.; Tom, S.K.; Szabo, P.A.; Cavanagh, P.C.; Dikeakos, J.D.; Haeryfar, S.M.M.; Barr, S.D. Interferon-induced HERC5 is evolving under positive selection and inhibits HIV-1 particle production by a novel mechanism targeting Rev/RRE-dependent RNA nuclear export. *Retrovirology* **2014**, *11*, 27. [CrossRef]

31. Woods, M.W.; Kelly, J.N.; Hattlmann, C.J.; Tong, J.G.K.; Xu, L.S.; Coleman, M.D.; Quest, G.R.; Smiley, J.R.; Barr, S.D. Human HERC5 restricts an early stage of HIV-1 assembly by a mechanism correlating with the ISGylation of Gag. *Retrovirology* **2011**, *8*, 95. [CrossRef] [PubMed]

32. Tang, Y.; Zhong, G.; Zhu, L.; Liu, X.; Shan, Y.; Feng, H.; Bu, Z.; Chen, H.; Wang, C. Herc5 attenuates influenza A virus by catalyzing ISGylation of viral NS1 protein. *J. Immunol.* **2010**, *184*, 5777–5790. [CrossRef] [PubMed]

33. Versteeg, G.A.; Hale, B.G.; van Boheemen, S.; Wolff, T.; Lenschow, D.J.; Garcia-Sastre, A. Species-specific antagonism of host ISGylation by the influenza B virus NS1 protein. *J. Virol.* **2010**, *84*, 5423–5430. [CrossRef] [PubMed]

34. Zhao, C.; Hsiang, T.Y.; Kuo, R.L.; Krug, R.M. ISG15 conjugation system targets the viral NS1 protein in influenza A virus-infected cells. *Proc. Natl. Acad. Sci. USA* **2010**, *107*, 2253–2258. [CrossRef]

35. Durfee, L.A.; Lyon, N.; Seo, K.; Huibregtse, J.M. The ISG15 conjugation system broadly targets newly synthesized proteins: Implications for the antiviral function of ISG15. *Mol. Cell* **2010**, *38*, 722–732. [CrossRef]

36. Paparisto, E.; Woods, M.W.; Coleman, M.D.; Moghadasi, S.A.; Kochar, D.S.; Tom, S.K.; Kohio, H.P.; Gibson, R.M.; Rohringer, T.J.; Hunt, N.R.; et al. Evolution-guided structural and functional analyses of the HERC family reveal an ancient marine origin and determinants of antiviral activity. *J. Virol.* **2018**, *92*, e0052818. [CrossRef]

37. Wong, J.J.Y.; Pung, Y.F.; Sze, N.S.; Chin, K.C. HERC5 is an IFN-induced HECT-type E3 protein ligase that mediates type I IFN-induced ISGylation of protein targets. *Proc. Natl. Acad. Sci. USA* **2006**, *103*, 10735–10740. [CrossRef]

38. Dastur, A.; Beaudenon, S.; Kelley, M.; Krug, R.M.; Huibregtse, J.M. Herc5, an interferon-induced HECT E3 enzyme, is required for conjugation of ISG15 in human cells. *J. Biol. Chem.* **2006**, *281*, 4334–4338. [CrossRef]

39. Watanabe, S.; Watanabe, T.; Noda, T.; Takada, A.; Feldmann, H.; Jasenosky, L.D.; Kawaoka, Y. Production of novel ebola virus-like particles from cDNAs: An alternative to ebola virus generation by reverse genetics. *J. Virol.* **2004**, *78*, 999–1005. [CrossRef]

40. Watt, A.; Moukambi, F.; Banadyga, L.; Groseth, A.; Callison, J.; Herwig, A.; Ebihara, H.; Feldmann, H.; Hoenen, T. A novel life cycle modeling system for Ebola virus shows a genome length-dependent role of VP24 in virus infectivity. *J. Virol.* **2014**, *88*, 10511–10524. [CrossRef]

41. Wendt, L.; Kämper, L.; Schmidt, M.L.; Mettenleiter, T.C.; Hoenen, T. Analysis of a putative late domain using an Ebola virus transcription and replication-competent virus-like particle system. *J. Infect. Dis.* **2018**, *218*, S355–S359. [CrossRef] [PubMed]

42. Côté, M.; Misasi, J.; Ren, T.; Bruchez, A.; Lee, K.; Filone, C.M.; Hensley, L.; Li, Q.; Ory, D.; Chandran, K.; et al. Small molecule inhibitors reveal Niemann–Pick C1 is essential for Ebola virus infection. *Nature* **2011**, *477*, 344–348. [CrossRef]

43. Chandran, K.; Sullivan, N.J.; Felbor, U.; Whelan, S.P.; Cunningham, J.M. Endosomal proteolysis of the Ebola virus glycoprotein is necessary for infection. *Science* **2005**, *308*, 1643–1645. [CrossRef]

44. Schmidt, M.L.; Hoenen, T. Characterization of the catalytic center of the Ebola virus L polymerase. *PLoS Negl. Trop. Dis.* **2017**, *11*, e0005996. [CrossRef] [PubMed]

45. Hoenen, T.; Watt, A.; Mora, A.; Feldmann, H. Modeling the lifecycle of Ebola virus under biosafety level 2 conditions with virus-like particles containing tetracistronic minigenomes. *J. Vis. Exp.* **2014**, 52381. [CrossRef]

46. Frick, C.; Ollmann-Saphire, E.; Stahelin, R. Live-cell imaging of Ebola virus matrix protein VP40. *FASEB J.* **2015**, *29*, 886.4. [CrossRef]

47. Noda, T.; Sagara, H.; Suzuki, E.; Takada, A.; Kida, H.; Kawaoka, Y. Ebola virus VP40 drives the formation of virus-like filamentous particles along with GP. *J. Virol.* **2002**, *76*, 4855–4865. [CrossRef]

48. Johnson, K.A.; Taghon, G.J.F.; Scott, J.L.; Stahelin, R.V. The Ebola Virus matrix protein, VP40, requires phosphatidylinositol extensive oligomerization at the plasma membrane and viral egress. *Sci. Rep.* **2016**, *6*, 19125. [CrossRef]

49. Bick, M.J.; Carroll, J.W.; Gao, G.; Goff, S.P.; Rice, C.M.; MacDonald, M.R. Expression of the zinc-finger antiviral protein inhibits alphavirus replication. *J. Virol.* **2003**, *77*, 11555–11562. [CrossRef]

50. Kerns, J.A.; Emerman, M.; Malik, H.S. Positive selection and increased antiviral activity associated with the PARP-containing isoform of human zinc-finger antiviral protein. *PLoS Genet.* **2008**, *4*, e21. [CrossRef]

51. Mao, R.; Nie, H.; Cai, D.; Zhang, J.; Liu, H.; Yan, R.; Cuconati, A.; Block, T.M.; Guo, J.-T.; Guo, H. Inhibition of Hepatitis B virus replication by the host zinc finger antiviral protein. *PLoS Pathog.* **2013**, *9*, e1003494. [CrossRef]

52. Muller, S.; Moller, P.; Bick, M.J.; Wurr, S.; Becker, S.; Gunther, S.; Kummerer, B.M. Inhibition of filovirus replication by the zinc finger antiviral protein. *J. Virol.* **2007**, *81*, 2391–2400. [CrossRef] [PubMed]

53. Wang, X.; Tu, F.; Zhu, Y.; Gao, G. Zinc-finger antiviral protein inhibits XMRV infection. *PLoS ONE* **2012**, *7*, e39159. [CrossRef]

54. Zhang, Y.; Burke, C.W.; Ryman, K.D.; Klimstra, W.B. Identification and characterization of interferon-induced proteins that inhibit alphavirus replication. *J. Virol.* **2007**, *81*, 11246–11255. [CrossRef]

55. Zhu, Y.; Chen, G.; Lv, F.; Wang, X.; Ji, X.; Xu, Y.; Sun, J.; Wu, L.; Zheng, Y.-T.; Gao, G. Zinc-finger antiviral protein inhibits HIV-1 infection by selectively targeting multiply spliced viral mRNAs for degradation. *Proc. Natl. Acad. Sci. USA* **2011**, *108*, 15834–15839. [CrossRef]

56. Li, M.M.H.; Aguilar, E.G.; Michailidis, E.; Pabon, J.; Park, P.; Wu, X.; de Jong, Y.P.; Schneider, W.M.; Molina, H.; Rice, C.M.; et al. Characterization of novel splice variants of zinc finger antiviral protein (ZAP). *J. Virol.* **2019**, *93*, e00715–e00719. [CrossRef] [PubMed]

57. Hayakawa, S.; Shiratori, S.; Yamato, H.; Kameyama, T.; Kitatsuji, C.; Kashigi, F.; Goto, S.; Kameoka, S.; Fujikura, D.; Yamada, T.; et al. ZAPS is a potent stimulator of signaling mediated by the RNA helicase RIG-I during antiviral responses. *Nat. Immunol.* **2011**, *12*, 37–44. [CrossRef]

58. Sánchez-Tena, S.; Cubillos-Rojas, M.; Schneider, T.; Rosa, J.L. Functional and pathological relevance of HERC family proteins: A decade later. *Cell. Mol. Life Sci.* **2016**, *73*, 1955–1968. [CrossRef]

59. Hochrainer, K.; Mayer, H.; Baranyi, U.; Binder, B.; Lipp, J.; Kroismayr, R. The human HERC family of ubiquitin ligases: Novel members, genomic organization, expression profiling, and evolutionary aspects. *Genomics* **2005**, *85*, 153–164. [CrossRef] [PubMed]

60. Jacquet, S.; Pontier, D.; Etienne, L. Rapid evolution of HERC6 and duplication of a chimeric HERC5/6 gene in rodents and bats suggest an overlooked role of HERCs in mammalian immunity. *Front. Immunol.* **2020**, *11*, 605270. [CrossRef]

61. Duggal, N.K.; Emerman, M. Evolutionary conflicts between viruses and restriction factors shape immunity. *Nat. Rev.* **2012**, *12*, 687–695. [CrossRef]

62. Daugherty, M.D.; Young, J.M.; Kerns, J.A.; Malik, H.S. Rapid evolution of PARP genes suggests a broad role for ADP-Ribosylation in host-virus conflicts. *PLoS Genet.* **2014**, *10*, e1004403. [CrossRef] [PubMed]

63. Xuan, Y.; Gong, D.; Qi, J.; Han, C.; Deng, H.; Gao, G. ZAP inhibits murine gammaherpesvirus 68 ORF64 expression and is antagonized by RTA. *J. Virol.* **2013**, *87*, 2735–2743. [CrossRef]

64. Guo, X.; Carroll, J.-W.N.; MacDonald, M.R.; Goff, S.P.; Gao, G. The zinc finger antiviral protein directly binds to specific viral mRNAs through the CCCH zinc finger motifs. *J. Virol.* **2004**, *78*, 12781–12787. [CrossRef] [PubMed]

65. Zhu, Y.; Wang, X.; Goff, S.P.; Gao, G. Translational repression precedes and is required for ZAP-mediated mRNA decay. *EMBO J.* **2012**, *31*, 4236–4246. [CrossRef]

66. Karki, S.; Li, M.M.H.; Schoggins, J.W.; Tian, S.; Rice, C.M.; MacDonald, M.R. Multiple interferon stimulated genes synergize with the zinc finger antiviral protein to mediate anti-alphavirus activity. *PLoS ONE* **2012**, *7*, e37398. [CrossRef]

67. Chen, G.; Guo, X.; Lv, F.; Xu, Y.; Gao, G. p72 DEAD box RNA helicase is required for optimal function of the zinc-finger antiviral protein. *Proc. Natl. Acad. Sci. USA* **2008**, *105*, 4352–4357. [CrossRef]

68. Guo, X.; Ma, J.; Sun, J.; Gao, G. The zinc-finger antiviral protein recruits the RNA processing exosome to degrade the target mRNA. *Proc. Natl. Acad. Sci. USA* **2007**, *104*, 151–156. [CrossRef]

69. Okumura, A.; Pitha, P.M.; Harty, R.N. ISG15 inhibits Ebola VP40 VLP budding in an L-domain-dependent manner by blocking Nedd4 ligase activity. *Proc. Natl. Acad. Sci. USA* **2008**, *105*, 3974–3979. [CrossRef]

70. Bray, M. The role of the Type I interferon response in the resistance of mice to filovirus infection. *J. Gen. Virol.* **2001**, *82*, 1365–1373. [CrossRef]

71. Ebihara, H.; Takada, A.; Kobasa, D.; Jones, S.; Neumann, G.; Theriault, S.; Bray, M.; Feldmann, H.; Kawaoka, Y. Molecular determinants of Ebola virus virulence in mice. *PLoS Pathog.* **2006**, *2*, e73. [CrossRef] [PubMed]

72. Guzzo, C.; Jung, M.; Graveline, A.; Banfield, B.W.; Gee, K. IL-27 increases BST-2 expression in human monocytes and T cells independently of type I IFN. *Sci. Rep.* **2012**, *2*, 974. [CrossRef]

73. Bailey, C.C.; Huang, I.-C.; Kam, C.; Farzan, M. Ifitm3 limits the severity of acute influenza in mice. *PLoS Pathog.* **2012**, *8*, e1002909. [CrossRef]

74. Kroismayr, R.; Baranyi, U.; Stehlik, C.; Dorfleutner, A.; Binder, B.R.; Lipp, J. HERC5, a HECT E3 ubiquitin ligase tightly regulated in LPS activated endothelial cells. *J. Cell Sci.* **2004**, *117*, 4749–4756. [CrossRef]

75. Kühl, A.; Banning, C.; Marzi, A.; Votteler, J.; Steffen, I.; Bertram, S.; Glowacka, I.; Konrad, A.; Stürzl, M.; Guo, J.-T.; et al. The Ebola virus glycoprotein and HIV-1 Vpu employ different strategies to counteract the antiviral factor Tetherin. *J. Infect. Dis.* **2011**, *204*, S850–S860. [CrossRef] [PubMed]

76. Lopez, L.A.; Yang, S.J.; Hauser, H.; Exline, C.M.; Haworth, K.G.; Oldenburg, J.; Cannon, P.M. Ebola virus glycoprotein counteracts BST-2/tetherin restriction in a sequence-independent manner that does not require tetherin surface removal. *J. Virol.* **2010**, *84*, 7243–7255. [CrossRef] [PubMed]

77. Lopez, L.A.; Yang, S.J.; Exline, C.M.; Rengarajan, S.; Haworth, K.G.; Cannon, P.M. Anti-tetherin activities of HIV-1 Vpu and Ebola virus glycoprotein do not involve removal of Tetherin from lipid rafts. *J. Virol.* **2012**, *86*, 5467–5480. [CrossRef]

78. González-Hernández, M.; Hoffmann, M.; Brinkmann, C.; Nehls, J.; Winkler, M.; Schindler, M.; Pöhlmann, S. A GXXXA motif in the transmembrane domain of the ebola virus glycoprotein is required for Tetherin antagonism. *J. Virol.* **2018**, *92*, e00403–e00418. [CrossRef]

79. Burgt, N.H.V.; Kaletsky, R.L.; Bates, P. Requirements within the Ebola Viral Glycoprotein for tetherin antagonism. *Viruses* **2015**, *7*, 5587–5602. [CrossRef]

80. Gustin, J.K.; Bai, Y.; Moses, A.V.; Douglas, J.L. Ebola virus glycoprotein promotes enhanced viral egress by preventing Ebola VP40 from associating with the host restriction factor BST2/Tetherin. *J. Infect. Dis.* **2015**, *212*, S181–S190. [CrossRef]

81. Brinkmann, C.; Nehlmeier, I.; Walendy-Gnirß, K.; Nehls, J.; González Hernández, M.; Hoffmann, M.; Qiu, X.; Takada, A.; Schindler, M.; Pöhlmann, S. The Tetherin antagonism of the Ebola virus glycoprotein requires an intact receptor-binding domain and can be blocked by GP1-specific antibodies. *J. Virol.* **2016**, *90*, 11075–11086. [CrossRef]

82. Gnirß, K.; Fiedler, M.; Krämer-Kühl, A.; Bolduan, S.; Mittler, E.; Becker, S.; Schindler, M.; Pöhlmann, S. Analysis of determinants in filovirus glycoproteins required for tetherin antagonism. *Viruses* **2014**, *6*, 1654–1671. [CrossRef]

83. Wang, M.K.; Lim, S.-Y.; Lee, S.M.; Cunningham, J.M. Biochemical basis for increased activity of Ebola Glycoprotein in the 2013-16 epidemic. *Cell Host Microbe* **2017**, *21*, 367–375. [CrossRef] [PubMed]

84. Marzi, A.; Chadinah, S.; Haddock, E.; Sow, S.; Massaquoi, M.; Feldmann, H. Recently identified mutations in the Ebola Virus-makona genome do not alter pathogenicity in animal models. *Cell Rep.* **2018**, *23*, 1806–1816. [CrossRef]

85. Sullivan, N.J.; Peterson, M.; Yang, Z.; Kong, W.; Duckers, H.; Nabel, E.; Nabel, G.J. Ebola virus glycoprotein toxicity is mediated by a dynamin-dependent protein-trafficking pathway. *J. Virol.* **2005**, *79*, 547–553. [CrossRef] [PubMed]

86. Iampietro, M.; Younan, P.; Nishida, A.; Dutta, M.; Lubaki, N.M.; Santos, R.I.; Koup, R.A.; Katze, M.G.; Bukreyev, A. Ebola virus glycoprotein directly triggers T lymphocyte death despite of the lack of infection. *PLoS Pathog.* **2017**, *13*, e1006397. [CrossRef] [PubMed]

87. Stewart, C.M.; Phan, A.; Bo, Y.; LeBlond, N.D.; Smith, T.K.T.; Laroche, G.; Giguère, P.M.; Fullerton, M.D.; Pelchat, M.; Kobasa, D.; et al. Ebola virus triggers receptor tyrosine kinase-dependent signaling to promote the delivery of viral particles to entry-conducive intracellular compartments. *PLoS Pathog.* **2021**, *17*, e1009275. [CrossRef]

Review

Complex Roles of Neutrophils during Arboviral Infections

Abenaya Muralidharan and St Patrick Reid *

Department of Pathology and Microbiology, University of Nebraska Medical Center, Omaha, NE 68198-5900, USA; abenaya.muralidharan@unmc.edu
* Correspondence: patrick.reid@unmc.edu; Tel.: +402-559-3644

Abstract: Arboviruses are known to cause large-scale epidemics in many parts of the world. These arthropod-borne viruses are a large group consisting of viruses from a wide range of families. The ability of their vector to enhance viral pathogenesis and transmission makes the development of treatments against these viruses challenging. Neutrophils are generally the first leukocytes to be recruited to a site of infection, playing a major role in regulating inflammation and, as a result, viral replication and dissemination. However, the underlying mechanisms through which neutrophils control the progression of inflammation and disease remain to be fully understood. In this review, we highlight the major findings from recent years regarding the role of neutrophils during arboviral infections. We discuss the complex nature of neutrophils in mediating not only protection, but also augmenting disease pathology. Better understanding of neutrophil pathways involved in effective protection against arboviral infections can help identify potential targets for therapeutics.

Keywords: neutrophils; arboviruses; mosquito; inflammation; pathology

Citation: Muralidharan, A.; Reid, S.P. Complex Roles of Neutrophils during Arboviral Infections. *Cells* **2021**, *10*, 1324. https://doi.org/10.3390/cells 10061324

Academic Editors: Thomas Hoenen and Allison Groseth

Received: 30 March 2021
Accepted: 21 May 2021
Published: 26 May 2021

Publisher's Note: MDPI stays neutral with regard to jurisdictional claims in published maps and institutional affil-iations.

1. Introduction

Neutrophils are the most abundant leukocytes in the blood. They serve as the first line of defense against incoming pathogens, quickly mobilizing to the site of infection [1]. While neutrophils can have protective immunostimulatory activities, they can also have debilitating immunosuppressive activities by inhibiting T cell functions [1–3]. In addition, some viruses such as influenza, specifically H5N1, and West Nile virus are known to infect and replicate within neutrophils, using these cells as reservoirs for dissemination, although mechanisms involved in this phenomenon remain unclear [4,5].

Arthropod-borne viruses or arboviruses have a unique effect on neutrophil function since viral factors as well as vector factors can affect the activity of these cells. Arboviruses are a diverse group of viruses [6–8]. They are transmitted through blood-feeding insects such as mosquitoes and ticks, and include viruses such as chikungunya virus (CHIKV), dengue virus (DENV), West Nile virus (WNV), Zika virus (ZIKV), yellow fever virus (YFV), and eastern equine encephalitis virus (EEEV). This group of viruses have been known to cause significant morbidity and mortality around the world, with the potential to spread quickly and expand their geographical range due to their distinct mode of transmission using arthropods. Furthermore, the changes in climate and increase in urbanization help augment transmission and infectivity of these viruses [9–15].

The heterogeneity among this exceptionally large group of viruses makes developing therapeutics challenging. Neutrophils are generally the first to infiltrate infected sites. However, the role of these cells during viral infection is not yet fully understood [16]. Targeting neutrophil pathways or proteins that activate or suppress neutrophils can serve as a useful strategy for drug and vaccine development. In this review, we highlight the recent advancements in understanding the beneficial and detrimental nature of neutrophils during arboviral infections.

2. Mosquitoes

Ticks and mosquitoes enhance disease severity as well as transmission of many viruses [17–21]. Mosquitoes are the most common type of arthropods that spread diseases including malaria, dengue, West Nile, Zika, and chikungunya fevers. Mosquito-transmitted diseases affect hundreds of millions of people each year, resulting in about 750,000 deaths every year [22–25]. During a blood meal, an adult female mosquito punctures the skin of a vertebrate and ingests the fluid. If the vertebrate is infected, the virus is also ingested along with the blood. The virus can then replicate, cross the midgut barrier, and reside in the salivary gland of the mosquito at high titers [26–29]. During the subsequent blood meal, the contents of the salivary gland are released below the skin to counteract the host's hemostasis and inflammatory responses, allowing the virus to enter the epidermis and dermis [30–32].

Mosquitoes' innate immune responses allow them to survive infections, making them effective carriers. They use the RNA interference pathway for protection against viral infections, including arboviruses [33,34]. In addition, *Aedes aegypti* mosquitoes have been shown to use the JAK/STAT pathway in response to WNV, DENV, and YFV [35]. Some species of mosquitoes also utilize the immediate response of apoptosis in the salivary glands and midgut to control viral load [36,37].

The mosquito plays a major role in creating an ideal environment for virus entry and replication. Indeed, the vector secretes anti-hemostatic, angiogenic, and vasodilatory molecules through its saliva to maintain optimum blood flow during feeding [31,32,38,39]. This microenvironment allows for enhanced infection and pathogenicity of the virus by controlling the initial replication of the virus and the potential for the infection to become systemic [38,40–45]. While some groups have hypothesized suppression of anti-viral immune responses by mosquito saliva during infection, Pingen et al. (2016) showed that mosquito bites facilitate infection by triggering a cellular influx that is inadvertently beneficial for the virus [31,32,38,39,46].

The saliva of a mosquito has been shown to contain highly active molecules involved in modulating early viral infection. Indeed, while infecting mice with arboviruses through a mosquito bite or a needle accompanied by an uninfected mosquito bite, the former resulted in more severe disease [17–20,47]. Furthermore, in chickens infected with West Nile virus (WNV) via mosquito bite, significantly high viral titers were observed in the serum compared to the group infected using a needle [48]. Similar augmentation of viremia was seen in mice infected with WNV via bite accompanied by faster neuro-invasion compared to needle-inoculated animals [42].

There are many effects a mosquito bite has on immune cells [49]. Some species of mosquitoes, namely *Anopheles stephensi* and *Anopheles gambiae*, secrete saliva that can result in chemotactic activity. The vascular permeabilization and mast cell degranulation in the skin caused by the saliva were shown to recruit dendritic cells to the feeding site and neutrophils to the draining lymph node [50–52]. Another study using humanized mice showed that seven days post-mosquito bite, there was a decrease in IL-8, a neutrophil chemoattractant, in the serum correlating to a decrease in circulating neutrophils. This corresponded to increased neutrophils in the skin [49].

Pingen et al. used mice infected with aedine mosquito-borne Semliki Forest virus (SFV), an alphavirus shown to replicate efficiently in immune-competent mice, and Bunyamwera virus (BUNV), a genetically unrelated RNA virus [6,46,53,54]. In their study, mosquito bites induced an influx of inflammatory neutrophils, which, in turn, promoted myeloid cell entry into the bite site in a CCR2-dependent manner. This augmented viral infection since myeloid cells are permissive to the virus. Interestingly, viral infection via bite synergistically enhanced CXCL2 and IL-1β expression, and neutrophil influx compared to bite alone [46]. The researchers further confirmed the role of neutrophils by depleting them and blocking inflammasome activity. This resulted in decreased inflammation and a suppressive environment for the viral infection. Depleting neutrophils also significantly reduced edema by further enhancing the vascular leakage caused by the bite. In addition, neutrophil influx into the bite site at earlier stages of infection was required for the induction of vital bite-associated genes such as IL-1β, CCL2, CCL7, and CCL12. Importantly, neutrophil depletion did not affect virus-induced genes, while neutrophils expressing IL-1β were necessary for establishing cutaneous inflammatory responses to mosquito bites [46]. Therefore, factors secreted by the mosquito augment infection by increasing neutrophil-mediated inflammation at the bite site during early stages of infection, which later determines the systemic course of the infection in mice. At later stages, however, neutrophils were required to effectively resolve the infection and decrease mortality in mice. Indeed, higher number of the neutrophil-depleted mice infected with the more virulent SFV6 succumbed to infection compared to neutrophil-sufficient mice [46].

3. Zika Virus

Zika virus (ZIKV) is a flavivirus transmitted mainly by *Aedes* species mosquitoes. ZIKV generally causes fever, cutaneous rash, headache, and malaise [55,56]. However, in the most recent 2015–2016 epidemic in Latin America and the Caribbean that affected more than 1.5 million people [57], ZIKV caused severe congenital malformations in the fetus, commonly known as Congenital Zika Syndrome, [58,59] and Guillain-Barre syndrome [60,61].

Recently, Hastings et al. conducted a study to identify specific antigenic salivary gland proteins in the *Aedes aegypti* mosquito that promotes ZIKV pathogenesis [62]. They used yeast display to identify a molecule in the saliva of the mosquito that can activate neutrophils in the host. The authors named this previously undescribed protein as neutrophil-stimulating factor 1 (NeSt1). When mice were treated with NeSt1-blocking antibodies before being bitten by ZIKV-infected mosquitoes, they had a 50% higher survival rate compared to untreated mice. Furthermore, NeSt1 was shown to activate neutrophils inducing their expression of IL-1β, and monocyte/macrophage-attracting chemokines CXCL2 and CCL2. The recruited macrophages may then be infected by the virus increasing the viral load [62]. Overall, NeSt1 stimulated neutrophils at the bite site augmenting early viral infection and ZIKV pathogenesis (Figure 1).

In contrast, another study using adult AG129 interferon α/β receptor knockout mice infected with a recent strain of ZIKV showed the protective effects of neutrophils [63]. In this mouse model, ZIKV has been shown to infect astrocytes and neurons in the brain and spinal cord. Zukor et al. observed that this infection resulted in astrogliosis along with T cell and neutrophil infiltration. The neutrophil recruitment inversely correlated with the virus-induced paresis protecting infected mice from motor deficits, indicating that neutrophils may be required for controlling ZIKV-induced disease [63]. Mechanisms underlying this protection need to be further explored. It is important to note that the differences in neutrophil activity observed in this study compared to Hastings et al. may be attributed to the absence of a mosquito vector or mosquito salivary components during infection of mice.

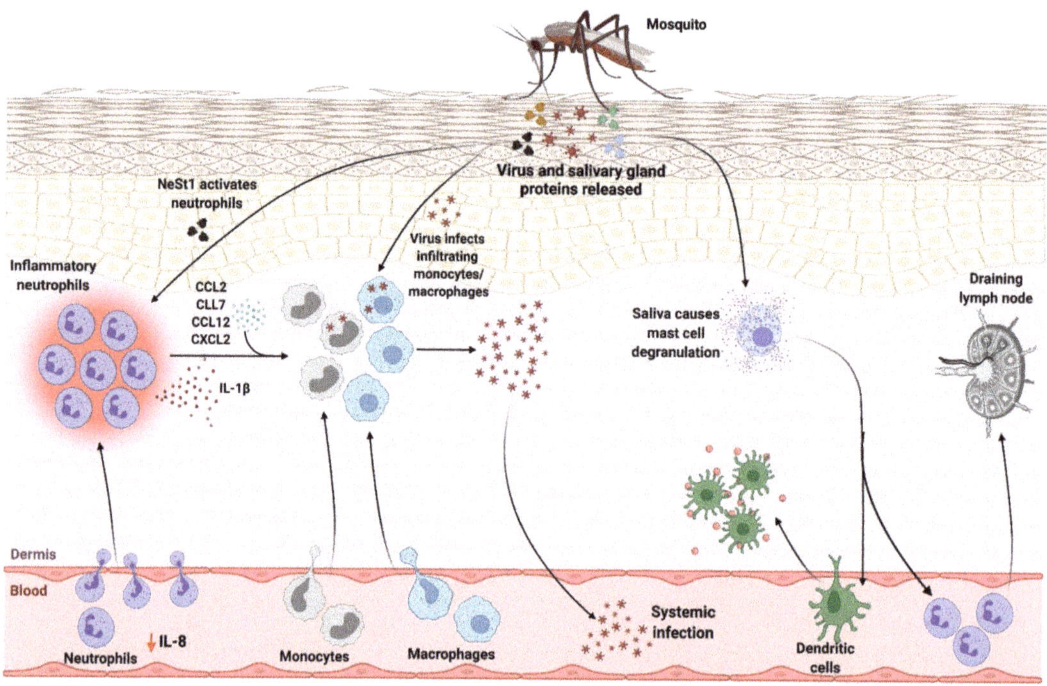

Figure 1. Neutrophil-mediated viral replication and dissemination induced by mosquito saliva. During a blood meal, mosquito carrying an arbovirus injects the virus along with its salivary gland proteins below the skin of the host. There is a decrease in IL-8 levels in the serum correlating to lower number of circulating neutrophils and higher number in the skin. One of the proteins in the saliva, neutrophil-stimulating factor 1 (NeSt1), activates the neutrophils in the dermis, the deepest layer of the skin, which houses the immune cells. IL-1β is secreted by these inflammatory neutrophils to establish cutaneous response to the bite. Additionally, bite-associated monocyte/macrophage-attracting chemokines, CCL2, CCL7, CCL12, and CXCL2, are upregulated. The infiltrating monocytes and macrophages are permissive to infection enhancing viral replication and increasing the potential for systemic spread. The mosquito saliva also causes vascular permeabilization and mast cell degranulation in the skin recruiting dendritic cells to the bite site, contributing to the inflammation, and neutrophils to the draining lymph nodes.

4. Dengue Virus

Dengue virus (DENV) can cause clinical outcomes that range from mild febrile illness to dengue fever to dengue hemorrhagic fever to life-threatening dengue shock syndrome [64]. With approximately 2.5 billion people at risk globally, DENV is the most common arbovirus [65,66]. Clinical studies of adult dengue patients showed severe neutropenia with lowest levels occurring five days post-infection [67,68]. The neutropenia, however, was not predictive of severe virus-induced disease or associated with prolonged hospital stay or death [67]. Interestingly, the low level of neutrophils was not for a lack of activation signals. In fact, neutrophil-activating cytokines, IL-8 and TNF-α, were high during DENV infections [69], while neutrophil-associated genes such as DEF4A, CEACAM8, BPI, and ELA2 were upregulated in the blood during severe DENV infection [70]. Neutrophil elastase levels were also increased in DENV-infected patients compared to uninfected controls, with higher elastase activity in patients with dengue hemorrhagic fever compared to dengue fever patients [71]. This suggests that enhanced neutrophil activation can be associated with severe disease.

In another study, researchers observed the formation of neutrophil extracellular traps (NETs) in vitro induced by DENV [72]. NET formation or NETosis consists of nuclear decondensation and delobulation, plasma membrane rupture, and release of DNA fibers that have anti-microbial peptides [73]. Although NETs play a crucial role when fighting infections, excessive NETosis and/or ineffective NET clearance can contribute to development of autoimmune diseases and inflammatory disorders [74,75]. Indeed, several NET-associated molecules, such as double-stranded DNA, histones, etc., are known to be autoantigens in systemic autoimmune diseases [74]. For instance, autoantibodies against NET components have been seen in systemic lupus erythematosus patients as well as an imbalance between NET formation and clearance, making them more prone to NET-mediated tissue damage [76–79]. Furthermore, NETs have also been implicated in the pathogenesis of inflammatory conditions including, but not limited to, small vessel vasculitis, psoriasis, and gout [74].

Examining the phenotypic and functional responses of neutrophils in adult dengue patients, Opasawatchai et al. observed an upregulation of CD66b on neutrophils and early stages of NET formation, indicating an activated state, during acute DENV infection [80]. CD66b is a granulocyte activation marker involved in degranulation and production of reactive oxygen species (ROS), which is essential for antiviral activity [81,82]. Interestingly, higher levels of NET components, IL-8, and TNF-α were found in patients diagnosed with the more severe dengue hemorrhagic fever compared to patients with dengue fever or healthy controls [80]. A study by Lien et al. identified the viral factor crucial for inducing NETosis in vitro and in mice to be DENV envelope protein domain III (EIII). This NET formation was alleviated in neutrophils from NLRP3 inflammasome-deficient mice, decreasing inflammation. Blocking EIII-neutrophil interactions also suppressed the NETosis [83].

The most severe disease caused by DENV comprises of systemic inflammation and increased vascular permeability. Many studies have also shown the activation of macrophages and platelets leading to an increase in proinflammatory cytokines and extracellular vesicles (EVs) [84–87] that transport proteins, peptides, and nucleic acids from one cell to another to modulate cell functions [88]. Indeed, DENV-induced release of IL-1β-containing EVs by platelets increased vascular permeability [87].

In addition, DENV enhanced release of EVs by activated platelets, which further activated CLEC5A, a spleen tyrosine kinase (Syk)-coupled C-type lectin receptor, and toll-like receptor 2 (TLR2) on neutrophils and macrophages. This induced NET formation and proinflammatory cytokine release [89]. Activation of CLEC5A is known to trigger NALP3 inflammasome activation and proinflammatory cytokine response [85,86,90], which augments systemic vascular permeability and hemorrhagic shock [86,91]. While blocking CLEC5A did not fully protect mice infected with a lethal dose of DENV [91], simultaneous blockade of CLEC5A and TLR2 significantly alleviated virus-induced inflammation and improved survival [89]. Together, these studies highlight the complex ways in which neutrophils mediate disease during the different stages of DENV infection.

5. West Nile Virus

Belonging to the same Flaviviridae family as DENV, West Nile virus (WNV) is a neuroinvasive pathogen [92]. WNV infection is typically only symptomatic in the elderly and immunocompromised individuals causing life-threatening neurological disease such as meningitis and encephalitis [92–95]. Strikingly, high levels of neutrophils were found in the cerebrospinal fluid collected from patients with WNV-induced disease, suggesting a major role of neutrophils in viral pathogenesis [96,97].

In mice infected with WNV, a rapid influx of neutrophils was seen at the site of infection promoting viral replication. Indeed, the expression of CXCL1 and CXCL2, neutrophil-attracting chemokines, was significantly upregulated in macrophages upon infection [5]. Interestingly, neutrophil-depletion studies revealed a dual role of these leukocytes during infection. Neutrophils were required for effective clearance of WNV and survival shown by higher viremia and death rate in mice depleted of neutrophils after infection. However, these cells were detrimental to the mice during early stages of infection since neutrophil depletion before WNV infection reduced viral burden and enhanced survival [5]. Overall, neutrophils can serve as reservoirs for WNV replication and dissemination as well as help defend against the virus at different stages of infection.

6. Alphaviruses

The alphavirus genus consists of many arthropod-borne viruses that are typically divided into two main groups, New World and Old World alphaviruses. New World alphaviruses such as eastern equine encephalitis virus (EEEV), western equine encephalitis virus (WEEV), and Venezuelan equine encephalitis virus (VEEV) cause encephalomyelitis in humans and are found in North and South America [98]. Old World alphaviruses that include chikungunya virus (CHIKV), Ross River virus (RRV), Mayaro virus, and o'nyong-nyong virus, are now found in Europe, Africa, Asia, and Oceania and generally induce fever, rash, and arthritis [8,99].

Although New World alphaviruses, such as EEEV, can have mortality rates as high as 70%, while Old World alphaviruses rarely cause death, the latter has caused many epidemics in the past, resulting in high infection rates [100]. A RRV epidemic in 1979–1980 in the South Pacific involved more than 60,000 patients [101] while the o'nyong nyong virus infected approximately 2 million people in Africa in the 1959–1962 epidemic [102]. CHIKV has caused reoccurring epidemics in numerous countries around the Indian Ocean since 2004 with millions of confirmed cases [103] and a surprising emergence in Europe and the Pacific Region for the first time in 2007 and 2011, respectively [104–107].

Humans and horses infected with New World alphaviruses show changes in the central nervous system characterized by high levels of neutrophil infiltration during early stages of disease, which is replaced by lymphocytes as the disease progresses [108,109]. Due to the lack of literature on the roles of neutrophils during New World alphavirus infection, we will focus on Old World alphaviruses in this section.

Old World alphaviruses can cause musculoskeletal inflammatory disease in humans that can be significantly debilitating. Infection with arthritis/myositis-associated alphaviruses can present with fever, joint pain, myalgia, and impaired movement [101,110]. Importantly, the musculoskeletal pain induced by arthritogenic alphaviruses can last for months to years in RRV- or CHIKV-infected individuals [111–117]. Many studies have been conducted to determine the cause of such chronic pain. In one study by Stoermer et al., RRV infection in mice with specific deletion of arginase 1 (Arg1) in neutrophils and macrophages was well controlled at later stages of infection enhancing viral clearance from musculoskeletal tissues and improving skeletal muscle tissue pathology [118]. Arg1 is expressed by monocytes/macrophages, neutrophils, and myeloid-derived suppressor cells (MDSCs) and plays an important role in regulating immune responses [119–121]. Although LysMCre Arg1$^{f/f}$ mice, with conditional deletion of Arg1 in macrophages and neutrophils, had no change in the disease outcomes during the acute phase of infection, significantly enhanced protection was observed in the late stages of RRV infection [118]. Furthermore, conditional knockout of Arg1 substantially reduced Arg1 expression in musculoskeletal tissues following CHIKV and RRV infection, suggesting that macrophages and neutrophils are the predominant cells at the inflammatory sites following arthritogenic alphavirus infection [118]. Overall, the study highlighted the crucial role of Arg1 in contributing to disease severity. Specific neutrophil depleting methods such as Ly6G antibody treatments could help further narrow down the responsible cell type.

CHIKV infection is primarily characterized by macrophage and monocyte infiltration into the primary sites of virus replication, which are typically the skin, muscle, and joints. However, an influx of neutrophils, dendritic cells, natural killer cells, and lymphocytes has also been observed [122]. Indeed, resident cells at the site of infection produce neutrophil-attracting chemokines, CXCL1 and CXCL2, following other viral infections [123,124]. This chemokine production by resident cells remains to be seen during CHIKV infection.

The recruited neutrophils produce ROS and other cytotoxic mediators to decrease viral replication [125]. In non-mammalian models of CHIKV infection such as zebrafish, the neutrophils also serve as an important source of type I interferon for eliminating the virus and alleviating disease [126]. Even in the absence of active viral replication during chronic phases of infection, CHIKV-induced arthritis may progress due to increased cytokine expression and immune cell infiltration [122,127].

A recent study found the role of CXCL10, a chemoattractant for monocytes/macrophages and T cells, during alphaviral infections using CHIKV and o'nyong nyong mouse models. At the peak of arthritic disease, which occurs 6 to 8 days post infection in mice, CXCL10$^{-/-}$ mice had decreased levels of immune infiltration as well as viral loads at the site of viral inoculation, the footpad, compared to wild-type mice [128]. The predominant populations in the infiltrates were macrophages and neutrophils in the wild-type mice following infection but this influx was significantly reduced in the CXCL10$^{-/-}$ mice. Interestingly, viral RNA was detected in these immune cells in wild-type mice, which was also significantly decreased in the knockout mice [128].

In another study, the role of NETs during CHIKV infection was explored. Ex vivo stimulation and infection of mouse-isolated neutrophils induced the release of NETs in a TLR7- and ROS-dependent manner neutralizing CHIKV [129]. The researchers used TLR3/7/9 triple knockout mice with TLR3$^{-/-}$ and TLR9$^{-/-}$ mice as controls due to the unavailability of TLR7$^{-/-}$ mice. Although knockout of TLR3 and TLR9 did not affect NET production after CHIKV infection, there may be some synergistic effects of the triple knockout affecting virus-mediated NET release [129]. In vivo infection of IFNAR$^{-/-}$ mice following NET inhibition enhanced susceptibility of the mice to an acute CHIKV infection confirming a crucial antiviral role of NETs. Moreover, clinical data also showed a correlation between the level of NETs in the blood and systemic viral loads in CHIKV infected patients [129–131]. Even though the role of NETs has been established during an acute CHIKV infection, they may also play a part during chronic infection. Indeed, neutrophils infiltrate the synovium and release NETs leading to damage of the joint tissues in rheumatoid arthritis [132].

CCR2 has been implicated in playing a protective role during CHIKV infection by preventing neutrophil-mediated pathology. CCR2$^{-/-}$ mice infected with CHIKV in the hind feet showed decreased levels of monocyte/macrophage infiltration with substantial increase in neutrophil infiltration, followed by eosinophils, compared to wild-type mice [133]. This change in cellular influx was associated with increased levels of CXCL1, CXCL2, G-CSF, and IL-1β with a decrease in IL-10, promoting neutrophil recruitment and exacerbating inflammation [134–140]. The eosinophil infiltration may be promoted by neutrophil-induced tissue damage to help control the inflammation in infected CCR2$^{-/-}$ mice [141]. CCR2 deficiency also led to cartilage damage in mice following CHIKV infection, which is normally not a symptom of alphaviral arthritis [133]. In fact, elevated macrophage and neutrophil infiltrates in CCR2$^{-/-}$ mice with collagen-induced arthritis is accompanied by more severe disease [134,142]. Interestingly, Poo et al. attempted neutrophil depletion in CCR2$^{-/-}$ mice after CHIKV infection, which resulted in new pathology characterized by increased foot swelling along with widespread hemorrhage and edema [133,143]. This, once again, may be suggestive of a dual role of neutrophils, where they are detrimental during certain stages of infection while protective during others.

Another group delineated the role of neutrophils during a pathogenic CHIKV infection on B cell maturation and lymphocyte influx. McCarthy et al. found that mice infected with a wild-type, not acutely cleared, strain of CHIKV had recruitment of monocytes and neutrophils to the draining lymph node (dLN). This aberrantly affected lymphocyte accumulation, lymph node organization, and virus-specific B cell responses, which was reversed by blocking the influx [144]. Interestingly, only pathogenic CHIKV decreased germinal center formation in the dLN, resulting in lower neutralizing antibodies in the serum compared to infection with an attenuated strain [145,146]. These diminished B cell responses were improved upon depletion of monocytes and neutrophils during early stages of infection [144].

Depleting either monocytes or neutrophils did not restore lymphocyte counts in the dLN, indicating that one of the two cell types is sufficient to block lymphocyte infiltration [144]. Furthermore, mice lacking type I interferon signaling (IFNAR$^{-/-}$) had higher percentage of neutrophils in the dLN compared to wild-type mice following pathogenic CHIKV infection. In contrast, MyD88-deficient mice and wild-type mice treated with IL-1 receptor (IL-1R) blocking antibody at the time of infection had reduced the percentage of neutrophils [144]. Together, MyD88-IL-1R signaling plays a crucial role in promoting the accumulation of neutrophils in the dLN while type I interferon signaling inhibits the recruitment during pathogenic CHIKV infection [144].

While IFN-α was observed to inhibit neutrophil influx into the dLN, IFN-β was found to inhibit neutrophil infiltration into the musculoskeletal tissues during CHIKV infection [147]. Following CHIKV inoculation in the foot of IFN-$\beta^{-/-}$ mice, although no change was observed in viral load in the foot or the blood compared to wild-type mice, there were increased levels of neutrophils in the foot [147]. Neutrophil depletion in IFN-$\beta^{-/-}$ mice alleviated musculoskeletal disease induced by CHIKV observed through significantly reduced foot swelling. On the other hand, IFN-$\alpha^{-/-}$ mice had higher viral burdens at the site of infection and in circulation [147]. This indicates that IFN-α helps limit viral replication whereas IFN-β modulates neutrophil recruitment to the site of infection that is necessary for exacerbation of disease pathology (Figure 2). Curiously, neither neutrophil-attracting chemokines nor proinflammatory cytokines were upregulated in the IFN-$\beta^{-/-}$ mice to accompany the neutrophil-mediated inflammation making the mechanism through which IFN-β regulates neutrophil infiltration during acute CHIKV infection unclear [147].

It is important to note that most studies deplete neutrophils in vivo to understand their function. All the studies involving neutrophil depletion referenced in this review use Ly6G antibody treatments in mice. While Ly6G may be transiently expressed on many myeloid cells in the bone marrow including monocytes and other granulocytes, neutrophils that are circulating and recruited to the site of inflammation typically have higher Ly6G expression [148]. Basophils and eosinophils are thought to be Ly6G$^-$ or Ly6G$^{low/intermediate}$ [148]. Additionally, some studies showed that Ly6G-mediated neutrophil depletion reduced only the Ly6C^{intermediate} neutrophil population and not the Ly6C^{high} monocyte population [144].

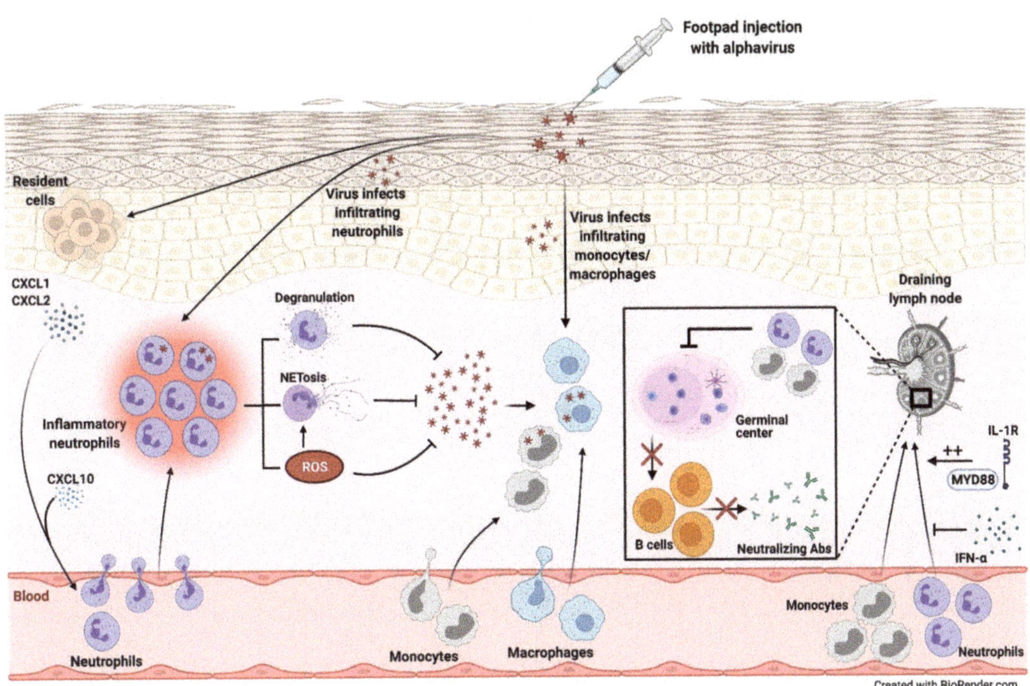

Figure 2. Alphavirus infection-induced neutrophil recruitment and inflammation. Following footpad injection of mice with alphavirus, levels of neutrophil-attracting chemokines, CXCL1 and CXCL2, increase. These chemokines and CXCL10 recruit neutrophils, which release reactive oxygen species (ROS), neutrophil extracellular traps (NETs), and other cytotoxic mediators through degranulation, promoting viral clearance. The infiltrating neutrophils can also be directly infected, triggering NET release in a ROS-dependent manner. Furthermore, monocytes/macrophages infiltrating the site of infection can be directly infected. On the other hand, alphaviral infections induce monocyte and neutrophil recruitment into the draining lymph node (dLN) that inhibit germinal center formation decreasing B cell maturation and neutralizing antibody (Ab) production. MyD88-IL-1R signaling promotes the accumulation of neutrophils in the dLN, while IFN-α inhibits this influx.

7. Conclusions

Neutrophils are key players in the immune system, being the most abundant leukocytes. They are one of the first responders to the site of infection. However, the heterogeneity of their roles and the variability from one infection to another makes it difficult to determine if the effects will be beneficial or detrimental to the host (Table 1). Arboviruses not only induce neutrophil-mediated inflammation using viral factors but also through factors in their vector, adding another level of complexity. Their mode of transmission through arthropods immensely increases the rate at which they spread, highlighting the need for better understanding of the underlying mechanisms involved in pathogenesis. During arboviral infections, the time and amount of neutrophil infiltration to the site of infection may have a significant effect on the outcome. Following infection, an early influx with a high number of hyperactivated neutrophils releasing high levels of IL-1β, ROS, and NETs may augment infection and disease. However, an influx at later stages of infection may be protective. Regularly causing epidemics in the vulnerable areas of the world, arboviral infections need to be controlled with unique therapeutics that can control the vector-mediated and virus-mediated symptoms. Neutrophils are implicated in disease pathology induced by arboviruses and their vectors, making them a potential therapeutic target. In-depth understanding of the neutrophil pathways involved may be crucial for successful treatment of arboviral infections.

Table 1. Summary of the roles of neutrophils during various arboviral infections.

Arbovirus	Beneficial Roles of Neutrophils	Detrimental Roles of Neutrophils	Replication in Neutrophils
Zika virus (ZIKV)	• In AG129 IFN-α/β receptor knockout mice, ZIKV induced neutrophil infiltration, which inversely correlated with virus-induced paresis protecting mice from motor deficits [63]	• Neutrophil-stimulating factor 1 (NeSt1) in mosquito saliva activated neutrophils, inducing IL-1β, CXCL2 and CCL2 expression, which augmented early viral infection in mice [62]	
Dengue virus (DENV)		Clinical studies: • Neutrophil-associated genes such as DEF4A, CEACAM8, BPI, and ELA2 were upregulated in the blood during severe DENV infection [70] • Patients with the more severe dengue hemorrhagic fever had higher neutrophil elastase activity [71] and higher levels of NET components [80] than dengue fever patients or healthy controls • In adult patients with acute DENV, CD66b was upregulated on neutrophils and there were early stages of NET formation [80] • Mice: • DENV enhanced release of extracellular vesicles by activated platelets activating CLEC5A and TLR2 on neutrophils and macrophages inducing NET formation and proinflammatory cytokine release decreasing survival [89]	
West Nile virus (WNV)	• Neutrophil depletion after infection increased viremia and death rate in mice [5]	• Neutrophil depletion before infection reduced viral burden and enhanced survival in mice [5]	Yes [5]
Semliki Forest virus (SFV)	Mice: • At later stages of infection, neutrophils were required to resolve the infection and decrease mortality [46]	Mice: • Mosquito bite induced an influx of inflammatory neutrophils which promoted CCR2-dependent myeloid cell entry augmenting viral infection [46] • Mosquito bite and virus synergistically enhanced CXCL2 and IL-1β expression, and neutrophil influx [46] • Depleting neutrophils and blocking inflammasome activity in the early stages of infection decreased inflammation, suppressed viral infection, and reduced edema [46]	
Ross River virus (RRV)		• Deletion of Arg1 in macrophages and neutrophils enhanced viral clearance and improved skeletal muscle tissue pathology in late stages of infection in mice, with no effect in the acute phase of infection [118]	

Table 1. *Cont.*

Arbovirus	Beneficial Roles of Neutrophils	Detrimental Roles of Neutrophils	Replication in Neutrophils
Chikungunya virus (CHIKV)	Zebrafish: • Neutrophils served as a major source of type I interferon for eliminating the virus and alleviating disease [126] Mice: • Neutrophils induced release of NETs in a TLR- and ROS-dependent manner, neutralizing CHIKV during an acute infection [129] • Neutrophil depletion in CCR2$^{-/-}$ mice after CHIKV infection increased foot swelling along with widespread hemorrhage and edema [133]	Mice: • CCR2 deficiency increased neutrophil infiltration at the site of infection exacerbating inflammation [133] • Recruitment of neutrophils to the draining lymph node (dLN) following pathogenic CHIKV infection affected lymphocyte accumulation, lymph node organization, and decreased germinal center formation resulting in lower virus-specific neutralizing antibodies in the serum [144] • Neutrophil depletion in IFN-β$^{-/-}$ mice alleviated CHIKV-induced musculoskeletal disease with reduced foot swelling [147]	
O'nyong nyong virus (ONNV)		• CXCL10$^{-/-}$ mice had reduced influx of macrophages and neutrophils, which was associated with decreased viral loads and foot swelling [128]	Yes [128]

Author Contributions: Conceptualization, A.M. and S.P.R.; writing—original draft preparation, A.M.; writing—review and editing, S.P.R. All authors have read and agreed to the published version of the manuscript.

Funding: This work was funded by startup funds for S.P.R.

Institutional Review Board Statement: Not applicable.

Informed Consent Statement: Not applicable.

Data Availability Statement: Not applicable.

Acknowledgments: All figures were created using Biorender.com (accessed on 21 May 2021).

Conflicts of Interest: The authors declare no conflict of interest.

References

1. Leliefeld, P.H.C.; Koenderman, L.; Pillay, J. How neutrophils shape adaptive immune responses. *Front. Immunol.* **2015**, *6*, 471. [CrossRef] [PubMed]
2. Geerdink, R.J.; Pillay, J.; Meyaard, L.; Bont, L. Neutrophils in respiratory syncytial virus infection: A target for asthma prevention. *J. Allergy Clin. Immunol.* **2015**, *136*, 838–847. [CrossRef] [PubMed]
3. Li, Y.; Wang, W.; Yang, F.; Xu, Y.; Feng, C.; Zhao, Y. The regulatory roles of neutrophils in adaptive immunity. *Cell Commun. Signal.* **2019**, *17*, 147. [CrossRef] [PubMed]
4. Zhao, Y.; Lu, M.; Lau, L.T.; Lu, J.; Gao, Z.; Liu, J.; Yu, A.C.H.; Cao, Q.; Ye, J.; McNutt, M.A.; et al. Neutrophils may be a vehicle for viral replication and dissemination in human h5n1 avian influenza. *Clin. Infect. Dis.* **2008**, *47*, 1575–1578. [CrossRef] [PubMed]
5. Bai, F.; Kong, K.F.; Dai, J.; Qian, F.; Zhang, L.; Brown, C.R.; Fikrig, E.; Montgomery, R.R. A paradoxical role for neutrophils in the pathogenesis of West Nile virus. *J. Infect. Dis.* **2010**, *202*, 1804–1812. [CrossRef]
6. Elliott, R.M. Orthobunyaviruses: Recent genetic and structural insights. *Nat. Rev. Microbiol.* **2014**, *12*, 673–685. [CrossRef] [PubMed]
7. Gould, E.; Solomon, T. Pathogenic flaviviruses. *Lancet* **2008**, *371*, 500–509. [CrossRef]
8. Powers, A.M.; Brault, A.C.; Shirako, Y.; Strauss, E.G.; Kang, W.; Strauss, J.H.; Weaver, S.C. Evolutionary relationships and systematics of the alphaviruses. *J. Virol.* **2001**, *75*, 10118–10131. [CrossRef]
9. Durbin, A.P.; Mayer, S.V.; Rossi, S.L.; Amaya-Larios, I.Y.; Ramos-Castaneda, J.; Eong Ooi, E.; Jane Cardosa, M.; Munoz-Jordan, J.L.; Tesh, R.B.; Messer, W.B.; et al. Emergence potential of sylvatic dengue virus type 4 in the urban transmission cycle is restrained by vaccination and homotypic immunity. *Virology* **2013**, *439*, 34–41. [CrossRef] [PubMed]
10. Huang, Y.J.S.; Higgs, S.; Horne, K.M.E.; Vanlandingham, D.L. Flavivirus-Mosquito interactions. *Viruses* **2014**, *6*, 4703–4730. [CrossRef] [PubMed]
11. Lambrechts, L.; Scott, T.W.; Gubler, D.J. Consequences of the expanding global distribution of aedes albopictus for dengue virus transmission. *PLoS Negl. Trop. Dis.* **2010**, *4*, e646. [CrossRef]
12. Miller, M.J.; Loaiza, J.R. Geographic expansion of the invasive mosquito aedes albopictus across Panama—Implications for control of dengue and chikungunya viruses. *Plos Negl. Trop. Dis.* **2015**, *9*. [CrossRef] [PubMed]
13. Guagliardo, S.A.; Barboza, J.L.; Morrison, A.C.; Astete, H.; Vazquez-Prokopec, G.; Kitron, U. Patterns of geographic expansion of aedes aegypti in the Peruvian Amazon. *PLoS Negl. Trop. Dis.* **2014**, *8*. [CrossRef] [PubMed]
14. Díaz-Nieto, L.M.; Maciá, A.; Perotti, M.A.; Berón, C.M. Geographical limits of the southeastern distribution of aedes aegypti (diptera, culicidae) in Argentina. *Plos Negl. Trop. Dis.* **2013**, *7*. [CrossRef] [PubMed]
15. MacKenzie, J.S.; Williams, D.T. The zoonotic flaviviruses of southern, South-Eastern and eastern Asia, and Australasia: The potential for emergent viruses. *Zoonoses Public Health* **2009**, *56*, 338–356. [CrossRef]
16. Gabriel, C.; Her, Z.; Ng, L.F.P. Neutrophils: Neglected players in viral diseases. *DNA Cell Biol.* **2013**, *32*, 665–675. [CrossRef]
17. Cox, J.; Mota, J.; Sukupolvi-Petty, S.; Diamond, M.S.; Rico-Hesse, R. Mosquito bite delivery of dengue virus enhances immunogenicity and pathogenesis in humanized mice. *J. Virol.* **2012**, *86*, 7637–7649. [CrossRef]
18. Edwards, J.F.; Higgs, S.; Beaty, B.J. Mosquito feeding-induced enhancement of cache valley virus (Bunyaviridae) infection in mice. *J. Med. Entomol.* **1998**, *35*, 261–265. [CrossRef] [PubMed]
19. Limesand, K.H.; Higgs, S.; Pearson, L.D.; Beaty, B.J. Potentiation of vesicular stomatitis New Jersey virus infection in mice by mosquito saliva. *Parasite Immunol.* **2000**, *22*, 461–467. [CrossRef]
20. Schneider, B.S.; Soong, L.; Girard, Y.A.; Campbell, G.; Mason, P.; Higgs, S. Potentiation of West Nile encephalitis by mosquito feeding. *Viral Immunol.* **2006**, *19*, 74–82. [CrossRef]
21. Dessens, J.T.; Nuttall, P.A. Mx1-based resistance to thogoto virus in A2G mice is bypassed in tick-mediated virus delivery. *J. Virol.* **1998**, *72*, 8362–8364. [CrossRef]
22. Bhatt, S.; Gething, P.W.; Brady, O.J.; Messina, J.P.; Farlow, A.W.; Moyes, C.L.; Drake, J.M.; Brownstein, J.S.; Hoen, A.G.; Sankoh, O.; et al. The global distribution and burden of dengue. *Nature* **2013**, *496*, 504–507. [CrossRef] [PubMed]
23. Weaver, S.C.; Lecuit, M. Chikungunya virus and the global spread of a mosquito-borne disease. *N. Engl. J. Med.* **2015**, *372*, 1231–1239. [CrossRef]

24. WHO. *A Global Brief on Vector-Borne Diseases*; WHO: Geneva, Switzerland, 2014.
25. WHO. *WHO Factsheet: Vector-Borne Diseases, Factsheet #387*; WHO: Geneva, Switzerland, 2014.
26. Luplertlop, N.; Surasombatpattana, P.; Patramool, S.; Dumas, E.; Wasinpiyamongkol, L.; Saune, L.; Hamel, R.; Bernard, E.; Sereno, D.; Thomas, F.; et al. Induction of a peptide with activity against a broad spectrum of pathogens in the aedes aegypti salivary gland, following infection with dengue virus. *PLoS Pathog.* **2011**, *7*, e1001252. [CrossRef]
27. Salazar, M.I.; Richardson, J.H.; Sánchez-Vargas, I.; Olson, K.E.; Beaty, B.J. Dengue virus type 2: Replication and tropisms in orally infected Aedes aegypti mosquitoes. *BMC Microbiol.* **2007**, *7*, 9. [CrossRef]
28. Vazeille, M.; Mousson, L.; Martin, E.; Failloux, A.-B. Orally co-infected aedes albopictus from La Reunion Island, Indian Ocean, can deliver both dengue and chikungunya infectious viral particles in their saliva. *PLoS Negl. Trop. Dis.* **2010**, *4*, e706. [CrossRef] [PubMed]
29. Ziegler, S.A.; Nuckols, J.; McGee, C.E.; Huang, Y.J.S.; Vanlandingham, D.L.; Tesh, R.B.; Higgs, S. In vivo imaging of chikungunya virus in mice and aedes mosquitoes using a Renilla luciferase clone. *Vector Borne Zoonotic Dis.* **2011**, *11*, 1471–1477. [CrossRef]
30. Heath, W.R.; Carbone, F.R. The skin-resident and migratory immune system in steady state and memory: Innate lymphocytes, dendritic cells and T cells. *Nat. Immunol.* **2013**, *14*, 978–985. [CrossRef] [PubMed]
31. Ribeiro, J.M.C.; Francischetti, I.M.B. Role of arthropod saliva in blood feeding: Sialome and post-sialome perspectives. *Annu. Rev. Entomol.* **2003**, *48*, 73–88. [CrossRef]
32. Schneider, B.S.; Higgs, S. The enhancement of arbovirus transmission and disease by mosquito saliva is associated with modulation of the host immune response. *Trans. R. Soc. Trop. Med. Hyg.* **2008**, *102*, 400–408. [CrossRef]
33. Olson, K.E.; Blair, C.D. Arbovirus-mosquito interactions: RNAi pathway. *Curr. Opin. Virol.* **2015**, *15*, 119–126. [CrossRef]
34. Ahlers, L.R.H.; Goodman, A.G. The immune responses of the animal hosts of West Nile virus: A comparison of insects, birds, and mammals. *Front. Cell. Infect. Microbiol.* **2018**, *8*, 96. [CrossRef]
35. Colpitts, T.M.; Cox, J.; Vanlandingham, D.L.; Feitosa, F.M.; Cheng, G.; Kurscheid, S.; Wang, P.; Krishnan, M.N.; Higgs, S.; Fikrig, E. Alterations in the Aedes aegypti Transcriptome during Infection with West Nile, Dengue and Yellow Fever Viruses. *PLoS Pathog.* **2011**, *7*, e1002189. [CrossRef] [PubMed]
36. Vaidyanathan, R.; Scott, T.W. Apoptosis in mosquito midgut epithelia associated with West Nile virus infection. *Apoptosis* **2006**, *11*, 1643–1651. [CrossRef]
37. Girard, Y.A.; Schneider, B.S.; McGee, C.E.; Wen, J.; Han, V.C.; Popov, V.; Mason, P.W.; Higgs, S. Salivary gland morphology and virus transmission during long-term cytopathologic west Nile virus infection in Culex mosquitoes. *Am. J. Trop. Med. Hyg.* **2007**, *76*, 118–128. [CrossRef]
38. Patramool, S.; Choumet, V.; Surasombatpattana, P.; Sabatier, L.; Thomas, F.; Thongrungkiat, S.; Rabilloud, T.; Boulanger, N.; Biron, D.G.; Missé, D. Update on the proteomics of major arthropod vectors of human and animal pathogens. *Proteomics* **2012**, *12*, 3510–3523. [CrossRef]
39. Briant, L.; Després, P.; Choumet, V.; Missé, D. Role of skin immune cells on the host susceptibility to mosquito-borne viruses. *Virology* **2014**, *464–465*, 26–32. [CrossRef]
40. Le Coupanec, A.; Babin, D.; Fiette, L.; Jouvion, G.; Ave, P.; Misse, D.; Bouloy, M.; Choumet, V. Aedes Mosquito Saliva Modulates Rift Valley Fever Virus Pathogenicity. *PLoS Negl. Trop. Dis.* **2013**, *7*, e2237. [CrossRef]
41. Schneider, B.S.; Soong, L.; Coffey, L.L.; Stevenson, H.L.; McGee, C.E.; Higgs, S. Aedes aegypti saliva alters leukocyte recruitment and cytokine signaling by antigen-presenting cells during West Nile Virus Infection. *PLoS ONE* **2010**, *5*, e11704. [CrossRef]
42. Styer, L.M.; Lim, P.-Y.; Louie, K.L.; Albright, R.G.; Kramer, L.D.; Bernard, K.A. Mosquito saliva causes enhancement of West Nile virus infection in mice. *J. Virol.* **2011**, *85*, 1517–1527. [CrossRef]
43. Surasombatpattana, P.; Ekchariyawat, P.; Hamel, R.; Patramool, S.; Thongrungkiat, S.; Denizot, M.; Delaunay, P.; Thomas, F.; Luplertlop, N.; Yssel, H.; et al. Aedes aegypti saliva contains a prominent 34-kDa protein that strongly enhances dengue virus replication in human keratinocytes. *J. Investig. Derm.* **2014**, *134*, 281–284. [CrossRef] [PubMed]
44. Thangamani, S.; Higgs, S.; Ziegler, S.; Vanlandingham, D.; Tesh, R.; Wikel, S. Host immune response to mosquito-transmitted chikungunya virus differs from that elicited by needle inoculated virus. *PLoS ONE* **2010**, *5*, e12137. [CrossRef]
45. McCracken, M.K.; Christofferson, R.C.; Chisenhall, D.M.; Mores, C.N. Analysis of early dengue virus infection in mice as modulated by aedes aegypti probing. *J. Virol.* **2014**, *88*, 1881–1889. [CrossRef]
46. Pingen, M.; Bryden, S.R.; Pondeville, E.; Schnettler, E.; Kohl, A.; Merits, A.; Fazakerley, J.K.; Graham, G.J.; McKimmie, C.S. Host inflammatory response to mosquito bites enhances the severity of arbovirus infection. *Immunity* **2016**, *44*, 1455–1469. [CrossRef] [PubMed]
47. Schneider, B.S.; Soong, L.; Zeidner, N.S.; Higgs, S. Aedes aegypti salivary gland extracts modulate anti-viral and T H1/TH2 cytokine responses to sindbis virus infection. *Viral Immunol.* **2004**, *17*, 565–573. [CrossRef] [PubMed]
48. Styer, L.M.; Bernard, K.A.; Kramer, L.D. Enhanced early West Nile virus infection in young chickens infected by mosquito bite: Effect of viral dose. *Am. J. Trop. Med. Hyg.* **2006**, *75*, 337–345. [CrossRef]
49. Vogt, M.B.; Lahon, A.; Arya, R.P.; Kneubehl, A.R.; Spencer Clinton, J.L.; Paust, S.; Rico-Hesse, R. Mosquito saliva alone has profound effects on the human immune system. *PLoS Negl. Trop. Dis.* **2018**, *12*, e0006439. [CrossRef]
50. Choumet, V.; Attout, T.; Chartier, L.; Khun, H.; Sautereau, J.; Robbe-Vincent, A.; Brey, P.; Huerre, M.; Bain, O. Visualizing non infectious and infectious anopheles gambiae blood feedings in naive and saliva-immunized mice. *PLoS ONE* **2012**, *7*, e50464. [CrossRef]

51. Demeure, C.E.; Brahimi, K.; Hacini, F.; Marchand, F.; Péronet, R.; Huerre, M.; St.-Mezard, P.; Nicolas, J.-F.; Brey, P.; Delespesse, G.; et al. Anopheles mosquito bites activate cutaneous mast cells leading to a local inflammatory response and lymph node hyperplasia. *J. Immunol.* **2005**, *174*, 3932–3940. [CrossRef]

52. Owhashi, M.; Harada, M.; Suguri, S.; Ohmae, H.; Ishii, A. The role of saliva of Anopheles stephensi in inflammatory response: Identification of a high molecular weight neutrophil chemotactic factor. *Parasitol. Res.* **2001**, *87*, 376–382. [CrossRef]

53. Ferguson, M.C.; Saul, S.; Fragkoudis, R.; Weisheit, S.; Cox, J.; Patabendige, A.; Sherwood, K.; Watson, M.; Merits, A.; Fazakerley, J.K. Ability of the encephalitic arbovirus semliki forest virus to cross the blood-brain barrier is determined by the charge of the E2 glycoprotein. *J. Virol.* **2015**, *89*, 7536–7549. [CrossRef] [PubMed]

54. Rodriguez-Andres, J.; Rani, S.; Varjak, M.; Chase-Topping, M.E.; Beck, M.H.; Ferguson, M.C.; Schnettler, E.; Fragkoudis, R.; Barry, G.; Merits, A.; et al. Phenoloxidase activity acts as a mosquito innate immune response against infection with semliki forest virus. *PLoS Pathog.* **2012**, *8*. [CrossRef] [PubMed]

55. Simpson, D. Zika virus infection in man. *Trans. R. Soc. Trop. Med. Hyg.* **1964**, *58*, 335–338. [CrossRef]

56. Bearcroft, W. Zika virus infection experimentally induced in a human volunteer. *Trans. R. Soc. Trop. Med. Hyg.* **1956**, *50*, 442–448. [CrossRef]

57. Bogoch, I.I.; Brady, O.J.; Kraemer, M.U.G.; German, M.; Creatore, M.I.; Kulkarni, M.A.; Brownstein, J.S.; Mekaru, S.R.; Hay, S.I.; Groot, E.; et al. Anticipating the international spread of Zika virus from Brazil. *Lancet* **2016**, *387*, 335–336. [CrossRef]

58. Schuler-Faccini, L.; Ribeiro, E.M.; Feitosa, I.M.L.; Horovitz, D.D.G.; Cavalcanti, D.P.; Pessoa, A.; Doriqui, M.J.R.; Neri, J.I.; de Neto, J.M.P.; Wanderley, H.Y.C.; et al. Possible association between zika virus infection and microcephaly—Brazil. *MMWR Morb. Mortal. Wkly. Rep.* **2016**, *65*, 59–62. [CrossRef] [PubMed]

59. Ventura, C.V.; Maia, M.; Bravo-Filho, V.; Góis, A.L.; Belfort, R. Zika virus in Brazil and macular atrophy in a child with microcephaly. *Lancet* **2016**, *387*, 228. [CrossRef]

60. Oehler, E.; Watrin, L.; Larre, P.; Leparc-Goffart, I.; Lastãre, S.; Valour, F.; Baudouin, L.; Mallet, H.P.; Musso, D.; Ghawche, F. Zika virus infection complicated by guillain-barré syndrome—Case report, French Polynesia, December. *Eurosurveillance* **2014**, *19*. [CrossRef]

61. Ioos, S.; Mallet, H.P.; Leparc Goffart, I.; Gauthier, V.; Cardoso, T.; Herida, M. Current Zika virus epidemiology and recent epidemics. *Med. Mal. Infect.* **2014**, *44*, 302–307. [CrossRef]

62. Hastings, A.K.; Uraki, R.; Gaitsch, H.; Dhaliwal, K.; Stanley, S.; Sproch, H.; Williamson, E.; MacNeil, T.; Marin-Lopez, A.; Hwang, J.; et al. Aedes aegypti NeSt1 protein enhances zika virus pathogenesis by activating neutrophils. *J. Virol.* **2019**, *93*. [CrossRef]

63. Zukor, K.; Wang, H.; Siddharthan, V.; Julander, J.G.; Morrey, J.D. Zika virus-induced acute myelitis and motor deficits in adult interferon $\alpha\beta/\gamma$ receptor knockout mice. *J. Neurovirol.* **2018**, *24*, 273–290. [CrossRef]

64. WHO. *Handbook for Clinical Management of Dengue*; WHO: Geneva, Switzerland, 2012.

65. Wilder-Smith, A.; Schwartz, E. Dengue in travelers. *N. Engl. J. Med.* **2005**, *353*, 924–932. [CrossRef]

66. Mackenzie, J.S.; Gubler, D.J.; Petersen, L.R. Emerging flaviviruses: The spread and resurgence of Japanese encephalitis, west Nile and dengue viruses. *Nat. Med.* **2004**, *10*, S98–S109. [CrossRef]

67. Thein, T.L.; Lye, D.C.; Leo, Y.S.; Wong, J.G.X.; Hao, Y.; Wilder-Smith, A. Short report: Severe neutropenia in dengue patients: Prevalence and significance. *Am. J. Trop. Med. Hyg.* **2014**, *90*, 984–987. [CrossRef] [PubMed]

68. Shourick, J.; Dinh, A.; Matt, M.; Salomon, J.; Davido, B. Severe neutropenia revealing a rare presentation of dengue fever: A case report. *BMC Res. Notes* **2017**, *10*, 415. [CrossRef] [PubMed]

69. Screaton, G.; Mongkolsapaya, J.; Yacoub, S.; Roberts, C. New insights into the immunopathology and control of dengue virus infection. *Nat. Rev. Immunol.* **2015**, *15*, 745–759. [CrossRef] [PubMed]

70. Hoang, L.T.; Lynn, D.J.; Henn, M.; Birren, B.W.; Lennon, N.J.; Le, P.T.; Duong, K.T.H.; Nguyen, T.T.H.; Mai, L.N.; Farrar, J.J.; et al. The early whole-blood transcriptional signature of dengue virus and features associated with progression to dengue shock syndrome in vietnamese children and young adults. *J. Virol.* **2010**, *84*, 12982–12994. [CrossRef]

71. Kunder, M.; Lakshmaiah, V.; Moideen Kutty, A.V. Plasma neutrophil elastase, α1-antitrypsin, α2-macroglobulin and neutrophil elastase–α1-antitrypsin complex levels in patients with dengue fever. *Indian J. Clin. Biochem.* **2018**, *33*, 218–221. [CrossRef] [PubMed]

72. Yost, C.C.; Schwertz, H.; Cody, M.J.; Wallace, J.A.; Campbell, R.A.; Vieira-De-Abreu, A.; Araujo, C.V.; Schubert, S.; Harris, E.S.; Rowley, J.W.; et al. Neonatal NET-inhibitory factor and related peptides inhibit neutrophil extracellular trap formation. *J. Clin. Investig.* **2016**, *126*, 3783–3798. [CrossRef] [PubMed]

73. Brinkmann, V.; Reichard, U.; Goosmann, C.; Fauler, B.; Uhlemann, Y.; Weiss, D.S.; Weinrauch, Y.; Zychlinsky, A. Neutrophil extracellular traps kill bacteria. *Science.* **2004**, *303*, 1532–1535. [CrossRef]

74. Kaplan, M.J.; Radic, M. Neutrophil extracellular traps: Double-edged swords of innate immunity. *J. Immunol.* **2012**, *189*, 2689–2695. [CrossRef]

75. Jorch, S.K.; Kubes, P. An emerging role for neutrophil extracellular traps in noninfectious disease. *Nat. Med.* **2017**, *23*, 279–287. [CrossRef] [PubMed]

76. Garcia-Romo, G.S.; Caielli, S.; Vega, B.; Connolly, J.; Allantaz, F.; Xu, Z.; Punaro, M.; Baisch, J.; Guiducci, C.; Coffman, R.L.; et al. Netting neutrophils are major inducers of type I IFN production in pediatric systemic lupus erythematosus. *Sci. Transl. Med.* **2011**, *3*, 73ra20. [CrossRef] [PubMed]

77. Lande, R.; Ganguly, D.; Facchinetti, V.; Frasca, L.; Conrad, C.; Gregorio, J.; Meller, S.; Chamilos, G.; Sebasigari, R.; Riccieri, V.; et al. Neutrophils activate plasmacytoid dendritic cells by releasing self-DNA-peptide complexes in systemic lupus erythematosus. *Sci. Transl. Med.* **2011**, *3*, 73ra19. [CrossRef] [PubMed]

78. Villanueva, E.; Yalavarthi, S.; Berthier, C.C.; Hodgin, J.B.; Khandpur, R.; Lin, A.M.; Rubin, C.J.; Zhao, W.; Olsen, S.H.; Klinker, M.; et al. Netting neutrophils induce endothelial damage, infiltrate tissues, and expose immunostimulatory molecules in systemic lupus erythematosus. *J. Immunol.* **2011**, *187*, 538–552. [CrossRef] [PubMed]

79. Hakkim, A.; Furnrohr, B.G.; Amann, K.; Laube, B.; Abed, U.A.; Brinkmann, V.; Herrmann, M.; Voll, R.E.; Zychlinsky, A. Impairment of neutrophil extracellular trap degradation is associated with lupus nephritis. *Proc. Natl. Acad. Sci. USA* **2010**, *107*, 9813–9818. [CrossRef]

80. Opasawatchai, A.; Amornsupawat, P.; Jiravejchakul, N.; Chan-in, W.; Spoerk, N.J.; Manopwisedjaroen, K.; Singhasivanon, P.; Yingtaweesak, T.; Suraamornkul, S.; Mongkolsapaya, J.; et al. Neutrophil activation and early features of NET formation are associated with dengue virus infection in human. *Front. Immunol.* **2019**, *9*, 3007. [CrossRef]

81. Lacy, P. Mechanisms of degranulation in neutrophils. *Allergy Asthma Clin. Immunol.* **2006**, *2*, 98. [CrossRef]

82. Yoon, J.; Terada, A.; Kita, H. CD66b regulates adhesion and activation of human eosinophils. *J. Immunol.* **2007**, *179*, 8454–8462. [CrossRef]

83. Lien, T.S.; Sun, D.S.; Hung, S.C.; Wu, W.S.; Chang, H.H. Dengue virus envelope protein domain III induces nlrp3 inflammasome-dependent NETosis-mediated inflammation in mice. *Front. Immunol.* **2021**, *12*, 618577. [CrossRef]

84. Ogura, H.; Kawasaki, T.; Tanaka, H.; Koh, T.; Tanaka, R.; Ozeki, Y.; Hosotsubo, H.; Kuwagata, Y.; Shimazu, T.; Sugimoto, H. Activated platelets enhance microparticle formation and platelet-leukocyte interaction in severe trauma and sepsis. *J. Trauma.* **2001**, *50*, 801–809. [CrossRef]

85. Wu, M.F.; Chen, S.T.; Hsieh, S.L. Distinct regulation of dengue virus-induced inflammasome activation in human macrophage subsets. *J. Biomed. Sci.* **2013**, *20*, 36. [CrossRef]

86. Wu, M.F.; Chen, S.T.; Yang, A.H.; Lin, W.W.; Lin, Y.L.; Chen, N.J.; Tsai, I.S.; Li, L.; Hsieh, S.L. CLEC5A is critical for dengue virus-induced inflammasome activation in human macrophages. *Blood* **2013**, *121*, 95–106. [CrossRef]

87. Hottz, E.D.; Lopes, J.F.; Freitas, C.; Valls-De-Souza, R.; Oliveira, M.F.; Bozza, M.T.; Da Poian, A.T.; Weyrich, A.S.; Zimmerman, G.A.; Bozza, F.A.; et al. Platelets mediate increased endothelium permeability in dengue through NLRP3-inflammasome activation. *Blood* **2013**, *122*, 3405–3414. [CrossRef] [PubMed]

88. Van Niel, G.; D'Angelo, G.; Raposo, G. Shedding light on the cell biology of extracellular vesicles. *Nat. Rev. Mol. Cell Biol.* **2018**, *19*, 213–228. [CrossRef]

89. Sung, P.S.; Huang, T.F.; Hsieh, S.L. Extracellular vesicles from CLEC2-activated platelets enhance dengue virus-induced lethality via CLEC5A/TLR2. *Nat. Commun.* **2019**, *10*, 1–13. [CrossRef]

90. Huang, Y.L.; Chen, S.T.; Liu, R.S.; Chen, Y.H.; Lin, C.Y.; Huang, C.H.; Shu, P.Y.; Liao, C.L.; Hsieh, S.L. CLEC5A is critical for dengue virus-induced osteoclast activation and bone homeostasis. *J. Mol. Med.* **2016**, *94*, 1025–1037. [CrossRef] [PubMed]

91. Chen, S.T.; Lin, Y.L.; Huang, M.T.; Wu, M.F.; Cheng, S.C.; Lei, H.Y.; Lee, C.K.; Chiou, T.W.; Wong, C.H.; Hsieh, S.L. CLEC5A is critical for dengue-virus-induced lethal disease. *Nature* **2008**, *453*, 672–676. [CrossRef] [PubMed]

92. Gubler, D.J. The continuing spread of West Nile virus in the Western Hemisphere. *Clin. Infect. Dis.* **2007**, *45*, 1039–1046. [CrossRef]

93. Shrestha, B.; Samuel, M.A.; Diamond, M.S. CD8+ T cells require perforin to clear west nile virus from infected neurons. *J. Virol.* **2006**, *80*, 119–129. [CrossRef]

94. Bai, F.; Town, T.; Qian, F.; Wang, P.; Kamanaka, M.; Connolly, T.M.; Gate, D.; Montgomery, R.R.; Flavell, R.A.; Fikrig, E. IL-10 signaling blockade controls murine West Nile virus infection. *PLoS Pathog.* **2009**, *5*. [CrossRef]

95. Town, T.; Bai, F.; Wang, T.; Kaplan, A.T.; Qian, F.; Montgomery, R.R.; Anderson, J.F.; Flavell, R.A.; Fikrig, E. Toll-like receptor 7 mitigates lethal west Nile encephalitis via interleukin 23-dependent immune cell infiltration and homing. *Immunity* **2009**, *30*, 242–253. [CrossRef] [PubMed]

96. Rawal, A.; Gavin, P.J.; Sturgis, C.D. Cerebrospinal fluid cytology in seasonal epidemic West Nile Virus meningo-encephalitis. *Diagn. Cytopathol.* **2006**, *34*, 127–129. [CrossRef] [PubMed]

97. Tyler, K.L.; Pape, J.; Goody, R.J.; Corkill, M.; Kleinschmidt-DeMasters, B.K. CSF findings in 250 patients with serologically confirmed West Nile virus meningitis and encephalitis. *Neurology* **2006**, *66*, 361–365. [CrossRef]

98. Baxter, V.K.; Heise, M.T. Genetic control of alphavirus pathogenesis. *Mamm. Genome* **2018**, *29*, 408–424. [CrossRef] [PubMed]

99. Hollidge, B.S.; González-Scarano, F.; Soldan, S.S. Arboviral encephalitides: Transmission, emergence, and pathogenesis. *J. Neuroimmune Pharmacol.* **2010**, *5*, 428–442. [CrossRef] [PubMed]

100. Steele, K.; Reed, D.; Glass, P. Medical aspects of biological warfare. In *Alphavirus Encephalitides*; Office of the Surgeon General, US Army Medical Department Center and School, Borden Institute: Washington, DC, USA, 2007; pp. 241–270.

101. Harley, D.; Sleigh, A.; Ritchie, S. Ross river virus transmission, infection, and disease: A cross-disciplinary review. *Clin. Microbiol. Rev.* **2001**, *14*, 909–932. [CrossRef]

102. Williams, M.C.; Woodall, J.P.; Gillett, J.D. O'Nyong-Nyong fever: An epidemic virus disease in East Africa. *Trans. R. Soc. Trop. Med. Hyg.* **1965**, *59*, 186–197. [CrossRef]

103. Weaver, S.C.; Reisen, W.K. Present and future arboviral threats. *Antivir. Res.* **2010**, *85*, 328–345. [CrossRef]

104. Chikungunya Fever in EU/EEA. Available online: https://www.ecdc.europa.eu/en/chikungunya/threats-and-outbreaks/chikungunya-fever-eueea (accessed on 28 March 2021).

105. Roth, A.; Hoy, D.; Horwood, P.F.; Ropa, B.; Hancock, T.; Guillaumot, L.; Rickart, K.; Frison, P.; Pavlin, B.; Souares, Y. Preparedness for threat of chikungunya in the pacific. *Emerg. Infect. Dis.* **2014**, *20*, e130696. [CrossRef]

106. Grandadam, M.; Caro, V.; Plumet, S.; Thiberge, J.M.; Souarès, Y.; Failloux, A.B.; Tolou, H.J.; Budelot, M.; Cosserat, D.; Leparc-Goffart, I.; et al. Chikungunya virus, Southeastern France. *Emerg. Infect. Dis.* **2011**, *17*, 910–913. [CrossRef]

107. Rezza, G.; Nicoletti, L.; Angelini, R.; Romi, R.; Finarelli, A.; Panning, M.; Cordioli, P.; Fortuna, C.; Boros, S.; Magurano, F.; et al. Infection with chikungunya virus in Italy: An outbreak in a temperate region. *Lancet* **2007**, *370*, 1840–1846. [CrossRef]

108. Greenlee, J.E. The equine encephalitides. In *Handbook of Clinical Neurology*; Elsevier B.V.: Amsterdam, The Netherlands, 2014; Volume 123, pp. 417–432.

109. Hatanpaa, K.J.; Kim, J.H. Neuropathology of viral infections. In *Handbook of Clinical Neurology*; Elsevier B.V.: Amsterdam, The Netherlands, 2014; Volume 123, pp. 193–214.

110. Staples, J.E.; Breiman, R.F.; Powers, A.M. Chikungunya fever: An epidemiological review of a re-emerging infectious disease. *Clin. Infect. Dis.* **2009**, *49*, 942–948. [CrossRef] [PubMed]

111. Sissoko, D.; Malvy, D.; Ezzedine, K.; Renault, P.; Moscetti, F.; Ledrans, M.; Pierre, V. Post-epidemic Chikungunya disease on reunion island: Course of rheumatic manifestations and associated factors over a 15-month period. *PLoS Negl. Trop. Dis.* **2009**, *3*, 389. [CrossRef] [PubMed]

112. Borgherini, G.; Poubeau, P.; Jossaume, A.; Gouix, A.; Cotte, L.; Michault, A.; Arvin-Berod, C.; Paganin, F. Persistent arthralgia associated with chikungunya virus: A study of 88 adult patients on reunion island. *Clin. Infect. Dis.* **2008**, *47*, 469–475. [CrossRef] [PubMed]

113. Larrieu, S.; Pouderoux, N.; Pistone, T.; Filleul, L.; Receveur, M.-C.; Sissoko, D.; Ezzedine, K.; Malvy, D. Factors associated with persistence of arthralgia among chikungunya virus-infected travellers: Report of 42 French cases. *J. Clin. Virol.* **2010**, *47*, 85–88. [CrossRef] [PubMed]

114. Simon, F.; Parola, P.; Grandadam, M.; Fourcade, S.; Oliver, M.; Brouqui, P.; Hance, P.; Kraemer, P.; Mohamed, A.A.; de Lamballerie, X.; et al. Chikungunya infection. *Medicine* **2007**, *86*, 123–137. [CrossRef]

115. Kularatne, S.A.M.; Weerasinghe, S.C.; Gihan, C.; Wickramasinghe, S.; Dharmarathne, S.; Abeyrathna, A.; Jayalath, T. Epidemiology, clinical manifestations, and long-term outcomes of a major outbreak of Chikungunya in a Hamlet in Sri Lanka, in 2007: A longitudinal cohort study. *J. Trop. Med.* **2012**, *2012*, 639178. [CrossRef]

116. Harley, D.; Bossingham, D.; Purdie, D.M.; Pandeya, N.; Sleigh, A.C. Ross river virus disease in tropical Queensland: Evolution of rheumatic manifestations in an inception cohort followed for six months. *Med. J. Aust.* **2002**, *177*, 352–355. [CrossRef]

117. Mylonas, A.D.; Brown, A.M.; Carthew, T.L.; Purdie, D.M.; Pandeya, N.; Collins, L.G.; Suhrbier, A.; McGrath, B.; Reymond, E.J.; Vecchio, P.C.; et al. Natural history of Ross River virus-induced epidemic polyarthritis. *Med. J. Aust.* **2002**, *177*, 356–360. [CrossRef]

118. Stoermer, K.A.; Burrack, A.; Oko, L.; Montgomery, S.A.; Borst, L.B.; Gill, R.G.; Morrison, T.E. Genetic ablation of arginase 1 in macrophages and neutrophils enhances clearance of an arthritogenic alphavirus. *J. Immunol.* **2012**, *189*, 4047–4059. [CrossRef] [PubMed]

119. Bronte, V.; Zanovello, P. Regulation of immune responses by L-arginine metabolism. *Nat. Rev. Immunol.* **2005**, *5*, 641–654. [CrossRef] [PubMed]

120. Munder, M. Arginase: An emerging key player in the mammalian immune system: REVIEW. *Br. J. Pharmacol.* **2009**, *158*, 638–651. [CrossRef] [PubMed]

121. Rodríguez, P.C.; Ochoa, A.C. Arginine regulation by myeloid derived suppressor cells and tolerance in cancer: Mechanisms and therapeutic perspectives. *Immunol. Rev.* **2008**, *222*, 180–191. [CrossRef]

122. Silva, L.A.; Dermody, T.S. Chikungunya virus: Epidemiology, replication, disease mechanisms, and prospective intervention strategies. *J. Clin. Investig.* **2017**, *127*, 737–749. [CrossRef] [PubMed]

123. Smith, P.; Wang, S.-Z.; Dowling, K.; Forsyth, K. Leucocyte populations in respiratory syncytial virus-induced bronchiolitis. *J. Paediatr. Child Health* **2001**, *37*, 146–151. [CrossRef]

124. Wojtasiak, M.; Pickett, D.L.; Tate, M.D.; Londrigan, S.L.; Bedoui, S.; Brooks, A.G.; Reading, P.C. Depletion of Gr-1+, but not Ly6G+, immune cells exacerbates virus replication and disease in an intranasal model of herpes simplex virus type 1 infection. *J. Gen. Virol.* **2010**, *91*, 2158–2166. [CrossRef]

125. Agraz-Cibrian, J.M.; Giraldo, D.M.; Mary, F.M.; Urcuqui-Inchima, S. Understanding the molecular mechanisms of NETs and their role in antiviral innate immunity. *Virus Res.* **2017**, *228*, 124–133. [CrossRef]

126. Palha, N.; Guivel-Benhassine, F.; Briolat, V.; Lutfalla, G.; Sourisseau, M.; Ellett, F.; Wang, C.-H.; Lieschke, G.J.; Herbomel, P.; Schwartz, O.; et al. Real-time whole-body visualization of chikungunya virus infection and host interferon response in zebrafish. *PLoS Pathog.* **2013**, *9*, e1003619. [CrossRef]

127. Chang, A.Y.; Martins, K.A.O.; Encinales, L.; Reid, S.P.; Acuña, M.; Encinales, C.; Matranga, C.B.; Pacheco, N.; Cure, C.; Shukla, B.; et al. Chikungunya arthritis mechanisms in the Americas. *Arthritis Rheumatol.* **2018**, *70*, 585–593. [CrossRef]

128. Lin, T.; Geng, T.; Harrison, A.G.; Yang, D.; Vella, A.T.; Fikrig, E.; Wang, P. CXCL10 signaling contributes to the pathogenesis of arthritogenic alphaviruses. *Viruses* **2020**, *12*, 1252. [CrossRef]

129. Hiroki, C.H.; Toller-Kawahisa, J.E.; Fumagalli, M.J.; Colon, D.F.; Figueiredo, L.T.M.; Fonseca, B.A.L.D.; Franca, R.F.O.; Cunha, F.Q. Neutrophil extracellular traps effectively control acute chikungunya virus infection. *Front. Immunol.* **2020**, *10*. [CrossRef]

130. Raftery, M.J.; Lalwani, P.; Krautkrämer, E.; Peters, T.; Scharffetter-Kochanek, K.; Krüger, R.; Hofmann, J.; Seeger, K.; Krüger, D.H.; Schönrich, G. β2 integrin mediates hantavirus-induced release of neutrophil extracellular traps. *J. Exp. Med.* **2014**, *211*, 1485–1497. [CrossRef] [PubMed]

131. Cortjens, B.; de Boer, O.J.; de Jong, R.; Antonis, A.F.; Sabogal Piñeros, Y.S.; Lutter, R.; van Woensel, J.B.; Bem, R.A. Neutrophil extracellular traps cause airway obstruction during respiratory syncytial virus disease. *J. Pathol.* **2016**, *238*, 401–411. [CrossRef] [PubMed]

132. Apel, F.; Zychlinsky, A.; Kenny, E.F. The role of neutrophil extracellular traps in rheumatic diseases. *Nat. Rev. Rheumatol.* **2018**, *14*, 467–475. [CrossRef]

133. Poo, Y.S.; Nakaya, H.; Gardner, J.; Larcher, T.; Schroder, W.A.; Le, T.T.; Major, L.D.; Suhrbier, A. CCR2 deficiency promotes exacerbated chronic erosive neutrophil-dominated chikungunya virus arthritis. *J. Virol.* **2014**, *88*, 6862–6872. [CrossRef]

134. Quinones, M.P.; Ahuja, S.K.; Jimenez, F.; Schaefer, J.; Garavito, E.; Rao, A.; Chenaux, G.; Reddick, R.L.; Kuziel, W.A.; Ahuja, S.S. Experimental arthritis in CC chemokine receptor 2–null mice closely mimics severe human rheumatoid arthritis. *J. Clin. Investig.* **2004**, *113*, 856–866. [CrossRef]

135. Sawanobori, Y.; Ueha, S.; Kurachi, M.; Shimaoka, T.; Talmadge, J.E.; Abe, J.; Shono, Y.; Kitabatake, M.; Kakimi, K.; Mukaida, N.; et al. Chemokine-mediated rapid turnover of myeloid-derived suppressor cells in tumor-bearing mice. *Blood* **2008**, *111*, 5457–5466. [CrossRef] [PubMed]

136. Fujii, H.; Baba, T.; Yamagishi, M.; Kawano, M.; Mukaida, N. The role of a chemokine receptor, CCR2, in suppressing the development of arthritis in IL-1 receptor antagonist-deficient mice. *Inflamm. Regen.* **2012**, *32*, 124–131. [CrossRef]

137. Montgomery, R.R.; Booth, C.J.; Wang, X.; Blaho, V.A.; Malawista, S.E.; Brown, C.R. Recruitment of macrophages and polymorphonuclear leukocytes in lyme carditis. *Infect. Immun.* **2007**, *75*, 613–620. [CrossRef]

138. Eyles, J.L.; Hickey, M.J.; Norman, M.U.; Croker, B.A.; Roberts, A.W.; Drake, S.F.; James, W.G.; Metcalf, D.; Campbell, I.K.; Wicks, I.P. A key role for G-CSF induced neutrophil production and trafficking during inflammatory arthritis. *Blood* **2008**, *112*, 5193–5201. [CrossRef]

139. Sadik, C.D.; Kim, N.D.; Iwakura, Y.; Luster, A.D. Neutrophils orchestrate their own recruitment in murine arthritis through C5aR and FcγR signaling. *Proc. Natl. Acad. Sci. USA* **2012**, *109*, E3177–E3185. [CrossRef] [PubMed]

140. Ajuebor, M.; Das, A.; Virag, L.; Flower, R.; Szabo, C.; Perretti, M. Role of resident peritoneal macrophages and mast cells in chemokine production and neutrophil migration in acute inflammation: Evidence for an inhibitory loop involving endogenous IL-10. *J. Immunol.* **1999**, *162*, 1685–1691. [PubMed]

141. Lotfi, R.; Herzog, G.I.; DeMarco, R.A.; Beer-Stolz, D.; Lee, J.J.; Rubartelli, A.; Schrezenmeier, H.; Lotze, M.T. Eosinophils oxidize damage-associated molecular pattern molecules derived from stressed cells. *J. Immunol.* **2009**, *183*, 5023–5031. [CrossRef] [PubMed]

142. Rampersad, R.R.; Tarrant, T.K.; Vallanat, C.T.; Quintero-Matthews, T.; Weeks, M.F.; Esserman, D.A.; Clark, J.; Di Padova, F.; Patel, D.D.; Fong, A.M.; et al. Enhanced Th17-cell responses render CCR2-deficient mice more susceptible for autoimmune arthritis. *PLoS ONE* **2011**, *6*, e25833. [CrossRef] [PubMed]

143. Rudd, P.A.; Wilson, J.; Gardner, J.; Larcher, T.; Babarit, C.; Le, T.T.; Anraku, I.; Kumagai, Y.; Loo, Y.-M.; Gale, M.; et al. Interferon response factors 3 and 7 protect against chikungunya virus hemorrhagic fever and shock. *J. Virol.* **2012**, *86*, 9888–9898. [CrossRef] [PubMed]

144. McCarthy, M.K.; Reynoso, G.V.; Winkler, E.S.; Mack, M.; Diamond, M.S.; Hickman, H.D.; Morrison, T.E. MyD88-dependent influx of monocytes and neutrophils impairs lymph node B cell responses to chikungunya virus infection via Irf5, Nos2 and Nox2. *PLoS Pathog.* **2020**, *16*, e1008292. [CrossRef]

145. Hawman, D.W.; Fox, J.M.; Ashbrook, A.W.; May, N.A.; Schroeder, K.M.S.; Torres, R.M.; Crowe, J.E.; Dermody, T.S.; Diamond, M.S.; Morrison, T.E. Pathogenic chikungunya virus evades B cell responses to establish persistence. *Cell Rep.* **2016**, *16*, 1326–1338. [CrossRef]

146. McCarthy, M.K.; Davenport, B.J.; Reynoso, G.V.; Lucas, E.D.; May, N.A.; Elmore, S.A.; Tamburini, B.A.; Hickman, H.D.; Morrison, T.E. Chikungunya virus impairs draining lymph node function by inhibiting HEV-mediated lymphocyte recruitment. *JCI Insight* **2018**, *3*. [CrossRef]

147. Cook, L.E.; Locke, M.C.; Young, A.R.; Monte, K.; Hedberg, M.L.; Shimak, R.M.; Sheehan, K.C.F.; Veis, D.J.; Diamond, M.S.; Lenschow, D.J. Distinct roles of interferon alpha and beta in controlling chikungunya virus replication and modulating neutrophil-mediated inflammation. *J. Virol.* **2019**, *94*. [CrossRef]

148. Lee, P.Y.; Wang, J.-X.; Parisini, E.; Dascher, C.C.; Nigrovic, P.A. Ly6 family proteins in neutrophil biology. *J. Leukoc. Biol.* **2013**, *94*, 585–594. [CrossRef] [PubMed]

 *cells*

Article

Dysregulated Host Responses Underlie 2009 Pandemic Influenza-Methicillin Resistant *Staphylococcus aureus* Coinfection Pathogenesis at the Alveolar-Capillary Barrier

Michaela E. Nickol [1], Sarah M. Lyle [1], Brendan Dennehy [1] and Jason Kindrachuk [1,2,*]

[1] Laboratory of Emerging and Re-Emerging Viruses, Department of Medical Microbiology, University of Manitoba, Winnipeg, MB R3E 0J9, Canada; nickolm@myumanitoba.ca (M.E.N.); lyles3@myumanitoba.ca (S.M.L.); dennehyb@myumanitoba.ca (B.D.)

[2] Vaccine and Infectious Disease Organization-International Vaccine Centre, University of Saskatchewan, Saskatoon, SK S7N 5E3, Canada

* Correspondence: Jason.Kindrachuk@umanitoba.ca; Tel.: +1-(204)-789-3807

Received: 17 October 2020; Accepted: 11 November 2020; Published: 13 November 2020

Abstract: Influenza viruses are a continual public health concern resulting in 3–5 million severe infections annually despite intense vaccination campaigns and messaging. Secondary bacterial infections, including *Staphylococcus aureus*, result in increased morbidity and mortality during seasonal epidemics and pandemics. While coinfections can result in deleterious pathologic consequences, including alveolar-capillary barrier disruption, the underlying mechanisms are poorly understood. We have characterized host- and pathogen-centric mechanisms contributing to influenza-bacterial coinfections in a primary cell coculture model of the alveolar-capillary barrier. Using 2009 pandemic influenza (pH1N1) and methicillin-resistant *S. aureus* (MRSA), we demonstrate that coinfection resulted in dysregulated barrier function. Preinfection with pH1N1 resulted in modulation of adhesion- and invasion-associated MRSA virulence factors during lag phase bacterial replication. Host response modulation in coinfected alveolar epithelial cells were primarily related to TLR- and inflammatory response-mediated cell signaling events. While less extensive in cocultured endothelial cells, coinfection resulted in changes to cellular stress response- and TLR-related signaling events. Analysis of cytokine expression suggested that cytokine secretion might play an important role in coinfection pathogenesis. Taken together, we demonstrate that coinfection pathogenesis is related to complex host- and pathogen-mediated events impacting both epithelial and endothelial cell regulation at the alveolar-capillary barrier.

Keywords: influenza; *Staphylococcus aureus*; alveolar-capillary barrier; coinfection; kinome; virulence factor

1. Introduction

Influenza A viruses (IAVs) infect approximately 10% of the global population each year, resulting in an estimated 3–5 million severe infections and 300,000–650,000 mortalities [1]. This occurs through both seasonal epidemics and sporadic pandemic outbreaks, despite an intensive vaccine program and the existence of antivirals [2–5]. Infections range from asymptomatic to severe or fatal [6,7] and generally manifest as acute, self-limiting infections in the upper or lower respiratory tract [8,9]. Clinical symptoms include high fever, headache, coryza, cough, myalgias, and general malaise [3,8–10]. In healthy adults, symptoms generally peak around 3–5 days post-infection and with convalescence at 7–10 days [8–10]. While influenza is generally mild in most of the population, infants <2 years of age, the elderly and individuals with underlying comorbidities (including respiratory, cardiac, neurological or immunosuppressive conditions) are at a high risk of severe disease [8,9,11,12].

There is an increasing appreciation that severe or fatal influenza infections are frequently complicated by bacterial coinfections [13]. The contribution of secondary bacterial infections has been well documented throughout prior influenza pandemics and most notably the 1918 H1N1 influenza pandemic [14]. Modern analyses of lung tissue and review of historical autopsy data from fatal 1918 influenza infections demonstrated that 95% of lethal cases were complicated by a bacterial coinfection, with *Staphylococcus aureus* and *Streptococcus pneumoniae* spp. most commonly identified [15–17]. Secondary bacterial pneumonia also resulted in significant morbidity and mortality during both the 1957 and 1968 influenza pandemics, with an estimated 44% of cases being complicated by *S. aureus* and *S. pneumoniae*. More recently, up to 55% of fatal cases during the 2009 H1N1 pandemic were complicated by bacterial coinfections, and methicillin-resistant *S. aureus* (MRSA) was commonly observed in coinfected patients [18–20]. Bacterial coinfections also complicate seasonal influenza infections where total influenza-related fatalities are estimated to be 65,000 (including both influenza- and pneumonia-related deaths) in the U.S. annually [18].

Previously, we characterized the host and pathogen molecular mechanisms that contribute to severe influenza-bacterial infections in the lower respiratory tract using a monolayer of an alveolar epithelial cell line [21]. We found that when respiratory epithelial cells were infected with MRSA during peak viral infection, host cell signaling responses shifted from viral- to bacterial-centric as infection moved from the early to late phase [21]. Further, a transition phase in host responses was identified at the mid-point of infection (8–12 h post-MRSA addition), which correlated with a loss of respiratory epithelial barrier function and integrity. While this prior investigation provided important insights into the molecular mechanisms underlying the pathogenesis of influenza-bacterial coinfections, it was limited to epithelial cells alone and did not account for the multicellular complexity of the alveolar-capillary barrier.

To address this, we established a coculture model of the alveolar-capillary barrier by using an in vitro coculture model using primary human alveolar epithelial cells and microvascular endothelial cells. This physiologically relevant model of the lower respiratory tract allowed us to elucidate the molecular mechanisms surrounding barrier dysfunction during IAV-MRSA coinfections. Based on our prior investigations, we hypothesized that secondary bacterial coinfections resulted in severe dysfunction of the alveolar-capillary barrier due to the modulation of bacterial virulence factor expression in the presence of IAV, thus leading to dysregulated host cell signaling responses in both epithelial and endothelial cells at the alveolar-capillary barrier. Our results suggest that the pathogenesis underlying severe influenza-bacterial coinfections in the lower respiratory tract results from both microbial- and host-centric activities.

2. Materials and Methods

2.1. Virus, Bacteria, and Cell Conditions

The 2009 pandemic H1N1 Influenza A/Mexico/4108/09 (pH1N1; GenBank GQ223112) was kindly provided by Dr. Kevin Coombs (University of Manitoba, Canada). Virus stocks were grown in Madin–Darby canine kidney cells (ATCC, Manassas, VA, USA) maintained in Dulbecco's modified Eagle medium (Gibco, Grand Island, NY, USA) with 1 µg/mL tosyl phenylalanyl chloromethyl ketone (TPCK)-treated trypsin, concentrated following ultracentrifugation on a 35% sucrose cushion, and kept at −80 °C. Viral titres were determined via plaque assay. CA-MRSA genotype CMRSA10 (USA300; herein referred to as MRSA) was kindly provided by Dr. George Zhanel (University of Manitoba, Winnipeg, MB, Canada). MRSA inocula were generated following growth to the mid-log phase in tryptic soy broth (Hardy Diagnostics, Santa Maria, CA, USA) at 37 °C with shaking. Bacterial titres were determined via the standard plate count. Human (HPAEpiC) were obtained from ScienCell Research Laboratories (Carlsbad, CA, USA). Cells were grown in the airway epithelial basal cell medium fully supplemented with the bronchial epithelial cells growth kit (ATCC, Manassas, VA, USA) at 37 °C and 5% CO_2. Human pulmonary microvascular endothelial cells (HPMECs) were obtained from ScienCell

Research Laboratories (Carlsbad, CA, USA). Cells were grown in the basal endothelial cell medium complete kit (ScienCell Research Laboratories, Carlsbad, CA, USA).

2.2. Coculture Model of the Alveolar-Capillary Barrier

The basal side of 0.4 µm transwell inserts (Corning Life Sciences, Montreal, QC, Canada) were coated with the GelTrex LDEV-free reduced growth factor basement membrane matrix (ThermoFisher Scientific, Mississauga, ON, Canada) and rested basal side up for 1 h at 37 °C and 5% CO_2. Transwell inserts were turned apical side up, the apical side coated with GelTrex, and rested for 1 h at 37 °C and 5% CO_2. Transwell inserts were turned basal side up, the basal side of the transwell inserts seeded with HPMEC at a concentration of 1.5×10^5 cells/mL (4.5×10^4 cells/cm^2) in a 1:1 mix of HPAEpiC and HPMEC media, and rested for 3 h at 37 °C and 5% CO_2. Transwell inserts were turned apical side up and HPAEpiC were seeded on the apical side at a concentration of 3×10^5 cells/mL (9×10^4 cells/cm^2) in a 1:1 mix of HPAEpiC and HPMEC media. After 24 h, media was removed from the upper compartment of the transwell insert, to allow primary epithelial cells to grow at the air-liquid interface. Media in the lower compartment was refreshed with a 1:1 mix of HPAEpiC and HPMEC media. Cells were permitted to grow to confluency for 14 days, with media in the lower compartment being refreshed every second day. An alveolar-capillary barrier coculture model schematic is presented in Figure S1.

2.3. Viral and Bacterial Infection of the Tissue Culture Model

Epithelial and endothelial cells were washed 2× with warm DPBS. Transwell infection media (a 1:1 mix of non-supplemented airway epithelial basal cell medium and basal endothelial cell medium without TPCK-trypsin) was added to the lower compartment and epithelial cells were infected by adding viral inocula to the upper compartment of the transwell insert. Cells were infected with pH1N1 at a multiplicity of infection (MOI) of 0.1 or mock with transwell infection media for 1 h at 37 °C and 5% CO_2. Following infection, viral inocula were aspirated from cells. Cells were rested for 24 h post-viral infection. Cells were infected with mid-log phase MRSA or mock 24 post-influenza addition with transwell infection media for 1 h. Bacterial MOIs of 0.1 were used and were achieved by serial dilution of mid-log phase culture in transwell infection media as described above. Bacterial inocula were aspirated from cells and both HPAEpiC and HPMEC were harvested at each time point by gentle scraping for further investigation of bacterial replication kinetics, virulence factor modulation, and kinome analysis.

2.4. Quantification of Bacterial and Viral Replication Kinetics

Quantification of the total number of adherent and internalized bacteria was determined at 1, 4, 8, 12, 16, 20, and 24 h post-bacterial infection. Respiratory epithelial HPAEpiCs were harvested for bacterial enumeration by washing 2× with DPBS followed by gentle scraping. Cells were pelleted by centrifugation at 5000 rpm for 10 min, supernatant removed, and cells resuspended in 0.025% TritonX-100. Colony forming units (CFU) were quantified by standard bacterial plating on tryptic soy agar (MP Biomedicals, LLC, Solon, OH, USA). Four biological replicates were performed at each time point for enumeration. RT-qPCR was used to quantify viral replication by collecting supernatant samples for IAV–MRSA infected alveolar epithelial cells. Total RNA was extracted from the supernatant using the PureLink Viral RNA/DNA Mini Kit (LifeTechnologies, Burlington, ON, Canada) according to the manufacturer's instructions. Reverse transcription of total RNA was performed using the Superscript IV first-strand cDNA synthesis kit (Life Technologies, Burlington, ON, Canada) using primers specific for the viral H1N1 HA sequence. Viral genome copy numbers were quantified by comparing RT-qPCR results to an established external viral genome copy number standard.

2.5. RNA Extraction, cDNA Synthesis, and Quantitative PCR

Three biological replicates with two technical replicates were collected at 1, 4, 8, 12, 16, 20, and 24 h post-bacterial infection to determine the modification of bacterial virulence factors in the presence of influenza. Following aspiration of media, HPAEpiCs were collected by gentle scraping, pelleting by centrifugation at 1200 rpm for 10 min, and stored at −80 °C until RNA extraction. Standard TRIzol-chloroform extraction (Ambion, Carlsbad, CA, USA) was performed to extract bacterial RNA, before concentration and purity of the RNA were assessed by A_{260}:A_{280} spectrophotometry. Total bacterial RNA was normalized to 35 ng and cDNA synthesized using random primers and the QuantiNova reverse transcription kit (Qiagen, Hilden, Germany). Of cDNA 10 ng was amplified in triplicate by RT-qPCR performed on the Applied Biosystems QuantStudio 6 Flex Real-Time PCR system (Life Technologies, Burlington, ON, Canada) using PowerUp SYBR Green Master Mix (Applied Biosystems, Austin, TX, USA) as a detection method and 8 μM of the appropriate primers (Table S1). Primers were designed and selected using PrimerQuest (https://www.idtdna.com/primerquest). Cycling conditions involved an initial 2 min incubation at 50 °C and a 2 min incubation at 95 °C for SYBR Green activation and polymerase activation, respectively. This was followed by 40 cycles of 15 s at 9 °C for denaturation and 1 min at 60 °C for annealing and extension. Bacterial gene expression was quantified through comparison to the MRSA housekeeping gene 16S, and relative fold change in expression was calculated using the $2^{-\Delta\Delta CT}$ method [22]. Relative fold change values represent IAV–MRSA (normalized to 16S)/MRSA-alone (normalized to 16S).

2.6. Determination of Barrier Integrity in a Coculture Model

The electric cell-substrate impedance sensing trans-epithelial/endothelial electrical resistance (ECIS TEER) 24, 24-well TEER 24 microplates, and common electrode array (Applied Biophysics, Troy, NY, USA) were employed to quantify barrier integrity in a coculture model during pH1N1-MRSA coinfection. Epithelial cells were infected with influenza (MOI 0.1) or mock with infection media (denoted as Time 0) for 1 h followed by resting for 24 h. Viral and mock cells were subsequently infected with mid-log phase MRSA (MOI 0.1) or mock with infection media for 1 h. Resistance measurements were acquired at 4000 Hz every 4 h for 48 h. At each time point, the upper compartment of each transwell insert was filled with 600 μL of infection media and resistance measured for 1 min. Infection media was removed from each transwell insert and the cells allowed to rest until the next time point. Control conditions included: (i) cells infected with influenza-alone (MOI 0.1); and (ii) cells infected with MRSA-alone (MOI 0.1). Three biological replicates were performed per time point and per condition.

2.7. Kinome Peptide Array Analysis

Kinome peptide array analysis was performed as previously described [23,24]. IAV-, MRSA-, IAV-MRSA-, and mock HPAEpiCs and HPMECs were collected at 4, 8, 12, and 24 h post-bacterial infection by gentle scraping. Cells were pelleted by centrifugation at 14,000 rpm for 10 min, treated with kinome lysis buffer (20 mM TrisHlC pH 7.5, 150 mM NaCl, 1 mM EDTA, 1 mM EGTA, 1% Triton X-100, 2.5 mM sodium pyrophosphate, and 1× Pierce Halt Protease and Phosphatase Inhibitor) incubated on ice for 10 min, and transferred to fresh microcentrifuge tubes. The Pierce BCA Protein Assay Kit (ThermoFisher Scientific, Mississauga, ON, Canada) was used to quantify total protein concentration. Activation mix (50% glycerol, 50 μM ATP, 60 mM $MgCl_2$, 0.05% Brij 35, and 0.25 mg/mL bovine serum albumin) was added to the equivalent amounts of the total protein (100 μg) for each sample, and total sample volumes were matched by the addition of kinome lysis buffer. Kinome peptide arrays (JPT Peptide Technologies GmnbH, Berling, Germany) were spotted with samples and incubated for 2 h at 37 °C and 5% CO_2. After incubation, arrays were rinsed once with 1% Triton X-100 and once with deionized H_2O. Arrays were stained using PRO-Q Diamond phosphoprotein stain (Invitrogen, Carlsbad, CA, USA) for 1 h with gentle agitation. Following staining, arrays were washed 3× with kinome destain (20% acetonitrile and 50 mM sodium acetate pH 4.0) for 10 min. Arrays were washed a final time with deionized water for 10 min and dried by centrifugation. A PowerScanner microarray scanner (Tecan, Morrisville, NC, USA) with a 580-nm filter was used to image arrays and Array-Pro Analyzer version 6.3 software (Media Cybernetics, Rockville, MD, USA) was used to collect signal intensity values. Intensity values for spots and background were collected for each array. The Platform for Integrated, Intelligent Kinome Analysis (PIIKA 2) software (available online: https://saphire.usask.ca/saphire/piika) was used to analyze kinome data as previously described [25]. Additional heatmaps were derived using the Heatmapper software suite [26].

2.8. Pathway Overrepresentation and Gene Ontology Analysis

Pathway overrepresentation and gene ontology analyses of differentially phosphorylated proteins were performed using InnateDB software as described previously [24,27]. Input data was limited to peptides that demonstrated statistically-significant changes in expression as compared to the respective time-matched mock controls, as described previously [28]. Protein identifiers, phosphorylation fold change values (>1), and *p*-values (<0.05) were uploaded to Innate DB.

2.9. Chemokine and Cytokine Measurement

Chemokine and cytokine levels were determined using the microbead array assay Milliplex MAP multiplex kit (Human Cytokine/Chemokine Magnetic Bead Panel 96 Well Plate Assay; Millipore, Billerica, MA, USA) and analyzed on the BioPlex-200 (Biorad, Mississauga, ON, Canada). Supernatants were collected at 4, 8, 12, and 24 h for mock-, pH1N1-, MRSA-, and pH1N1-MRSA-infected samples and stored at −80 °C until use. Supernatants were analyzed according to the manufacturer's overnight protocol. Lower detection limit was 2.09 pg/mL for EGF, 24.84 pg/mL for FGF-2, 1.73 pg/mL for IFN-α2, 2.02 pg/mL for IFN-γ, 1.84 pg/mL for GRO, 1.66 pg/mL for IL-1β, 1.61 pg/mL for IL-3, 2.47 pg/mL for IL-6, 2.26 pg/mL for IL-8, 1.72 pg/mL for IP-10, 1.95 pg/mL for MCP-1, 1.59 pg/mL for RANTES, 1.55 pg/mL for TNF-α, and 1.53 pg/mL for VEGF.

2.10. Statistical Analyses

All numerical data are presented as mean ± SEM. Statistical analyses were performed using ANOVA for comparisons of group means using Prism 8 for MacOS (version 8.2.1). This includes pathogen replication kinetics, RT-qPCR, ECIS, and Milliplex MAP multiplex kit. A *p* value of ≤ 0.05 was considered statistically significant for all analyses. *p* values less than 0.05 are summarized by a single asterisk (*), less than 0.01 are summarized by two asterisks (**), less than 0.001 are summarized by three asterisks (***), and less than 0.0001 are summarized by four asterisks (****).

3. Results

3.1. MRSA Replication Kinetics Are Similar during MRSA-Alone and pH1N1-MRSA Infection

We first sought to determine how pre-existing pH1N1 infection affects bacterial replication in an in vitro tissue culture model of the alveolar-capillary barrier. Primary HPMECs were seeded on the basal side of transwell inserts and HPAEpiCs were seeded on the apical side. Temporal enumeration of bacteria was investigated by adding MRSA to our mock or pH1N1-infected tissue culture model 24 h post-infection. The number of adherent and internalized bacteria in epithelial cells was quantified through standard bacterial plating (Figure 1). No bacteria were identified by plating from the endothelial cells.

While there appeared to be a trend towards faster bacterial replication in pH1N1-MRSA coinfection, no statistically significant differences were observed between either infection condition (*p* = 0.3258) or between either infection condition over time (*p* > 0.6000). This suggested that MRSA fitness within pulmonary respiratory epithelial cells at the alveolar-capillary barrier is not affected by the presence of pH1N1. In contrast, viral loads decreased over time post-MRSA infection (Supplementary Figure S1). These matched our previous observations [21]. No virus was identified by RT-qPCR in the endothelial cells.

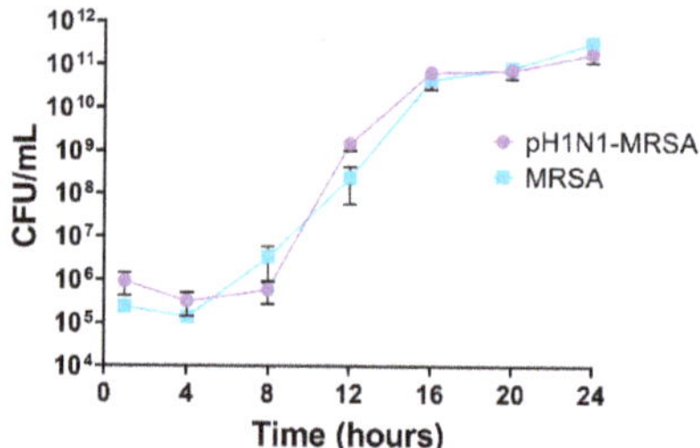

Figure 1. MRSA replication kinetics during MRSA infection and pH1N1-MRSA coinfection in primary alveolar epithelial cells. Human primary epithelial cells of the alveolar-capillary barrier were infected with pH1N1 (MOI 0.1) or mock followed by MRSA infection 24 h later (MOI 0.1). CFU were quantified by standard bacterial plating. Error bars represent SEM calculated from three biological replicates ($n = 3$). Statistical analyses were performed using ANOVA for comparisons of group means using Prism 8.

3.2. Modulation of Bacterial Virulence Factors in the Presence of pH1N1

As our results suggested that coinfection did not result in altered bacterial replication kinetics, we next sought to characterize how the modulation of bacterial virulence factors related to adhesion and invasion might be altered during coinfection. Our prior work with coinfection in A549 cells demonstrated that altered virulence factor expression was only found during early infection (1–4 h post-MRSA addition). Thus, here we focused on the same time points. The alveolar pulmonary cells of our coculture model were infected with pH1N1 at a MOI of 0.1 or mock and allowed to rest for 24 h prior to MRSA-infection (MOI 0.1). Cell lysates were collected at multiple time points post-infection and RT-qPCR employed to examine differential modulation of MRSA virulence factor gene expression in the presence or absence of pre-existing influenza virus infection. We studied 13 virulence factor genes directly related to adhesion and invasion: coa, ebpS, eno, fnbA, fnbB, hla, hlgA, icaA, icaB, sbi, sek, seq, and spA. Modulation of virulence factors was observed at 1 and 4 h post-MRSA infection, which mimicked our prior results in immortalized A549 cells [21].

At 1 h post-infection, *eno* ($p = 0.0120$), *icaB* ($p < 0.001$), *sek* ($p = 0.0146$), and *seq* ($p = 0.0135$) were significantly upregulated at 1 h in coinfected samples as compared to MRSA-alone (Figure 2). At 4 h post-MRSA infection, *coa* ($p < 0.0001$), *fnbB* ($p < 0.0001$), *hla* ($p = 0.0014$), *hlgA* ($p < 0.0001$), *icaA* ($p < 0.0001$), *icaB* ($p < 0.0001$), *sbi* ($p < 0.0001$), and *sek* ($p < 0.0001$) were significantly upregulated during coinfection. This data coincides with the lag phase of MRSA in the presence of pH1N1 at 1 and 4 h, suggesting that adhesion- and invasion-associated virulence factors may play a role in the initial stages of MRSA infection in primary alveolar cells previously infected with pH1N1.

3.3. Barrier Integrity of a Coculture Model of the Alveolar-Capillary Barrier during pH1N1-MRSA Coinfection

We next sought to characterize the effect of pH1N1-MRSA coinfection on barrier integrity in our coculture model by measuring temporal changes in resistance. Cells were either mock or infected with pH1N1 at a MOI of 0.1 (first arrow; designated as Time 0), allowed to rest for 24 h, and either mock or infected with MRSA at a MOI of 0.1 (second arrow). No change in resistance was observed following pH1N1-alone infection as compared with mock cells; the resistance of each of the observed conditions remained steadily at 110 ohms (Figure 3). Following bacterial addition, infection with MRSA-alone resulted in no changes in resistance. No significant differences in barrier integrity were observed at any time point between models infected with MRSA-alone and pH1N1-alone. Samples coinfected with pH1N1-MRSA resulted in a steady decrease in resistance beginning at 8 h post-MRSA addition (30 h). By 45 h, pH1N1-MRSA was significantly downregulated ($p = 0.0005$) as compared with the mock model. This decrease in barrier resistance beginning at 8 h post-MRSA infection coincided with the beginning of the exponential phase of MRSA in the presence of pH1N1.

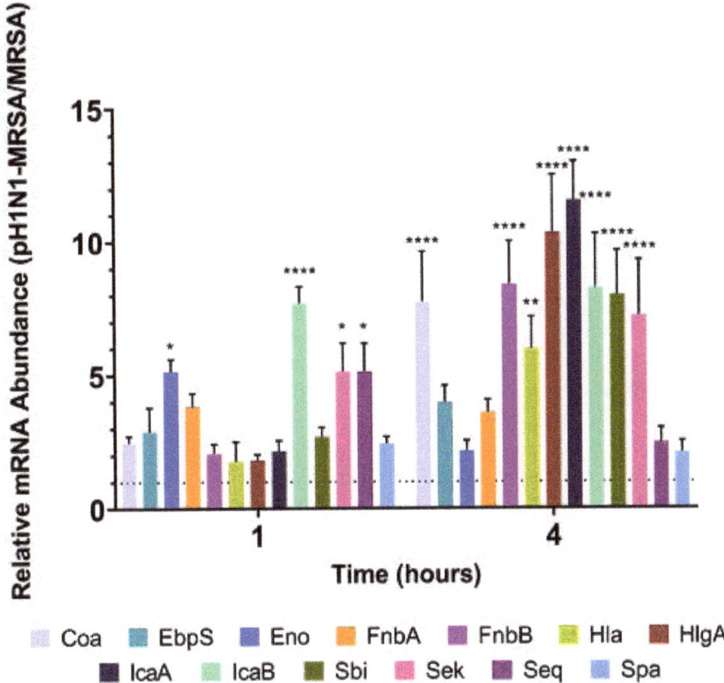

Figure 2. MRSA virulence factors are modulated during coinfection at the alveolar-capillary barrier. RT-qPCR was employed to examine differential modulation of relative MRSA virulence factor mRNA abundance at 1 and 4 h in infected primary epithelial cells of the alveolar-capillary barrier. Relative mRNA abundance fold changes represent pH1N1-MRSA vs. MRSA infection alone and were calculated by the $2^{-\Delta\Delta CT}$ method. The dashed line signifies a fold-change of 1. Error bars represent SEM calculated from three biological replicates ($n = 3$). Statistical analyses were performed using ANOVA for comparisons of group means using Prism 8. *: $p < 0.1$ **: $p < 0.01$ ****: $p < 0.0001$.

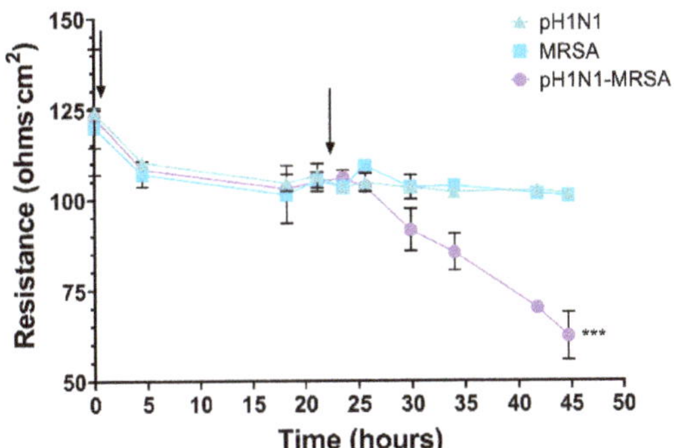

Figure 3. pH1N1-MRSA coinfection decreases barrier function in an alveolar-capillary coculture model. Human primary epithelial cells of the alveolar-capillary barrier were infected or mock with pH1N1 (MOI 0.1; first arrow) and MRSA (MOI 0.1; second arrow) was added to cells 24 h later. Error bars represent SEM calculated from three biological replicates ($n = 3$). Statistical analyses were performed using ANOVA for comparisons of group means using Prism 8. ***: $p < 0.001$.

3.4. Temporal Analysis of the Host Kinome Response in a Coculture Model of the Alveolar-Capillary Barrier during pH1N1-MRSA Coinfection

As our temporal analysis of barrier integrity suggested that pH1N1-MRSA coinfection results in more severe barrier dysregulation compared with either pathogen alone, we addressed whether aberrant cell-mediated immune responses contribute to coinfection pathogenesis. We performed temporal kinome analysis of pH1N1-, MRSA-, and pH1N1-MRSA-infected alveolar epithelial and microvascular endothelial cells of the alveolar-capillary cocultures. Time-matched mock controls cells served as controls. Alveolar-capillary barrier cocultures were initially infected with pH1N1 (MOI 0.1) or mock and rested for 24 h prior to bacterial infection. MRSA addition to cells (+ or − pH1N1) was designated as time 0. Epithelial and endothelial cells were harvested separately at 4, 8, 12, and 24 h post-MRSA infection. Both pH1N1-alone infected cells and mock control cells were treated with MRSA-free infection inoculum at time 0 to normalize cellular responses resulting from physical stress during inoculum addition. Time-matched pH1N1-, MRSA-, and mock control cells were collected throughout the duration of the experiment.

To gain insight into the host kinome response of pulmonary epithelial cells during pH1N1-MRSA coinfection as compared to infection with either pathogen alone, biological subtraction of the time-matched mock kinome datasets from their respective infected counterparts was performed. Respective hierarchical clustering analysis of the kinome data following mock background subtraction is presented in Figure 4. Notably, each of the time-matched samples from IAV-alone, MRSA-alone, and IAV–MRSA infected samples clustered together, resulting in four major clusters. From left to right, the first cluster (denoted as A) consisted of each of the 4 h time-matched samples, the second cluster (denoted as B) consisted of each of the 12 h time-matched samples, the third cluster (denoted as C) consisted of each of the 8 h time-matched samples, and the fourth cluster (denoted as D) consisted of each of the 24 h time-matched samples. Clusters B and C, consisting of the 8 and 12 h time points, respectively, clustered together more strongly than with the samples from 4 and 24 h post-MRSA infection. Moreover, the 24 h samples differentiated most strongly from each of the other time points. This data suggested that the modulation of the host kinome response were strongly related to post-infection time points, with intra-time point dependent differences in host responses to infection.

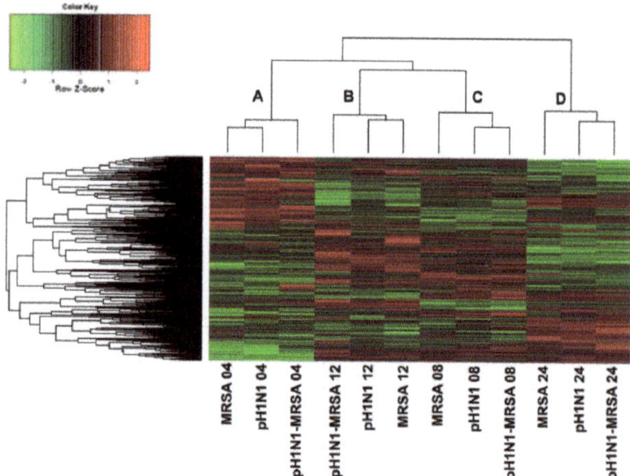

Figure 4. Temporal kinome responses of pH1N1, MRSA, and pH1N1-MRSA infection in epithelial cells within an alveolar-capillary barrier coculture model. Mock kinome responses were subtracted from time-matched infected samples. Fold change phosphorylation values are plotted for all kinase recognition sequences. Red depicts upregulation, while green depicts downregulation as compared with the background. A–D designate the four major dataset clusters as identified following hierarchical clustering.

We next sought to identify host cell signaling responses or biological networks in the pH1N1-MRSA-infected pulmonary alveolar cells that were selectively modulated at 24 h post-MRSA infection. Kinome analysis at 24 h post-MRSA addition demonstrated that pH1N1-MRSA coinfection resulted in the activation of numerous signaling pathways as compared with either pH1N1- or MRSA-infection alone (Table S2). There was an overrepresentation of numerous pathways related directly to the cell cycle, TLR-related signaling, interleukins, and interferon signaling. TLR pathways were not identified in pH1N1- or MRSA-alone datasets, while there was a unique over-representation of TLR signaling pathways and TLR-associated pathways during coinfection. Infection with pH1N1-alone resulted in a lower total number of pathways identified as compared to the coinfection data with the overrepresentation of IFN signaling pathways (lowest *p*-value), cytokine signaling, and apoptosis-associated pathways (Table S3). This suggests that IFN-mediated responses are muted during coinfection as compared to pH1N1 alone in alveolar epithelial cells. In contrast, MRSA infection alone resulted in relatively few upregulated pathways as compared with the mock control, namely IGF- and inflammasome-associated signaling pathways (Table S3). This data demonstrates that pH1N1-MRSA coinfection results in a unique cell response signature in primary differentiated alveolar epithelial cells grown in close proximity to pulmonary endothelial cells. This contrasts to our prior analysis of coinfection in A549 cells where kinome responses from pH1N1- and MRSA-alone infections largely overlapped.

Pulmonary endothelial cells from the cocultures were also isolated throughout the time course of infection. Biological subtraction of time-matched mock kinome datasets was performed as previous (Figure 5). Similar to the epithelial cells, biological subtraction revealed four major clusters matched by time point. From left to right, the 8 h time points of pH1N1-alone, MRSA-alone, and pH1N1-MRSA infected samples clustered together (denoted as A). The second cluster consisted of pH1N1-alone, MRSA-alone, and pH1N1-MRSA infected samples at 12 h post-bacterial infection (denoted as B). The third cluster was comprised of the 24 h time points of IAV-alone, MRSA-alone, and IAV–MRSA infected samples (denoted as C). Lastly, samples infected with IAV-alone, MRSA-alone, and IAV–MRSA at 4 h post-bacterial infection clustered together (denoted as D). A and B clustered together; C and D formed a separate cluster. This suggested that the modulation of the host kinome response related strongly to time post-infection as also seen in the epithelial cells.

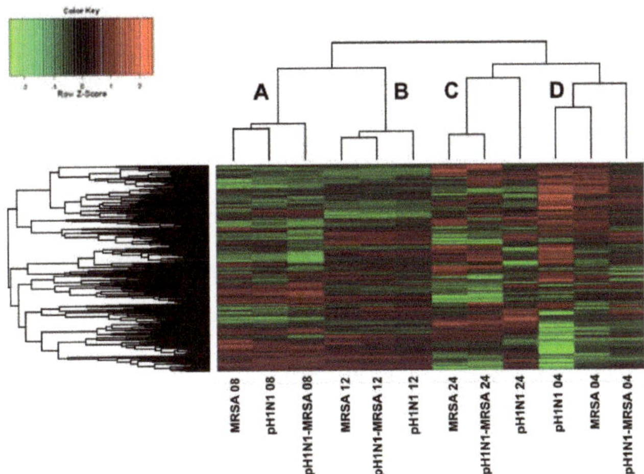

Figure 5. Temporal kinome responses of pH1N1, MRSA, and pH1N1-MRSA infection in pulmonary endothelial cells within an alveolar-capillary barrier coculture model. Mock kinome responses were subtracted from time-matched infected samples. Red depicts upregulation, while green depicts downregulation as compared with the background. A–D designate the four major dataset clusters as identified following hierarchical clustering.

Coinfection resulted in the overrepresentation of signaling pathways related to hedgehog-, proteasome-, cell stress-, and Wnt- -related pathways at 24 h (Table S5). In contrast, only a single signaling pathway was differentially modulated in the endothelial cells during pH1N1-alone infection, which was unsurprising given the lack of susceptibility of these cells to pH1N1 transmission from the virus-infected epithelial cells. MRSA-alone infection showed notable upregulation of hedgehog- and Wnt/β-catenin-associated signaling pathways as compared with time-matched mock controls (Table S6). Similar to the coinfected cells, MRSA infection alone resulted in over-representation of signaling pathways related to hedgehog signaling and Wnt/β-catenin-related signaling suggesting that endothelial cell host responses are largely driven by responses to the MRSA-infected epithelial cells during coinfection (Table S7).

3.5. Cytokine Expression Is Modulated during pH1N1-MRSA Coinfection

As our kinome data revealed that pathways related to interleukins and cytokine signaling were both overrepresented when alveolar epithelial respiratory cells were infected with pH1N1 and pH1N1-MRSA, we next sought to characterize the expression of proinflammatory cytokines by pH1N1-, MRSA-, and pH1N1-MRSA infected cells. Supernatants were collected at 4, 8, 12, and 24 h post-MRSA addition and cytokine secretion was assessed for each condition. Cytokines that were measured included the epidermal growth factor (EGF), fibroblast growth factor 2 (FGF-2), IL-6, IL-8, interferon-γ induced protein 10 (IP-10/CXCL10), monocyte chemoattractant protein-1 (MCP-1/CCL2), and VEGF. Significant downregulation of EGF expression was observed at 12 h in both pH1N1-alone ($p = 0.0003$) and pH1N1-MRSA ($p = 0.0002$) infection and was resolved by 24 h (Figure 6).

At 4 h, FGF-2 expression was significantly upregulated in pH1N1-alone infection ($p = 0.0052$) and also at 12 h and 24 h. FGF-2 was significantly upregulated in the case of pH1N1-alone ($p = 0.0035$) and pH1N1-MRSA coinfection ($p = 0.0140$) at 12 h, and in the case of pH1N1-alone ($p = 0.0008$) and pH1N1-MRSA ($p = 0.0002$) at 24 h. MRSA-alone infection resulted in significant repression at 12 h compared with mock-infection ($p = 0.0001$). Significant upregulation of IL-6 was observed at 24 h ($p < 0.0001$) in pH1N1-alone infections, and at 8 and 24 h ($p = 0.0303$ and $p < 0.0001$, respectively) in pH1N1-MRSA coinfections. IL-8 was significantly upregulated at 4, 8, 12, and 24 h in pH1N1-alone infection ($p = 0.0059$, $p = 0.0056$, $p = 0.0280$, and $p < 0.0001$, respectively). IL-8 was also significantly upregulated at 4, 8, and 24 h in pH1N1-MRSA coinfection ($p = 0.0002$, $p < 0.0001$, and $p < 0.0001$). In MRSA-alone infections, IL-8 was significantly upregulated at 24 h only ($p = 0.0016$).

IP-10 was significantly upregulated at 4, 8, 12, and 24 h in pH1N1-alone infection and pH1N1-MRSA coinfection ($p < 0.0001$ for each), suggesting that the presence of pH1N1 has an important effect on IP-10 secretion. MCP-1 was only significantly upregulated at 24 h in pH1N1-alone infection ($p < 0.0001$), but not at any other time point. VEGF was significantly repressed at 4 h in each infection condition as compared with the mock samples ($p < 0.0001$ for all) and remained repressed at the 8 h timepoint in the coinfection samples alone ($p = 0.0019$). Taken together, this data suggest that cytokine expression during coinfection was largely driven by pH1N1 infection and may play an important role in barrier disruption.

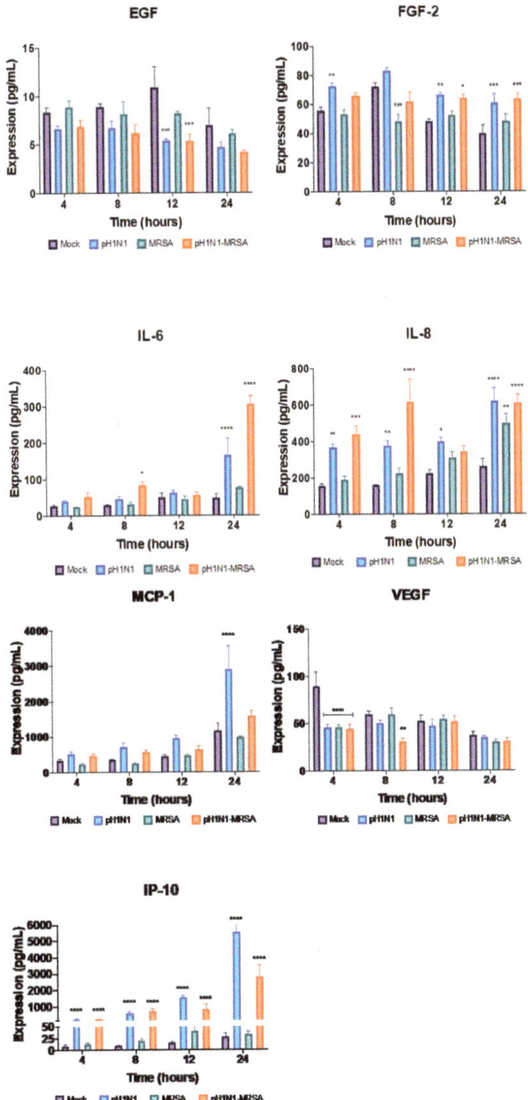

Figure 6. Induction of cytokine expression during pH1N1-, MRSA-, and pH1N1-MRSA infection in the alveolar-capillary barrier model. Primary epithelial cells of the alveolar-capillary barrier were mock-infected or infected with pH1N1 (MOI 0.1). MRSA (MOI 0.1) was added 24 h later. Cytokine levels were determined using the Milliplex MAP multiplex kit. Error bars represent SEM calculated from three biological replicates ($n = 3$). Statistical analyses were performed using ANOVA for comparisons of group means using Prism 8. *: $p < 0.1$ **: $p < 0.01$ ***: $p < 0.001$ ****: $p < 0.0001$.

4. Discussion

As severe influenza and influenza-bacterial coinfections within the lower respiratory tract can lead to disruption of the alveolar-capillary barrier with potentially deleterious effects on both gas exchange and normal lung function, we sought to examine influenza-bacterial coinfections in a physiologically relevant model of the alveolar-capillary barrier. No significant differences were found for MRSA replication kinetics at any time point across each of our infection conditions similar to our prior analysis in A549 cells [21]. This suggests that MRSA fitness is not altered by pre-existing pH1N1 infection at the alveolar-capillary barrier. Further, this suggests that the increased disease severity associated with influenza-bacterial coinfection is not simply due to increased bacterial burden within the lungs during coinfection. The expression of MRSA virulence factors related to adhesion and invasion were selectively upregulated at 1 and 4 h in our coculture model, which was similar to our prior findings [21]. The pattern of upregulated gene expression corresponded with the lag phase of MRSA at 1 and 4 h in the presence of pH1N1. At 1 h post-MRSA infection, significant upregulation was only observed in *eno*, *icaB*, *sek*, and *seq*. The *eno* protein codes for enolase, which binds to laminin in the basal lamina of the epithelia; previous studies have shown that the high prevalence of *eno* could play an important role in future MRSA vaccine design, which is further underlined by our results [29,30]. The ica locus, which codes for *icaB*, is involved in intracellular adhesion and biofilm formation [31–35]. Upregulation of *icaB* early in infection may be indicative of the increased lag phase of MRSA in the presence of pH1N1 during alveolar epithelial infection. This upregulation of *icaB* was not observed in our alveolar monolayer and may have been due to the production of surfactant by our primary differentiated alveolar cells [21]. Lastly, *seq* and *sek* encode secreted exotoxins that alter the host cell membranes, resulting in lysis [36–38]. The superantigen properties of *sek* and *seq* also contribute directly to MRSA virulence [39–41]. Lysis of neutrophils by exotoxins such as *sek* and *seq* result in reactive oxygen species release, which leads to damage and inflammation to surrounding lung tissue [39,41]. Further, mouse models have suggested that the *sek* and *seq* superantigens of MRSA play a role in T-cell signaling responsible for much of the early lung damage seen in *S. aureus* infection [40]. At 4 h post-MRSA addition, significant upregulation seen in *coa*, *fnbB*, *hla*, *hlgA*, *icaA*, *sbi*, and *sek*. The product of *coa*, coagulase, plays a role in initial adhesion to epithelial cells by cleaving fibrinogen and activating prothrombin [42–44]. The fnb locus, which codes for *fnbB*, is also involved in initial bacterial adhesion to epithelial cells for internalization [38,43,45–48]. As both *hla* and *hlgA* result in the lysis of infected cells, this may support the idea that bacterial toxins play a role in secondary bacterial pathogenesis early in infection of alveolar epithelial cells, as both virulence factors were also upregulated during coinfection of our alveolar monolayer [21]. Upregulation of *icaA*, which plays a similar role to *icaB*, may underlie the increased lag phase observed in pH1N1-MRSA coinfection [31,33–35]. Lastly, the upregulation of *sbi* may suggest that pre-existing pH1N1 infection in alveolar epithelial cells may indirectly facilitate a strong immune evasion response in MRSA through interaction of bacteria with secreted messengers or agonists from damaged epithelium resulting in the inhibition of antibody responses through the binding of IgG and C3, a novel immune evasion approach [49].

Coinfection had significant effects on alveolar-capillary barrier permeability as assessed by ECIS. As the presence of MRSA alone did not result in permeabilization, our data suggest that underlying pH1N1 infections at the alveolar-capillary barrier contribute to dysfunction following secondary bacterial coinfections. This is unsurprising as pH1N1-alone infection in healthy adults rarely results in severe disease [11,12]. Likewise, severe illness and death in otherwise healthy adults infected with *S. aureus* is often associated with prior influenza infection [50–52]. We also utilized kinome analysis to provide insights into host response modulation during pH1N1-MRSA coinfections. Our human kinome peptide arrays had an increased breadth of kinase recognition sequences as compared to our prior analysis in A549 cells (1294 targets compared to 309 targets) [21]. The use of a transwell coculture system provides for intercellular interactions between the epithelial and endothelial cells. Interestingly, in our primary epithelial cells, the 8 and 12 h time points still clustered together more strongly than with the 4 and 24 h time points.

Our pathway over-representation analysis of differentiated alveolar epithelial cells further supports our postulate that IAV-bacterial coinfections are able to specifically modulate host cellular responses independently of infection with pH1N1- or MRSA-alone. Most notably, pathways related to TLRs, including TLR-2, -4, -5, -7/8, -9, and -10 were overrepresented only in IAV–MRSA coinfected cells. TLRs are an important aspect of severe IAV- MRSA coinfections [53,54]. MyD88, an important mediator of

TLR signaling and NK-κB activation, was also strongly overrepresented in our coinfected epithelial cells [55–57]. Interestingly, *S. aureus* can dampen TLR-2 activation and thus NK-κB activation [57]. Taken together, our data supports the hypothesis that further characterization of TLR signaling in pH1N1-MRSA coinfection may provide new opportunities for targeted drug therapies. Cell cycle pathways were also overrepresented in coinfected epithelial cells. Many of these overrepresented pathways were involved in the G1 and G2 phases, and the transition between these two phases. Previous gene-expression analysis in patients infected with pH1N1 has shown that progression towards severe IAV infection is often characterized by abnormal deviations in cell cycle and apoptosis signaling pathways [58]. Specifically, progression to severe infection was characterized by increased aberrant DNA replication in the G1/S phase but delayed exit from the G2/M phase. Cytokine signaling was overrepresented at 24 h in the presence of pH1N1, whether MRSA was present or not. "Cytokine Signaling in Immune System" was highly upregulated (56 associated proteins from our kinome arrays) in both pH1N1-MRSA and MRSA-alone infection conditions in alveolar epithelial cells. Infection with pH1N1-alone also resulted in upregulation in "Signaling by Interleukins", with 44 associated proteins from our kinome arrays, in addition to upregulation of IL-1, IL-2, and IL-6. Various studies have suggested that severe lung inflammation seen in highly pathogenic influenza strains may be due to increased levels of proinflammatory cytokines [58–61].

In comparison to pH1N1-MRSA and pH1N1-alone infection, relatively few pathways were upregulated in alveolar epithelial cells infected with MRSA-alone. Notably, upregulation of pathways related to hedgehog, Wnt, and β-catenin signaling was observed. Our analysis demonstrated that coinfection resulted in a paucity of upregulated signaling pathways compared to the epithelial cells. Overrepresentation of pathways was most commonly found in pH1N1-MRSA infection, followed by MRSA-alone infection, and pH1N1-alone infection. Only one pathway was overrepresented in cells infected with pH1N1-alone. This was perhaps unsurprising as no evidence of pH1N1 infection was identified in the endothelial cells although the alveolar epithelial cells were susceptible to productive viral infection. Pathways related to hedgehog and Wnt/β-catenin signaling were most commonly overrepresented in endothelial cells in both our MRSA-alone and pH1N1-MRSA infection conditions. The Wnt signaling pathway regulates a number of genes involved in cell growth, differentiation, survival, and immune functions [62]. However, recent studies have revealed both a pro- and anti-inflammatory response from Wnt/β-catenin signaling, suggesting that the role of Wnt may be dependent on the stimulus, cell type, and crosstalk with other signaling pathways [62–66]. The majority of evidence notes that activation of the Wnt pathway is able to reduce inflammatory processes triggered by bacterial pathogens [62]. As MRSA is not present in the endothelial cells of the coculture model, overrepresentation of pathways related to Wnt signaling may be indicative of the ability of the alveolar epithelial cells to communicate with the underlying endothelial cells of the capillary.

Based on our kinome analysis we also assessed the concentration of cytokines that were previously implicated in severe pH1N1-MRSA infections. The proinflammatory response may be an important factor in disease outcome from secondary bacterial pneumonia, as various studies have suggested that severe lung inflammation in influenza infections may be related to increased levels of proinflammatory cytokines in the lung [58–61]. Specifically, pH1N1 is known to induce the expression of a number of interleukins in both the respiratory tract and central nervous system [67]. Significant downregulation of EGF was observed at 12 h in both pH1N1-alone and pH1N1-MRSA infection. EGF is a growth factor capable of stimulating proliferation of epithelial cells by activating cellular signaling through engagement of the EGF receptor [68]. A previous study reported that EGF was significantly higher in healthy patients compared to pH1N1 infection and the authors suggested that EGF was actively suppressed, in an effort to protect the lung from host or virus mediated damage [69]. This may explain the significant downregulation of EGF observed in our pH1N1-alone and pH1N1-MRSA infected cells. Significant upregulation of FGF-2, a member of the fibroblast growth factor family, was also observed in infection with pH1N1-alone and in pH1N1-MRSA coinfections. Conversely, FGF-2 was significantly downregulated at 8 h in cells infected with MRSA-alone. FGF-2 plays an important role in epithelial repair in the lung and in wound healing [70–72]. FGF-2 dysregulation is implicated in many inflammatory diseases, and a study in mice suggested that FGF-2 plays a vital role in IAV-induced lung injury [73]. While our data seems to suggest that pH1N1 infection results in a significant increase in FGF-2 expression regardless of the presence of MRSA, further investigation will need to be done to fully understand the role of FGF-2 in severe pH1N1-MRSA infections. IL-6 was significantly upregulated at multiple time points in pH1N1-MRSA infections, and at 24 h in pH1N1-alone infections. IL-6 levels have been shown to be significantly elevated in the presence of a clinically relevant secondary bacterial infection, which may make its upregulation in pH1N1-MRSA coinfection unsurprising [54,74]. It is enticing to speculate that IL-6 upregulation may be related to increased barrier permeability found in our pH1N1-MRSA ECIS results, as high levels of IL-6 are known to directly damage endothelial cells [75]. Further, elevated serum IL-6 levels have been implicated as a potential biomarker for disease severity in pH1N1-alone infections [76]. IL-8 was also significantly upregulated in pH1N1-alone and pH1N1-MRSA infections. IL-8 shows distinct target specificity for attracting and activating neutrophils to inflammatory regions [77]. IP-10 was significantly upregulated at all time points in pH1N1-MRSA and pH1N1-alone infections. This is perhaps unsurprising as previous studies have reported that interferon-related signaling, such as IP-10, were more abundant in cases of severe disease [74]. IP-10 is able to directly influence apoptosis in disease, which may explain the upregulation of apoptosis pathways in our epithelial cell kinome data from pH1N1-MRSA and pH1N1-alone infections [78]. Interestingly, IP-10 was decreased in expression in the coinfected samples as compared to pH1N1 alone. Prior analysis has demonstrated that *S. aureus* downregulates IP-10 production [79] and this difference likely represents the inhibitory activity imparted by MRSA during coinfection. Similarly, MCP-1 was significantly upregulated in pH1N1-infected cells at 24 h but not in MRSA-alone or coinfected cells. This is perhaps unsurprising as MCP-1 was not upregulated in response to *S. aureus* infection in airway epithelial cells [80] and may reduce the overall induction of this chemokine during coinfection. Lastly, VEGF was significantly downregulated across all conditions early during infection as compared to mock cells. VEGF is a regulator of cell growth, and is most abundant in the lung; transcripts are primarily localized in alveolar type II cells in the alveoli [81–83]. The downregulation observed in our model is perhaps surprising, as previous studies have reported that hypoxia, commonly seen in severe pH1N1-MRSA infection, results in increased induction of VEGF from ATII cells [81]. Further investigations in more advanced in vitro or in vivo models of severe pH1N1-MRSA and/or a longer course of infection may better describe the role VEGF plays at the alveolar-capillary barrier.

Overall, significant upregulation of cytokine expression is often observed in pH1N1-MRSA coinfection. Further understanding of how elevation of specific cytokines impact pH1N1-MRSA disease severity may reveal potential additional therapeutic targets to reduce the generation of a cytokine storm in infected patients. This strategy has previously shown promise as inhibition of certain cytokines, such as IL-10, have shown improved survival from bacterial pneumonia late after influenza infection [84,85].

5. Conclusions

Our investigations into influenza–bacterial coinfections in a primary coculture model of the alveolar-capillary barrier suggest that infection with both pH1N1 and MRSA appears to have a synergistic pathologic effect, as infection with either pathogen on its own did not result in the loss of barrier integrity nor strong dysregulation of host response during coinfection. Strikingly, dysregulation of the host response seems to be driven primarily by the response of alveolar epithelial cells to both pathogens. This is contrary to what was observed in an alveolar cell line, which suggested that MRSA-alone infection resulted in a similar dysregulation as pH1N1-MRSA coinfection [21]. This disparity could be due to a number of reasons. Notably, our results indicate that alveolar-capillary barrier cocultures are likely a more reliable surrogate for assessing lower respiratory tract pathogenesis than alveolar epithelial cells alone, though further investigations are needed to confirm this. Additionally, the coculture model is able to reflect the crosstalk that is able to occur between the alveolar epithelial and microvascular endothelial cells at the alveolar-epithelial barrier.

Supplementary Materials: The following are available online at http://www.mdpi.com/2073-4409/9/11/2472/s1, Figure S1: Alveolar-capillary barrier coculture model schematic, Table S1: MRSA Virulence Factor Primer Sequences, Table S2: Upregulated signaling pathways in pH1N1-MRSA coinfected alveolar epithelial cells at the alveolar-capillary barrier as determined by temporal kinome analysis, Table S3: Upregulated signaling pathways in pH1N1-infected alveolar epithelial cells at the alveolar-capillary barrier as determined by temporal kinome analysis, Table S4: Upregulated signaling pathways in MRSA-infected alveolar epithelial cells at the alveolar-capillary barrier, Table S5: Upregulated signaling pathways in pH1N1-MRSA-infected pulmonary endothelial cells at the alveolar-capillary barrier, Table S6: Upregulated signaling pathways in pH1N1-infected pulmonary endothelial cells at the alveolar-capillary barrier, Table S7: Upregulated signaling pathways in MRSA-infected pulmonary endothelial cells at the alveolar-capillary barrier.

Author Contributions: J.K. and M.E.N. conceived of the ideas presented herein. M.E.N., S.M.L., and B.D. performed the described experiments. M.E.N., S.M.L., B.D. and J.K. analyzed the data. M.E.N., S.M.L., B.D. and J.K. performed the drafting and revising of the manuscript. All authors have read and agreed to the published version of the manuscript.

Funding: M.E.N. was funded by a Research Manitoba Master's Studentship Award. J.K. is funded by a Tier 2 Canada Research Chair in the Molecular Pathogenesis of Emerging and Re-Emerging Viruses provided by the Canadian Institutes of Health Research (Grant No. 950-231498) and a Research Manitoba New Investigator Operating Grant (Grant No. 3531).

Conflicts of Interest: The authors declare no conflict of interest.

References

1. Iuliano, A.D.; Roguski, K.M.; Chang, H.H.; Muscatello, D.J.; Palekar, R.; Tempia, S.; Cohen, C.; Gran, J.M.; Schanzer, D.; Cowling, B.J.; et al. Estimates of global seasonal influenza-associated respiratory mortality: A modelling study. *Lancet* **2018**, *391*, 1285–1300. [CrossRef]

2. Krammer, F.; Smith, G.J.D.; Fouchier, R.A.M.; Peiris, M.; Kedzierska, K.; Doherty, P.C.; Palese, P.; Shaw, M.L.; Treanor, J.; Webster, R.G.; et al. Influenza. *Nat. Rev.* **2018**, *4*, 3. [CrossRef] [PubMed]

3. Centers for Disease Control and Prevention (CDC). Influenza (Flu). Available online: https://www.cdc.gov/flu/index.htm (accessed on 1 July 2019).

4. McKimm-Breschkin, J.L. Influenza meuraminidase inhibitors: Antiviral action and mechanisms of resistance. *Influenza Other Respir. Viruses* **2013**, *7*, 25–36. [CrossRef] [PubMed]

5. Barberis, I.; Myles, P.; Ault, S.K.; Bragazzi, N.L.; Martini, M. History and evolution of influenza control through vaccination: From the first monovalent vaccine to universal vaccines. *J. Prev. Med. Hyg.* **2016**, *57*, E115–E120.

6. Neumann, G.; Kawaoka, Y. Transmission of Influenza A Viruses. *Virology* **2015**, 234–246. [CrossRef]

7. Richard, M.; Fouchier, R.A.M. Influenza A virus transmission via respiratory aerosols or droplets as it relates to pandemic potential. *FEMS Microbiol. Rev.* **2016**, *40*, 68–85. [CrossRef]

8. Taubenberger, J.K.; Morens, D.M. The Pathology of Influenza Virus Infections. *Annu. Rev. Pathol. Mech. Dis.* **2008**, *3*, 499–522. [CrossRef]

9. Kash, J.C.; Taubenberger, J.K. The Role of Viral, Host, and Secondary Bacterial Factors in Influenza Pathogenesis. *Am. J. Pathol.* **2015**, *185*, 1528–1536. [CrossRef]

10. Honce, R.; Schultz-Cherry, S. Impact of Obesity on Influenza A Virus Pathogenesis, Immune Response, and Evolution. *Front. Immunol.* **2019**, *10*, 1071. [CrossRef]

11. Ghebrehewet, S.; MacPherson, P.; Ho, A. Influenza. *BMJ* **2016**, *355*. [CrossRef]

12. Moghadami, M. A Narrative Review of Influenza: A Seasonal and Pandemic Disease. *Iran. J. Med. Sci.* **2017**, *42*, 2–13. [PubMed]

13. McCullers, J.A. The co-pathogenesis of influenza viruses with bacteria in the lung. *Nat. Rev. Microbiol.* **2014**, *12*. [CrossRef] [PubMed]

14. Taubenberger, J.K.; Morens, D.M. 1918 Influenza: The mother of all pandemics. *Emerg. Infect. Dis.* **2006**, *12*, 15–22. [CrossRef] [PubMed]

15. Rudd, J.M.; Ashar, H.K.; Chow, V.T.K.; Teluguakula, N. Lethal Synergism between Influenza and *Streptococcus pneumoniae*. *J. Infect. Pulm. Dis.* **2016**, *2*. [CrossRef]

16. Morens, D.M.; Taubenberger, J.K.; Fauci, A.S. Predominant role of bacterial pneumonia as a cause of death in pandemic influenza: Implications for pandemic influenza preparedness. *J. Infect. Dis.* **2008**, *198*. [CrossRef]

17. Rynda-Apple, A.; Robinson, K.M.; Alcorn, J.F. Influenza and Bacterial Superinfection: Illuminating the Immunologic Mechanisms of Disease. *Infect. Immun.* **2015**, *83*. [CrossRef]

18. Chertow, D.S.; Memoli, M.J. Bacterial Coinfection in Influenza: A Grand Rounds Review. *JAMA* **2013**, *309*. [CrossRef]

19. Louie, J.; Jean, C.; Chen, T.H.; Park, S.; Ueki, R.; Harper, T.; Chmara, E.; Myers, J.; Stoppacher, R.; Catanese, C.; et al. Bacterial coinfection in lung tissue specimens from fatal cases of 2009 pandemic influenza A (H1N1)—United States, May–August 2009. *Morb. Mortal. Wkly. Rep.* **2009**, *58*, 1–4.

20. Gill, J.R.; Sheng, Z.M.; Ely, S.F.; Guinee, D.G., Jr.; Beasley, M.B.; Suh, J.; Deshpande, C.; Mollura, D.J.; Morens, D.M.; Bray, M.; et al. Pulmonary Pathologic Findings of Fatal 2009 Pandemic Influenza A/H1N1 Viral Infections. *Arch. Pathol. Lab. Med.* **2010**, *134*, 235–243. [CrossRef]

21. Nickol, M.E.; Ciric, J.; Falcinelli, S.; Chertow, D.S.; Kindrachuk, J. Characterization of Host and Bacterial Contributions to Lung Barrier Dysfunction Following Co-infection with 2009 Pandemic Influenza and Methicillin Resistant. *Staphylococcus Aureus Viruses* **2019**, *11*, 116. [CrossRef]

22. Atshan, S.S.; Shamsudin, M.N.; Karunandihi, A.; van Belkum, A.; Lung, L.T.; Sekawi, Z.; Nathan, J.J.; Ling, K.H.; Seng, J.S.; Ali, A.M.; et al. Quantitative PCR analysis of genes expressed during biofilm development of methicllin resistant *Staphylococcus Aureus* (MRSA). *Infect. Genet. Evol.* **2013**, *18*, 106–112. [CrossRef] [PubMed]

23. Kindrachuk, J.; Wahl-Jensen, V.; Safronetz, D.; Trost, B.; Hoenen, T.; Arsenault, R.; Feldmann, F.; Traynor, D.; Postnikova, E.; Kusalik, A.; et al. Ebola virus modulates transforming growth factor β signaling and cellular markers of mesenchyme-like transition in hepatocytes. *J. Virol.* **2014**, *88*, 9877–9892. [CrossRef] [PubMed]

24. Kindrachuk, J.; Ork, B.; Hart, B.J.; Mazur, S.; Holbrook, M.R.; Frieman, M.B.; Traynor, D.; Johnson, R.F.; Dyall, J.; Kuhn, J.H.; et al. Antiviral Potential of ERK/MAPK and PI3K/AKT/mTOR Signaling Modulation for Middle East Respiratory Syndrom Coronavirus Infection as Identified by Temporal Kinome Analysis. *Antimicrob. Agents Chemother.* **2014**, *59*, 1088–1099. [CrossRef] [PubMed]

25. Trost, B.; Kindrachuk, J.; Maattanen, P.; Napper, S.; Kusalik, A. PIIKA 2: An expanded, web-based platform for analysis of kinome microarray data. *PLoS ONE* **2013**, *8*, e80837. [CrossRef]

26. Babicki, S.; Amdt, D.; Marcu, A.; Liang, Y.; Grant, J.R.; Maciejewski, A.; Wishart, D.S. Heatmapper: Web-enabled heat mapping for all. *Nucleic Acids Res.* **2016**, *44*, W147–W153. [CrossRef]

27. Lynn, D.J.; Winsor, G.L.; Chan, C.; Richard, N.; Laird, M.R.; Barsky, A.; Gardy, J.L.; Roche, F.M.; Chan, T.H.W.; Shah, N.; et al. InnateDB: Facilitating systems-level analyses of the mammalian innate immune response. *Mol. Syst. Biol.* **2008**, *4*, 218. [CrossRef]

28. Li, Y.; Arsenault, R.J.; Trost, B.; Slind, J.; Griebel, P.J.; Napper, S.; Kusalik, A. A Systemic Approach for Analysis of Peptide Array Kinome Data. *Sci. Signal.* **2012**, *5*, pl2. [CrossRef]

29. Ghasemian, A.; Peerayeh, S.N.; Bakhshi, B.; Mirzaee, M. The Microbial Surface Components Recognizing Adhesive Matrix Molecules (MSCRAMMs) Genes among Clinical Isolates of *Staphylococcus aureus* from Hospitalized Childre. *Iran. J. Pathol.* **2015**, *10*, 258–264.

30. Spellberg, B.; Daum, R. Development of a vaccine against *Staphylococcus aureus*. *Semin. Immunopathol.* **2012**, *34*, 335–348. [CrossRef]

31. Jenkins, A.; Diep, B.A.; Mai, T.T.; Vo, N.H.; Warrener, P.; Zuzich, J.; Stover, C.K.; Sellman, B.R. Differential Expression and Roles of *Staphylococcus aureus* Virulence Determinants during Colonization and Disease. *mBio* **2015**, *6*, e02272-14. [CrossRef]

32. Cramton, S.E.; Gerke, C.; Schnell, N.F.; Nichols, W.W.; Götz, F. The Intercellular Adhesion (*ica*) Locus Is Present in *Staphylococcus aureus* and Is Required for Biofilm Formation. *Infect. Immun.* **1999**, *67*, 5427–5433. [CrossRef] [PubMed]

33. Basanisi, M.G.; La Bella, G.; Nobili, G.; Franconieri, I.; La Salandra, G. Genotyping of methicillin-resistant Staphyloccous aureus (MRSA) isolated from milk and dairy products in South Italy. *Food Microbiol.* **2017**, *62*, 141–146. [CrossRef] [PubMed]

34. Ocal, D.N.; Dolapci, I.; Karahan, Z.C.; Tekeli, A. Investigation of biofilm formation properties of staphylococcus isolates. *Mikrobiyoloji Bulteni* **2017**, *51*, 10–19. [CrossRef]

35. Piechota, M.; Kot, B.; Frankowska-Maciejewska, A.; Grużewska, A.; Woźniak-Kosek, A. Biofilm formation by Methicillin-Resistant and Methicillin-Sensitive *Staphylococcus aureus* Strains from Hospitalized Patients in Poland. *BioMed. Res. Int.* **2018**, *2018*, 1–7. [CrossRef] [PubMed]

36. Aguilar, J.L.; Varshney, A.K.; Pechuan, X.; Dutta, K.; Nosanchuk, J.D.; Fries, B.C. Monoclonal antibodies protect from Staphylococcal Enterotoxin K (SEK) induced toxic shock and sepsis by USA300 *Staphylococcus aureus*. *Virulence* **2017**, *8*, 741–750. [CrossRef] [PubMed]

37. Yarwood, J.M.; McCormick, J.K.; Paustian, M.L.; Orwin, P.M.; Kapur, V.; Schlievert, P.M. Characterization and expression analysis of Staphylococcus aureus pathogenicity island 3. Implications for the evolution of staphylococcal pathogenicity islands. *J. Biol. Chem.* **2002**, *277*, 13138–13147. [CrossRef]

38. Chavakis, T.; Preissner, K.T.; Hermann, M. The anti-inflammatory activities of *Staphylococcus aureus*. *Trends Immunol.* **2007**, *28*, 408–418. [CrossRef]

39. Bloes, D.A.; Haasbach, E.; Hartmayer, C.; Hertlein, T.; Kingel, K.; Kretschmer, D.; Planz, O.; Peschel, A. Phenol-Soluble Modulin Peptides Contribute to Influenza A Virus-Associated *Staphylococcus aureus* Pneumonia. *Infect. Immun.* **2017**, *85*, e00620-17. [CrossRef]

40. Parker, D.; Ryan, C.L.; Alonzo III, F.; Torres, V.J.; Planet, P.J.; Prince, A.S. CD4+ T cells Promote the Pathogenesis of *Staphylococcus aureus* Pneumonia. *J. Infect. Dis.* **2015**, *211*, 835–845. [CrossRef]

41. Sibille, Y.; Marchandise, F.X. Pulmonary immune cells in health and disease: Polymorphonuclear neutrophils. *Eur. Respir. J.* **1993**, *6*, 6.

42. McAdow, M.; Missiakas, D.M.; Schneewind, O. *Staphylococcus aureus* secretes coagulase and von Willebrand factor binding protein to modify the coagulation cascade and establish host infections. *J. Innate Immun.* **2012**, *4*, 141–148. [CrossRef] [PubMed]

43. Peacock, S.J.; Moore, C.E.; Justice, A.; Kantzanou, M.; Story, L.; Mackie, K.; O'Neill, G.; Day, N.P.J. Virulent combinations of adhesin and toxin genes in natural populations of *Staphylococcus aureus*. *Infect. Immun.* **2002**, *70*, 4987–4996. [CrossRef] [PubMed]

44. Cheng, A.G.; McAdow, M.; Kim, H.K.; Bae, T.; Missiakas, D.M.; Schneewind, O. Contribution of Coagulases towards *Staphylococcus aureus* Disease and Protective Immunity. *PLoS Pathog.* **2010**, *6*, e1001036. [CrossRef] [PubMed]

45. Wardenburg, J.B.; Bae, T.; Otto, M.; DeLeo, F.R.; Schneewind, O. Poring over pores: α-hemolysin and Panton-Valentine leukocidin in *Staphylococcus aureus* pneumonia. *Nat. Med.* **2007**, *13*, 1405–1406. [CrossRef] [PubMed]

46. Speziale, P.; Pietrocola, G.; Rindi, S.; Provenzano, G.; Di Poto, A.; Visai, L.; Arciola, C.R. Structural and functional role of *Staphylococcus aureus* surface components recognizing adhesive matrix molecules of the host. *Future Microbiol.* **2009**, *4*, 1337–1352. [CrossRef] [PubMed]

47. Clarke, S.R.; Foster, S.J. Surface adhesins of *Staphylococcus aureus*. *Adv. Microb. Physiol.* **2006**, *51*, 187–224. [CrossRef] [PubMed]

48. Shinji, H.; Yosizawa, Y.; Tajima, A.; Iwase, T.; Sugimoto, S.; Seki, K.; Mizunoe, Y. Role of fibronectin-binding proteins A and B in in vitro cellular infections and in vivo septic infections by Staphylococcus aureus. *Infect. Immun.* **2011**, *79*, 2215–2223. [CrossRef]

49. Zhao, F.; Chong, A.S.; Montgomery, C.P. Importance of B Lymphocytes and the IgG-Binding Protein Sbi in *Staphylococcus aureus* Skin Infection. *Pathogens* **2016**, *5*, 12. [CrossRef]

50. Hageman, J.C.; Uyeki, T.M.; Francis, J.S.; Jernigan, D.B.; Wheeler, J.G.; Bridges, C.B.; Barenkamp, S.J.; Sievert, D.M.; Srinivasan, A.; Doherty, M.C.; et al. Severe Community-acquired Pneumonia Due to *Staphylococcus aureus*, 2003-04 Influenza Season. *Emerg. Infect. Dis* **2006**, *12*, 894–899. [CrossRef]

51. Chickering, H.T.; Park, J.H. *Staphylococcus aureus* pneumonia. *JAMA* **1919**, *72*, 617. [CrossRef]

52. Martin, C.M.; Kunin, C.M.; Gottlieb, L.S.; Finland, M. Asian influenza A in Boston, 1957-1958. II. Severe staphylococcal pneumonia complicating influenza. *AMA Arch. Intern. Med.* **1959**, *103*, 532–542. [CrossRef] [PubMed]

53. Kadioglu, A.; Andrew, P.W. The innate immune response to pneumococcal lung infection: The untold story. *Trends Immunol.* **2004**, *25*, 143–149. [CrossRef] [PubMed]

54. McCullers, J.A. Insights into the Interaction between Influenza Virus and Pneumococcus. *Clin. Microbiol. Rev.* **2006**, *19*, 571–582. [CrossRef] [PubMed]

55. Koedel, U.; Rupprecht, T.; Angele, B.; Heesemann, J.; Wagner, H.; Pfister, H.W.; Krischning, C.J. MyD88 is required for mounting a robust host immune response to Streptococcus pneumoniae in the CNS. *Brain* **2004**, *127*, 1437–1445. [CrossRef] [PubMed]

56. Yoshimura, A.; Lien, E.; Ingalis, R.R.; Tuomanen, E.; Dziarski, R.; Golenbock, D. Cutting edge: Recognition of Gram-positive bacterial cell wall components by the innate immune system occurs via Toll-like receptor 2. *J. Immunol.* **1999**, *163*, 163.

57. Askarian, F.; Wagner, T.; Johannessen, M.; Nizet, V. *Staphylococcus aureus* modulation of innate immune responses through Toll-like (TLR), (NOD)-like (NLR) and C-type lectin (CLR) receptors. *FEMS Microbiol. Rev.* **2018**, *42*, 656–671. [CrossRef]

58. Tumpey, T.M.; Garcia-Sastre, A.; Taubenberger, J.K.; Palese, P.; Swayne, D.E.; Pantin-Jackwood, M.J.; Schultz-Cherry, S.; Solórzano, A.; Van Rooijen, N.; Katz, J.M.; et al. Pathogenicity of influenza viruses with genes from the 1918 pandemic virus: Functional roles of alveolar macrophages and neutrophils in limiting virus replication and mortality in mice. *J. Virol.* **2005**, *79*, 14933–14944. [CrossRef]

59. Tumpey, T.M.; Basier, C.F.; Aguilar, P.V.; Zeng, H.; Solórzano, A.; Swayne, D.E.; Cox, N.J.; Katz, J.M.; Taubenberger, J.K.; Palese, P.; et al. Characterization of the reconstructed 1918 Spanish influenza pandemic virus. *Science* **2005**, *310*, 77–80. [CrossRef]

60. Beigel, J.H.; Farrar, J.; Han, A.M.; Hayden, F.G.; Hyer, R.; de Jong, M.D.; Lochindarat, S.; Nguyen, T.K.; Nguyen, T.H.; Tran, T.H.; et al. Avian influenza A (H5N1) infection in humans. *N. Engl. J. Med.* **2005**, *353*, 1374–1385. [CrossRef]

61. Cheung, C.Y.; Poon, L.L.; Lau, A.S.; Luk, W.; Lau, Y.L.; Shortridge, K.F.; Gordon, S.; Guan, Y.; Peiris, J.S. Induction of proinflammatory cytokines in human macrophages by influenza A (H5N1) viruses: A mechanism for the unusal severity of human disease? *Lancet* **2002**, *360*, 1831–1837. [CrossRef]

62. Silva-García, O.; Valdez-Alarcón, J.J.; Baizabal-Aguirre, V.M. The Wnt/ B-Catenin Signaling Pathway Controls the Inflammatory Response in Infections Caused by Pathogenic Bacteria. *Mediat. Inflamm.* **2014**, *2014*, 1–7. [CrossRef] [PubMed]

63. Gustafson, B.; Smith, U. Cytokines Promote Wnt SIgnaling and Inflammation and Impair the Normal Differentiation and Lipid Accumulation in 3T3-L1 Preadipocytes. *J. Biol. Chem.* **2006**, *281*, 9507–9516. [CrossRef] [PubMed]

64. Halleskog, C.; Mulder, J.; Dahlström, J.; Mackie, K.; Hortobágyi, T.; Tanila, H.; Kumar Puli, L.; Färber, K.; Harkany, T.; Schulte, G. WNT signaling in activated microglia is proinflammatory. *Glia* **2011**, *59*, 119–131. [CrossRef] [PubMed]

65. Liu, X.; Lu, R.; Wu, S.; Sun, J. Salmonella regulation of intestinal stem cells through the Wnt/beta-catenin pathway. *FEBS Lett.* **2010**, *584*, 911–916. [CrossRef] [PubMed]

66. Neumann, J.; Schaale, K.; Farhat, K.; Endermann, T.; Ulmer, A.J.; Ehlers, S.; Reiling, N. Frizzled1 is a marker of inflammatory macrophages, and its ligand Wnt3a is involved in reprogramming Mycobacterium tuberculosis-infected macrophages. *FASEB J.* **2010**, *24*, 4599–4612. [CrossRef]

67. Short, K.R.; Veeris, R.; Leitjen, L.M.; van den Brand, J.M.; Jong, V.L.; Stittelaar, K.; Osterhaus, A.D.M.E.; Andeweg, A.; van Riel, D. Proinflammatory Cytokine Responses in Extra-Respiratory Tissues During Severe Influenza. *J. Infect. Dis.* **2017**, *216*, 829–833. [CrossRef]

68. Federspiel, C.K.; Liu, K.D. *Critical Care Nephrology*, 3rd ed.; Elsevier: Amsterdam, The Netherlands, 2019.

69. Bradley-Stewart, A.; Jolly, L.; Adamson, W.; Gunson, R.; Frew-Gillespie, C.; Templeton, K.; Aitken, C.; Carman, W.; Cameron, S.; McSharry, C. Cytokine responses in patients with mild or severe influenza A(H1N1)pdm09. *J. Clin. Virol.* **2013**, *58*, 100–107. [CrossRef] [PubMed]

70. Guzy, R.D.; Stoilov, I.; Elton, T.J.; Mecham, R.P.; Omitz, D.M. Firboblast growth factor 2 is required for epithelial recovery, but not for pulmonary fibrosis, in response to bleomycin. *Am. J. Respir. Cell Mol. Biol.* **2015**, *52*, 116–128. [CrossRef]

71. Braun, S.; auf dem Keller, U.; Steiling, H.; Werner, S. Fibroblast growth factors in epithelial repair and cytoprotection. *Philos. Trans. R Soc. Lond B Biol. Sci.* **2004**, *359*, 753–757. [CrossRef]

72. Meyer, M.; Müller, A.K.; Yang, J.; Moik, D.; Ponzio, G.; Grose, R.; Werner, S. FGF receptors 1 and 2 are key regulators of keratinocyte migration in vitro and in wounded skin. *J. Cell Sci.* **2012**, *125*, 5690–5701. [CrossRef]

73. Wang, K.; Lai, C.; Li, T.; Wang, C.; Wang, W.; Ni, B.; Bai, C.; Zhang, S.; Han, L.; Gu, H.; et al. Basic fibroblast growth factor protects against influenza A virus-induced acute lung injury by recruiting neutrophils. *J. Mol. Cell Biol.* **2018**, *10*, 573–585. [CrossRef] [PubMed]

74. Dunning, J.; Blankley, S.; Hoang, L.T.; Cox, M.; Graham, C.M.; James, P.L.; Bloom, C.I.; Chaussabel, D.; Banchereau, J.; Brett, S.J.; et al. Progression of whole-blood transcriptional signatures from interferon-induced to neutrophil-associated patterns in severe influenza. *Nat. Immunol.* **2018**, *19*, 625–635. [CrossRef] [PubMed]

75. Jia, L.; Zhao, J.; Yang, C.; Liang, Y.; Long, P.; Liu, X.; Qiu, S.; Wang, L.; Xie, J.; Li, H.; et al. Severe Pneumonia Caused by Coinfection with Influenza Virus Followed by Methicillin-Resistant *Staphylococcus aureus* Induces Higher Mortality in Mice. *Front. Immunol.* **2019**, *9*, 3189. [CrossRef] [PubMed]

76. Paquette, S.G.; Banner, D.; Zhao, Z.; Fang, Y.; Huang, S.S.; León, A.J.; Ng, D.C.; Almansa, R.; Martin-Loaches, I.; Ramirez, P.; et al. Interleukin-6 is a potential biomarker for severe pandemic H1N1 influenza A infection. *PLoS ONE* **2012**, *7*, e38214. [CrossRef]

77. Bickel, M. The role of interleukin-8 in inflammation and mechanisms of regulation. *J. Periodontol.* **1993**, *64*, 64.

78. Liu, M.; Guo, S.; Hibbert, J.M.; Jain, V.; Singh, N.; Wilson, N.O.; Stiles, J.K. CXCL10/IP-10 in Infectious Diseases Pathogenesis and Potential Therapeutic Implications. *Cytokine Growth. Factor Rev.* **2011**. [CrossRef]

79. Li, Z.; Levast, B.; Madrenas, J. Staphylococcus aureus Downregulates IP-10 Production and Prevents Th1 Cell Recruitment. *J. Immunol.* **2017**, *198*, 1865–1874. [CrossRef]

80. Al Alam, D.; Deslee, G.; Tournois, C.; Lamkhioued, B.; Lebargy, F.; Merten, M.; Belaaouaj, A.; Guenounou, M.; Gangloff, S.C. Impaired interleukin-8 chemokine secretion by staphylococcus aureus-activated epithelium and T-cell chemotaxis in cystic fibrosis. *Am. J. Respir. Cell Mol. Biol.* **2010**, *42*, 644–650. [CrossRef]

81. Tuder, R.M.; Flook, B.E.; Voelkel, N.F. Increased gene expression for VEGF and the VEGF receptors KDR/Flk and Flt in lungs exposed to acute or to chronic hypoxia. Modulation of gene expression by nitric oxide. *J. Clin. Investig.* **1995**, *95*, 1798–1807. [CrossRef]

82. Monacci, W.; Merrill, M.; Oldfield, E. Expression of vascular permeability factor/vascular endothelial growth factor in normal rat tissues. *Am. J. Physiol. Cell Physiol.* **1993**, *264*, C995–C1002. [CrossRef]

83. Marti, H.H.; Risau, W. Systemic hypoxia changes the organ-specific distribution of vascular endothelial growth factor and its receptors. *Proc. Natl. Acad. Sci. USA* **1998**, *95*, 15809–15814. [CrossRef] [PubMed]

84. van der Poll, T.; Marchant, A.; Keogh, C.V.; Goldman, M.; Lowry, S.F. Interleukin-10 impairs host defense in murine pneumococcal pneumonia. *J. Infect. Dis.* **1996**, *174*, 994–1000. [CrossRef] [PubMed]

85. van der Sluijs, K.F.; van Elden, L.J.; Nijhuis, M.; Schuurman, R.; Pater, J.M.; Florquin, S.; Goldman, M.; Jansen, H.M.; Lutter, R.; van der Poll, T. IL-10 is an important mediator of the enhanced susceptibility to pneumococcal pneumonia after influenza infection. *J. Immunol.* **2004**, *172*, 7603–7609. [CrossRef] [PubMed]

Publisher's Note: MDPI stays neutral with regard to jurisdictional claims in published maps and institutional affiliations.

MDPI
St. Alban-Anlage 66
4052 Basel
Switzerland
Tel. +41 61 683 77 34
Fax +41 61 302 89 18
www.mdpi.com

Cells Editorial Office
E-mail: cells@mdpi.com
www.mdpi.com/journal/cells